Library of Davidson College

VOID

**THE CELLULAR AND MOLECULAR BIOLOGY
OF INVERTEBRATE DEVELOPMENT**

The Belle W. Baruch Library in Marine Science

1. ESTUARINE MICROBIAL ECOLOGY
2. SYMBIOSIS IN THE SEA
3. PHYSIOLOGICAL ECOLOGY OF MARINE ORGANISMS
4. BIOLOGICAL RHYTHMS IN THE MARINE ENVIRONMENT
5. THE MECHANISMS OF MINERALIZATION IN THE INVERTEBRATES AND PLANTS
6. ECOLOGY OF MARINE BENTHOS
7. ESTUARINE TRANSPORT PROCESSES
8. MARSH-ESTUARINE SYSTEMS SIMULATION
9. REPRODUCTIVE ECOLOGY OF MARINE INVERTEBRATES
10. ADVANCED CONCEPTS IN OCEAN MEASUREMENTS FOR MARINE BIOLOGY
11. MARINE BENTHIC DYNAMICS
12. PROCESSES IN MARINE REMOTE SENSING
13. MARINE POLLUTION AND PHYSIOLOGY: RECENT ADVANCES
14. SHALLOW-WATER MARINE BENTHIC MACROINVERTEBRATES OF SOUTH CAROLINA: SPECIES IDENTIFICATION, COMMUNITY COMPOSITION AND SYMBIOTIC ASSOCIATIONS.
15. THE CELLULAR AND MOLECULAR BIOLOGY OF INVERTEBRATE DEVELOPMENT

THE BELLE W. BARUCH LIBRARY IN MARINE SCIENCE NUMBER 15

The Cellular and Molecular Biology of Invertebrate Development

Edited by Roger H. Sawyer
and Richard M. Showman

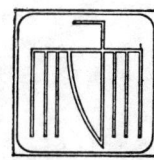

Published for the Belle W. Baruch Institute for Marine Biology and

Coastal Research by the

UNIVERSITY OF SOUTH CAROLINA PRESS

Copyright© University of South Carolina 1985

Published in Columbia, South Carolina by the
University of South Carolina Press

First Edition

Manufactured in the United States of America

Library of Congress Cataloging-in-Publication Data
Main entry under title:

The Cellular and molecular biology of invertebrate
   development.

   (The Belle W. Baruch library in marine science ;
no. 15)
   Proceedings of the Twenty-sixth Southeastern
Conference on Developmental Biology held in 1984 at the
Belle Baruch Laboratory, Georgetown, South Carolina in
honor of Ernest Everett Just.
   Includes index.
   1. Invertebrates--Development--Congresses.
2. Invertebrates--Cytology--Congresses. 3. Molecular
biology--Congresses. 4. Just, Ernest Everett, 1883-
1941. I. Sawyer, Roger H. II. Showman, Richard M.
III. Just, Ernest Everett, 1883-1941. IV. Bell W.
Baruch Institute for Marine Biology and Coastal
Research. V. Southeastern Conference on Developmental
Biology (26th : 1984 : Belle Baruch Laboratory)
VI. Series.
QL364.C45   1985   592'.03   85-22695
ISBN 0-87249-464-0

DEDICATION

Ernest Everett Just
1883–1941

One of the greatest honors for a scientist is a publication or an event dedicated to his life's work. The Twenty-sixth Southeastern Conference on Developmental Biology held at the Belle Baruch Laboratory in Georgetown, South Carolina, and this resulting publication of the proceedings represent both for Ernest Everett Just. One of this country's early, eminent embryologists who lived a sad life and died a tragic death, he would feel a deep appreciation for being recognized by today's leading biologists at a conference in his home state of South Carolina, and for being an inspiration for many young biologists, black and white alike. That the conference occurred as a celebration of the centennial of his birth was a special touch indeed. The recognition of and true sympathy for the social as well as scientific purpose of the event will mark the occasion and this publication as truly rare in the history of modern science. We owe special gratitude to the organizers and participants.

> Kenneth R. Manning
> Professor of the History of Science
> Massachusetts Institute of Technology

CONTENTS

Speakers — viii

Special Guests — x

Preface — xi

Opening Remarks: Preserving the Anarchy of Science
    J.D. Ebert — 1

A Scanning Electron Microscopical Overview of Cellular and Extracellular Patterns During Blastulation and Gastrulation in the Sea Urchin, Lytechinus variegatus
    J.B. Morrill and L. Santos — 3

Egg Plasma Membrane Changes at Fertilization
    F.J. Longo — 35

Physical Interactions Between Asters and the Cortex in Echinoderm Eggs
    T.E. Schroeder — 69

Translational Control in Echinoid Eggs and Early Embryos
    M.B. Hille, M.V. Danilchik, A.M. Colin, and R.T. Moon — 91

Patterns of Maternal mRNA Distribution and their Role in Early Development
    W.R. Jeffery — 125

Maternal Messenger RNA: Synthesis and Localization of Histone in the Sea Urchin Oocyte
    R. Showman, D.E. Wells, J.A. Anstrom, D.A. Hursh, D.S. Leaf, and R.A. Raff — 153

Expression and Appearance of Germ Layer-
Specific Antigens on the Surface of
Embryonic Sea Urchin Cells
    D.R. McClay, V. Matranga and G. Wessel   171

A Short Review of Germ Cell Determination
in Drosophila Melanogaster
    R.E. Boswell   187

The Many Motors of Morphogenesis: The
Role of Muscles, Cilia, and Micro-
filaments in the Metamorphosis of
Marine Bryozoans
    C.G. Reed   197

Tissue Architecture and Hydroid Morphogenesis:
The Role of Locomotory Traction in Shaping
the Tissue
    R.D. Campbell   221

Isolation and Characterization of
Blastoderm-Specific Genes of Drosophila
Melanogaster
    J.A. Lengyel, M. Roark, K. Kongsuwan,
    P.A. Mahoney, P.D. Boyer, and J.R. Merriam   239

Translational Control of Cell Cycle-Related
Proteins in Early Embryos
    J.V. Ruderman   259

The Sea Urchin Spec Family of Calcium
Binding Proteins: Characterization and
Consideration of Possible Role in
Larval Development
    W.H. Klein, C.D. Carpenter, L.E. Philpotts,
    and B.P. Brandhorst   275

The Origin of the Micromeres and Formation
of the Skeletal Spicules in Developing
Sea Urchin Embryos
    F.H. Wilt, S. Benson and J.A. Uzman   297

Aspects of Gene Expression in the Sea
Urchin Micromere - Primary Mesenchyme
Cell Line
    M.A. Harkey   311

Index   329

SPEAKERS

Dr. Everett Anderson
Laboratory of Human Reproduction
Harvard Medical School
Boston, MA   02115

Dr. Robert E. Boswell
NIEHS
PO Box 12233
Research Triangle Park, NC
   27709

Dr. Richard Campbell
Developmental Biology Center
University of California
Irvine, CA   92717

Dr. James Ebert, President
Carnegie Institute of
   Washington
1530 P Street, NW
Washington, DC   20005

Dr. Michael Harkey
Department of Zoology, Microbiology and Friday Harbor
   Labs
University of Washington
Seattle, WA   98195

Dr. Merrill B. Hille
Department of Zoology
University of Washington
Seattle, WA   98195

Dr. William R. Jeffrey
Center for Developmental
   Biology
Department of Zoology
University of Texas
Austin, TX   78712

Dr. William Klein
Program in Molecular,
   Cellular and
   Developmental Biology
Indiana University
Bloomington, IN   47405

Dr. Judith A. Lengyel
Department of Biology and
   Molecular Biology Institute
University of California
Los Angeles, CA   90024

Dr. Frank Longo
Department of Anatomy
University of Iowa
Iowa City, IA   52242

Dr. David R. McClay
Department of Zoology
Duke University
Durham, NC   27706

Dr. John B. Morrill
Divison of Natural Sciences
New College of USF
Sarasota, FL   33580

Dr. Clifton A. Poodry
Thimann Laboratories
University of California
Santa Cruz, CA   95064

Dr. Christopher Reed
Department of Biological
   Sciences
Dartmouth College
Hanover, NY   03755

Dr. Joan V. Ruderman
Department of Anatomy
Harvard Medical School
Boston, MA   02115

Dr. Thomas E. Schroeder
Friday Harbor Laboratories
University of Washington
Friday Harvor, WA 98250

Dr. Richard Showman
Department of Biology
University of South Carolina
Columbia, SC 29208

Dr. J. Richard Whittaker
Boston University Marine Program
Marine Biological Laboratory
Woods Hole, MA 02543

Dr. Fred Wilt
Department of Zoology
University of California
Berkeley, CA 94720

SPECIAL GUESTS

Dr. James H. Arrington, Chairman
Department of Natural Sciences
South Carolina State College
Orangeburg, SC   29117

Dr. George P. Fulton, Director
South Carolina Hall of Science and Technology
4740 Cedar Springs Road
Columbia, SC   29206

Dr. Kenneth R. Manning
Department of History and Science
Massachusetts Institute of Technology
Cambridge, MA   02139

Dr. Lewie C. Roache, Dean
School of Arts and Sciences
South Carolina State College
Orangeburg, SC   29117

Dr. Oswald F. Schuette
Department of Physics
University of South Carolina
Columbia, SC   29208

Mr. James L. Solomon, Jr., Commissioner
South Carolina Department of Social Services
P.O. Box 1520
Columbia, SC   29202

Dr. F. John Vernberg, Director
Baruch Institute for Marine Biology and Coastal Research
University of South Carolina
Columbia, SC   29208

Dr. Clemmie E. Webber
Professor Emeritus
Department of Education
South Carolina State College
Orangeburg, SC   29117

PREFACE

In the one hundred years since the birth of E.E. Just, the field of developmental biology has seen radical transformations in the way we approach our subject. Electron microscopes, acrylamide gels and molecular probes have become the tools of our trade. Even so, the basic principles and the way we view the complexity of the developmental process have changed little since Just's time. We still struggle with the concepts of determination and differentiation. We still search for the mechanisms that regulate developmental events. For Just, who had to struggle not only with the intricacies of a demanding science, but also with the vicissitudes of an intolerant society, the fruits of success often remained elusive. Yet he persevered, and in doing so he laid a part of the foundation upon which we stand today.

In bringing together the speakers for this Symposium in honor of Just, we sought to offer as much breadth as possible, both in terms of organisms and approaches to invertebrate development. This was done in the hope that somewhere at the core of all our work lie those common threads that tie the developmental process together as a unified whole. Only as we begin to integrate the cytology, molecular biology, and genetics of developing organisms will we begin to understand the "hows" and "whys" of the developmental process. We are a long way from achieving that goal but we are making progress, as the contents of this volume show, in many areas. It is our hope that in assembling these papers, we will stimulate new ideas and approaches that will increase our understanding of the process of development in all organisms.

Many individuals and organizations have made the Just Symposium and this volume possible. Financial support from the Belle W. Baruch Institute for Marine Biology and Coastal Research, the Department of Biology of the University of South Carolina, The Society for Developmental Biology, The American Society of Zoologists, the South Carolina Academy of Sciences, and the National Science Foundation have made it possible to hold these meetings. We wish to thank Drs. Lewis Bowman, Wallace Dawson, Michael Dewey and John Wourms for helping to organize and chair sessions. Particular thanks go to Dr. John Verberg, Director of the Belle W. Baruch Institute, and his staff including Dennis Allen, Douglas Baughman, Mary Beth Sawyer, and Virginia Smith who helped ease the logistic burden of the meeting and made available the facilities of the Baruch Institute Field Station at Georgetown, South Carolina. We also wish to thank James L. Solomon, Jr., Commissioner of the South Carolina Department of Social Services, George P. Fulton, Director of the South Carolina Hall of Science and Technology, and Professor Oswald F. Schuette of the University of South Carolina for their encouragement and support in organizing this meeting in honor of one of South Carolina's finest black scientists. Last and far from least we wish to thank Ms. Debra Chavis, Ms. Anne Miller, and Dr. L.T. Wimer, for their tireless efforts and assistance in the editing and processing of the text. Without them our job would have been much harder.

              Roger H. Sawyer
              Richard M. Showman
              University of South Carolina
              May, 1985

PRESERVING THE ANARCHY OF SCIENCE

Today, all of us -- individuals and institutions alike -- are concerned about our roots. Kenneth Manning's brilliant biography of Ernest Everett Just, Black Apollo of Science, does more than explore Just's roots; it is at once a penetrating history of ideas and an incisive analysis of the scientific and philanthropic institutions of Just's time.

This symposium dedicated to Just is timely, an occasion not only to celebrate the one-hundredth anniversary of his birth, but also, in his name, to dedicate ourselves to the re-creation of a national spirit of science.

In this brief introduction to Professor Manning's address, drawing especially from Just's experiences at the Marine Biological Laboratory, I shall explore some of the forces that shape a scholar, that nourish and winnow a field, and that appear again and again in our great teachers and scientists.

Scholarly pursuits above all require the expression of individuality. One must often be alone. Yet, except possibly for the most brilliant, the most gifted, there must be, at crucial points in a career, the most significant personal interactions between perceptor and student, who must develop a very special intellectual intimacy, a very special interdependence. In The World As I See It, Albert Einstein wrote, "A hundred times every day I remind myself that my inner and outer life depend on the labors of others, living and dead, and that I must exert myself in order to give in the same measure as I have received and am still receiving".

Too, there must be for most of us, a favorable environment, an intellectual and physical climate supportive of creativity, requiring the very best of those of us whose joys of scholarship are largely vicarious, but whose role it is to nourish our fields; and above all, to provide for our institutions the consistency and stability of support that are crucial to creative endeavors.

The operative "key words" -- to use a phrase in vogue -- are consistency and <u>stability</u> of support and the <u>flexibility</u> of the human mind.

The operative condition is an environment -- an intellectual climate -- favoring the free flow of ideas. Here the crucial phrase is the free flow of ideas. Over two decades ago, Jerome Wiesner, then Science Advisor to President Kennedy was asked by a reporter to define that position. He reflected a moment and offered the following job description. It was, he said, "To preserve the anarchy of science."

It is a lesson too many of us have forgotten; it is a lesson devoutly to be relearned. We must find ways of encouraging more Ernest Everett Justs, the bold, the unusual, the breakers of rules --- Wiesner's words, of preserving the anarchy of science.

James D. Ebert
Carnegie Institution of Washington

A SCANNING ELECTRON MICROSCOPICAL OVERVIEW OF CELLULAR AND EXTRACELLULAR PATTERNS DURING BLASTULATION AND GASTRULATION IN THE SEA URCHIN, *LYTECHINUS VARIEGATUS*

John B. Morrill and Laurinda L. Santos

ABSTRACT

SEM analysis of dry-fractured embryos fixed in an isotonic gluteraldehyde-osmium fixative containing Ruthenium red or cetylpyridinium chloride showed stage-specific changes in the morphological patterns of the cells of the blastocoelic wall and the colloidal, extracellular substances of the blastocoel and archenteron during the 8- to 10-h period (25°C) of blastulation and gastrulation. Regionalization of cellular patterns in the blastular wall were correlated with Horstadius' developmental fate map. Cellular and extracellular patterns associated with bilateral symmetry and dorsoventrality appeared in the blastula before ingression of the primary mesenchyme cells in normal embryos. In Li-treated embryos these patterns were markedly altered in radialized Li-exogastrulae.

INTRODUCTION

> "It is not birth, marriage, or death, but gastrulation which is truly the most important time in your life."
>   Lewis Wolpert (From J.M.W. Slack, 1983)

In spite of its importance and the plethora of original studies and reviews, the mechanisms of gastrulation in the sea urchin are incompletely characterized and understood. This is because of variations in the timing and morphologies of the developmental events of the various species used in different laboratories throughout the world, asynchronous cultures of embryos, variations in quality controls, variations in the treatment and preparation of embryos for biochemical and microscopical analyses, and frequently the study of only one or a few stages of the developmental continuum from early blastulation to the completion of gastrulation - a continuum that may span a period of 8 h at 25°C in Lytechinus variegatus or 18 h at 18°C in Psammechinus miliaris (Czihak, 1971).

During the last 25 years, beginning with the studies of Motomura (1960) on the secretion of a mucosubstance in the sea urchin gastrula, Immers (1961) on the incorporation of $^{35}SO_4$ in the blastula and gastrula and Okazaki and Nijima (1964) on the isolation of the "basement membrane" (basal lamina) in sea urchin larvae, there has been an increasing interest in the synthesis, composition and potential morphogenetic roles of extracellular substances during blastulation, gastrulation and formation of normal and abnormal larvae.

In this chapter we will outline the early development through gastrulation in Lytechinus variegatus and describe the temporal and spatial changes in cellular and extracellular patterns of normal embryos fractured to expose the inside (blastocoelic side) and the cellular walls of blastulae and gastrulae when seen with the scanning electron microscope (SEM). We will then describe the Li-exogastrula as seen with the SEM and finally explore the future potentials of SEM analysis of sea urchin development.

MATERIALS AND METHODS

"The individual experiment capable of being
repeated at will depends very strongly on
the adherence to some general rules."
E.E. Just (1939)

Lytechinus variegatus, referred to as Toxopnuestes variegatus in the early literature, is a variable species that ranges from North Carolina to Santos, Brazil, on the American Coast and to Bermuda and the Cape Verde Islands (Serafy, 1973). For our study we collected adults from local populations of the subspecies carolinus in Sarasota Bay, FL. In this region the natural breeding season extends from the end of September to mid-May.

Gametes of freshly collected adults were obtained by intracoelomic injection of 0.5 M KCl. Sperm were collected 'dry' in syracuse dishes; eggs were shed into 100-ml beakers of natural sea water (NSW) over a 15-min period; settled eggs were rinsed twice with NSW before insemination. Five drops of concentrated eggs suspended in 50 ml of NSW in an Erlenmeyer flask were fertilized by adding 0.1 ml of diluted sperm (1 drop dry sperm/50 ml NSW) with continuous mixing over one min. Only batches of eggs exhibiting 95% fertilization membranes and two cells at first cleavage were used. After fertilization, eggs were first sedimented with a hand centrifuge and then resuspended in fresh NSW containing 200 µg/ml of streptomycin sulfate (Sigma) and cultured as a sparse monolayer in 3-inch glass culture dishes at 25°C in a constant temperature incubator. Even with uniform culturing conditions we have observed temporal variations in the time of hatching from the fertilization membrane, mesenchyme cell ingression and the beginning and completion of gastrulation with batches of eggs collected throughout the breeding season, with batches of eggs from females collected on the same day, and within an individual batch of eggs. Therefore, we routinely used morphological criteria to stage embryos since morphological stages are reproducible regardless of the exact rate of development and are readily observed in living embryos with Nomarski differential interference contrast (DIC) optics.

Preparation of embryos for SEM analysis involved fixing concentrated embryos in one of three isotonic (1200 milliosmols) 'cocktail' fixatives concocted by Deni Galileo (DG) and Matthew Wahl (MW). DG-RR fixative was prepared by mixing ice cold solutions of 2X artificial sea water (2 ml), 10% E.M. grade gluteraldehyde (1 ml), 4% osmium tetroxide (2 ml) and 20 mg of 50% Ruthenium red (RR) (Sigma). DG-CPC fixative was the same as DG-RR except that 50 mg of cetylpyridinium chloride (CPC) replaced the RR. Embryos were fixed for 1 h at 4°C, rinsed five times with cold distilled water and dehydrated in an ethanol series (10-100%).

The MW fixative consisted of distilled water (4.5 ml), 8 M S-collidine (2.5 ml), 10% gluteraldehyde (3 ml), sucrose (2.05 g) and calcium chloride (10 mg). This fixative had a pH of 7.4 and was 1100 milliosmols. After the embryos were fixed for 12 to 18 h at 23°C, they were rinsed in the above S-collidine buffer (pH 7.4) that contained 2.74 g sucrose, 10 mg calcium chloride and 3 ml distilled water before they were dehydrated in an ethanol series (30-100%).

The DG-RR and DG-CPC fixatives resulted in a minimal shrinkage of the embryos and maximized the preservation of extracellular substances such as the hyaline layer, basal lamina, extracellular vesicles and fibers, and variable amounts of acid mucopolysaccharides, glycosaminoglycans (GAGs). Microtubule- and microfilament-dependent cellular processes were probably distorted to some degree because of the low temperature of fixation and osmium in the fixatives. Additional subtle artifacts undoubtedly occurred because of the multi-compartmental nature of the embryo and the fact that there is no universally perfect fixative.

Dehydrated specimens were critical point dried using liquid CO_2 as the transitional fluid. The dried embryos were then transferred to double sided Scotch tape on aluminum stubs with the aid of a fine needle before they were sputter coated with gold or gold-palladium. The mounted embryos were fractured either by cutting them with a No. 11 scalpel blade mounted in a micromanipulator, pricking them with a fine needle, or plucking off the upper side of the embryo with a piece of double sided tape on a second stub that was first lowered with a micromanipulator until the tape touched the embryo and then retracting the stub. The plucking method produced the cleanest fractures and frequently resulted in both fractured halves of an embryo being preserved. Interestingly, in embryos fixed in DG-RR and MW fixatives the fractures occurred between the cells, while in DG-CPC fixed embryos the fractures passed through the cells of the blastular and gastrular walls. We have observed that dry fracturing can result in several artifacts. The embryo may be distorted from the physical pressures and tensions of fracturing; dried cells, pieces of cells and extracellular entities may be displaced; and the extracellular matrix (ECM) in the blastocoels of fractured embryos may be partially or totally lost during the pump down period in the sputter coater.

The specimens were observed with an ISI SS-40 scanning electron microscope at a beam voltage of 10kV. Because this microscope is equipped with two S.E. detectors, one in the stage chamber and one in the column, it is possible to collect secondary electron signals from what normally appear as dark areas on the image of a specimen. Thus, by varying the several operational parameters on the microscope - the working distance, stage tilt, size of the final aperture, gamma attenuation, and the two S.E. detectors - we were able to view and photograph the inner recesses of the fractured embryos. With the HR (high resolution) mode on this microscope, we were able to view such features as cell surfaces and processes, the basal lamina, and other extracellular entities at magnifications up to 50,000X.

Stereo pair micrographs were taken routinely on Polaroid P/N 55 film at stage tilt angles of 4°, 6°, or 10° depending on the magnification of the image. Stereo pairs were viewed with a Stereoaids bench top viewer in order to determine Z axis spatial relationships because, as Russell Steere has emphasized, "If you haven't seen it in stereo, you haven't seen it!" Since we did not have an analytical stereo viewer, the dimensions of various structures taken from mono images are only an approximation. Furthermore, the diameters of ultrastructural features such as vesicles, fibers, cilia, and cellular processes are increased by approximately 20 nm because of the 10nm coating of gold or gold-palladium.

NORMAL DEVELOPMENT

Cleavage

During the fourth cleavage cycle the four cells at the animal pole of the eight-cell egg divide by vertical furrows to form a loose plate of eight mesomeres; four larger macromeres and four small micromeres are formed by an unequal cleavage of the cells at the vegetal pole. In the resulting 16-cell embryo the cells are first round and loosely associated (Fig. 1a). Upon compaction during the 16-cell stage, the interblastomeric sides of the cells flatten; individual cells may have up to seven flattened faces (Fig. 1b) as well as cellular processes that extend from the faces and edges of individual cells and attach to the surfaces of adjoining cells (Figs. 1c & d). In addition, small granules are abundant on the cleavage cavity surfaces of the cells (Fig. 1d). In some eggs fractured at this stage, the incipient cleavage cavity is filled with an ECM (Fig. 1e) that is composed of a meshwork of fibers coated with small granules (Fig. 1f). These granules may be one or more kinds of GAGs. Thus, as early as the 16-cell stage, probably earlier, the positioning of cells relative to one another is maintained not only by their attachment to the hyaline layer and to one another via cellular junctions (Spiegel & Howard, 1983) but also by filopodial-like attachment processes. Coincident with each cleavage cycle there may be an active formation and retraction of these attachment processes as well as a programmed

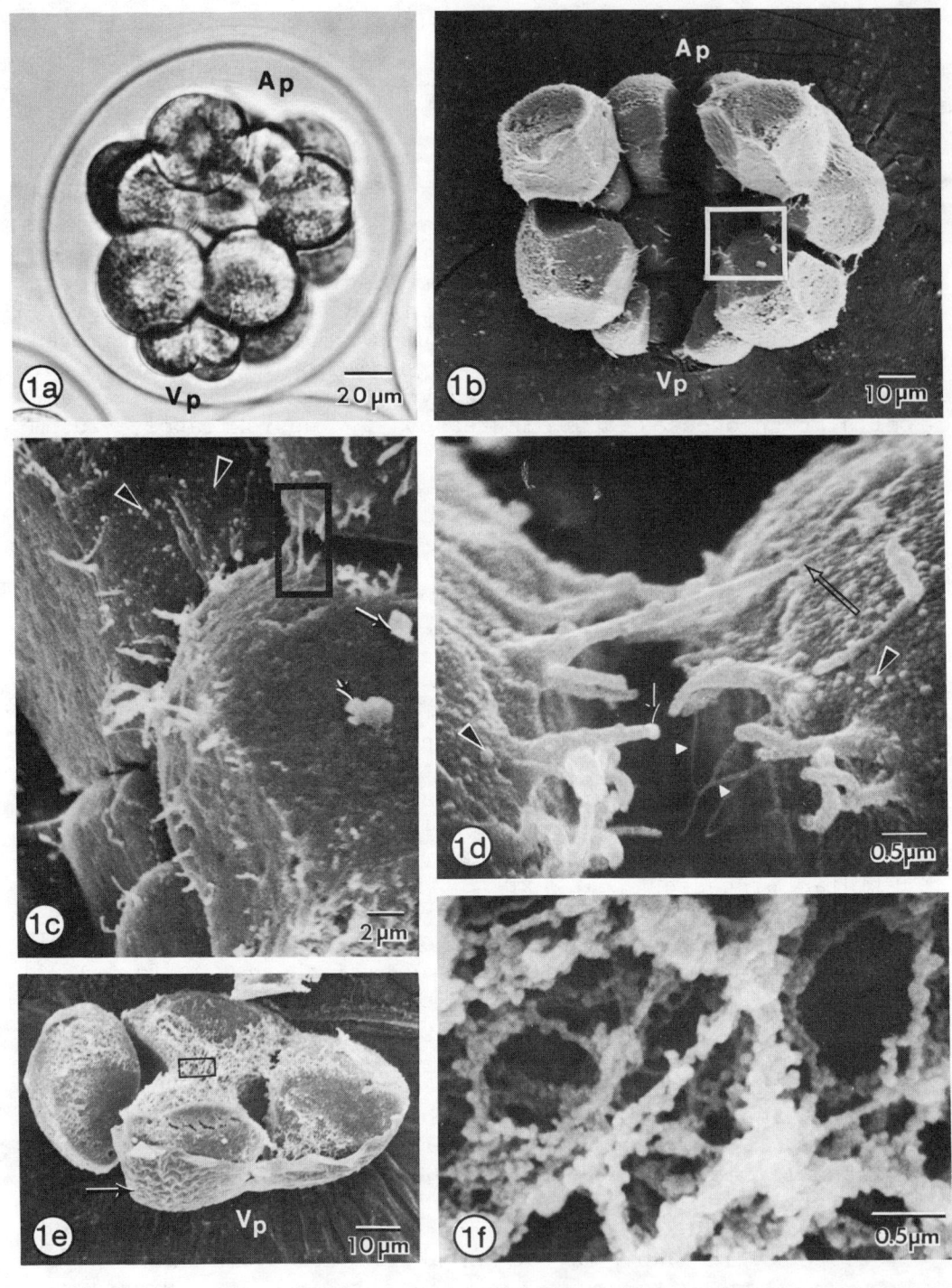

Fig. 1. Shapes of blastomeres and morphological features of cell surfaces and ECM of cleavage cavity of 16-cell embryos. (a) DIC micrograph of live embryo within its fertilization membrane shortly after completion of the fourth cleavage. The spherical images of the cells may be an optical illusion because of the limited depth of field and the edges of the cells being out of focus. (b) SEM (700X) of a fractured egg with one mesomere and one micromere missing. The edges and flattened faces of the blastomeres have short cellular processes that extend between pairs of cells. The hyaline layer is partially lost as a result of the fertilization membrane having been removed by treatment with 1 M urea for 30 sec at 2 min after insemination. M.W. fixative. The square in (b) shows the location of (c). Many of the cellular processes appear to be broken near their tips by the fracturing. Small granules (arrowheads) are present on the surfaces of some cells. Two pieces of debris (arrows) are on the surface of one cell. In (d) (15,000X) additional features of the boxed area in (c) become visible. The tip of one cell process (outline arrow) appears to extend into what may be a surface coating of the adjoining cell. Other cell processes have bulbous tips (arrow). The surfaces of the cells have numerous granules (arrowheads). Thin extracellular fibrils (white triangles) are also present. (e) In a second egg fractured horizontally, an ECM is abundant in the intercellular spaces between the four macromeres. Portions of intact hyaline layer (arrow) are present. In (f) (30,000X) the ECM in the box area in (e) consists of a meshwork of fibers coated with granules. Ap, animal pole; Vp, vegetal pole.

secretion and degradation or physical alteration of the colloidal ECM in the cleavage cavity.

From the 16-cell stage on, cleavage of the initial four tiers of cells is asynchronous. Table 1 from Guthrie and Hibbard (1919) summarizes the temporal sequence of the early cleavages in L. variegatus up to the 76-cell stage. Between the 16- and 32-cell stage, the four mesomeres divide vertically and the four macromeres divide vertically to form two tiers of eight cells. The four micromeres divide unequally in a horizontal plane to form four smaller micromeres at the vegetal pole and four larger micromeres that surround the crowded smaller micromeres. Interestingly, this period is particularly sensitive to pulse treatments with LiCl (Mitsunaga et al., 1983; Santos, unpubl.).

At the next cleavage the eight macromeres divide horizontally to form an upper Veg_1 tier of macromeres and a lower Veg_2 tier of macromeres. Twenty minutes later the Veg_1 macromeres divide vertically to form a tier of 16 cells; the mesomeres divide vertically to form 32 cells. Next the upper micromeres divide vertically into eight cells that form a ring around the four lower micromeres. Finally, the Veg_2 macromeres divide vertically to form a tier of 16 cells.

Thus at the 76-cell stage there are in the animal hemisphere 32 mesomeres (An_1 and An_2) arranged as a hemispherical sheet of cells, a tier of 16 Veg_1 cells, a tier of 16 Veg_2 cells and 8 larger micromeres that encircle the 4 small micromeres at the vegetal pole. Both the temporal and spatial patterns of cleavage during this period have an orderliness analogous to the classical spiralian eggs where subtle alterations in the sequence of cleavages are related to predictable developmental abnormalities.

The developmental fates of the several tiers of cells in L. variegatus probably are similar to those in Paracentrotus lividus originally traced by Horstadius. According to Horstadius (1939), descendents of the An_1 blastomeres give rise to the ciliated apical tuft cells at the animal pole, the oral lobe, the dorsal oral arms and the stomodaeum. Descendents of An_2 form the lateral ectoderm; and those from Veg_1 form the ventral, anal ectoderm of the larva. Descendents of Veg_2 invaginate to form the endodermal wall of the archenteron, while descendents of the four small micromeres are first localized at the invaginated tip of the archenteron from where they may differentiate into exploratory secondary mesenchyme cells (SMC) during gastrulation. Descendents of the ring of larger micromeres ingress into the blastocoel during blastulation and differentiate into the primary mesenchyme cells (PMC).

Table 1. Sequence of early cleavages of Lytechinus variegatus at 28°C[1].

TIME (hours) (minutes)	40	1 10	1 35	1 50	1 50	2 10	2 20	2 35	2 55	2 55	3 5	3 10
Mesomeres			4	4	8	16	16	16	32	32	32	32
Blastomeres	2	4										
Macromeres			4	4	4	8	8	16	16	24	24	32
Micromeres				4	4	4	4	4	4	4	8	8
							4	4	4	4	4	4
TOTAL CELLS	2	4	8	12	16	28	32	40	56	64	68	76

[1]Adapted from Guthrie and Hibbard (1919). The two columns headed 1 h 50 min and the two columns headed 2 h 55 min indicate that division of the four micromeres is completed before the four mesomeres and that the division of the sixteen mesomeres is completed before that of the upper (vegetal 1 tier) eight macromeres.

To this general outline of cell lineage of the sea urchin, one additional phenomenon is worth mentioning. Young (1958) traced the origins of the echinochrome-containing pigment cells (echinophores) that appear at early gastrulation in L. variegatus to the Veg_1 tier of cells at the 64-cell stage. In the early gastrula the echinophores appear as large, irregular cells located in the region of invagination.

When invagination is nearly completed, these cells are dark red and have increased in number. They also become amoeboid and begin to migrate within the ectodermal layer. Young observed that in Li-exogastrulae echinophores were in the blastocoelic cavity and even in the lumen of the everted gut. In his operative and chemical treatment experiments, Young observed that pigment formation occurred only in embryos that gastrulated and concluded that pigment cell differentiation was related to gastrulation in some way. Young's observations raise a number of intriguing questions regarding the synthesis and regulation of proteins involved in the differentiation of these pigment cells, their invasive, amoeboid migration throughout the ectodermal wall of the gastrula, and their migration through the basal lamina of the Li-exogastrula.

Blastulation

Between 6 to 7 h (25°C) post-fertilization and 1 h before hatching, the round, hollow, ciliated blastula rotates slowly in the perivitelline fluid within the fertilization membrane. SEMs of fractured embryos at this stage (Fig. 2) show that the cells of the blastular wall are flat at their outer, hyaline ends and round at their inner, basal ends. Lenticular spaces occur between the cells that are attached to one another just below their basal ends by short, spike-like cellular processes (Fig. 2a) that differ morphologically from the attachment processes at the 16-cell stage. We observed no ECM in the blastocoel and no basal lamina lining the blastular wall. The rough texture of the basal ends of the cells suggests that individual cells have a surface coating that is more pronounced in embryos fixed in DG-CPC. Extracellular vesicles (0.1 to 1.0 μm diameter) first appear on the blastular wall at the animal pole (Fig. 2b). Then the inner ends of the cells in this region begin to spread, flatten, and form filopodia that extend over the surfaces of neighboring cells (Figs. 2c & d). The aerial extent of this cellular activity may be a morphological correlate associated with the early differentiation of the apical plate derived from the An_1 cells.

After it hatches (7 to 8 h), the blastula begins to elongate along the animal-vegetal (A-V) axis. The embryo becomes oval shaped with the narrow end of the oval at the vegetal pole. At the same time the blastular wall thickens in the vegetal hemisphere and becomes thinner in the animal

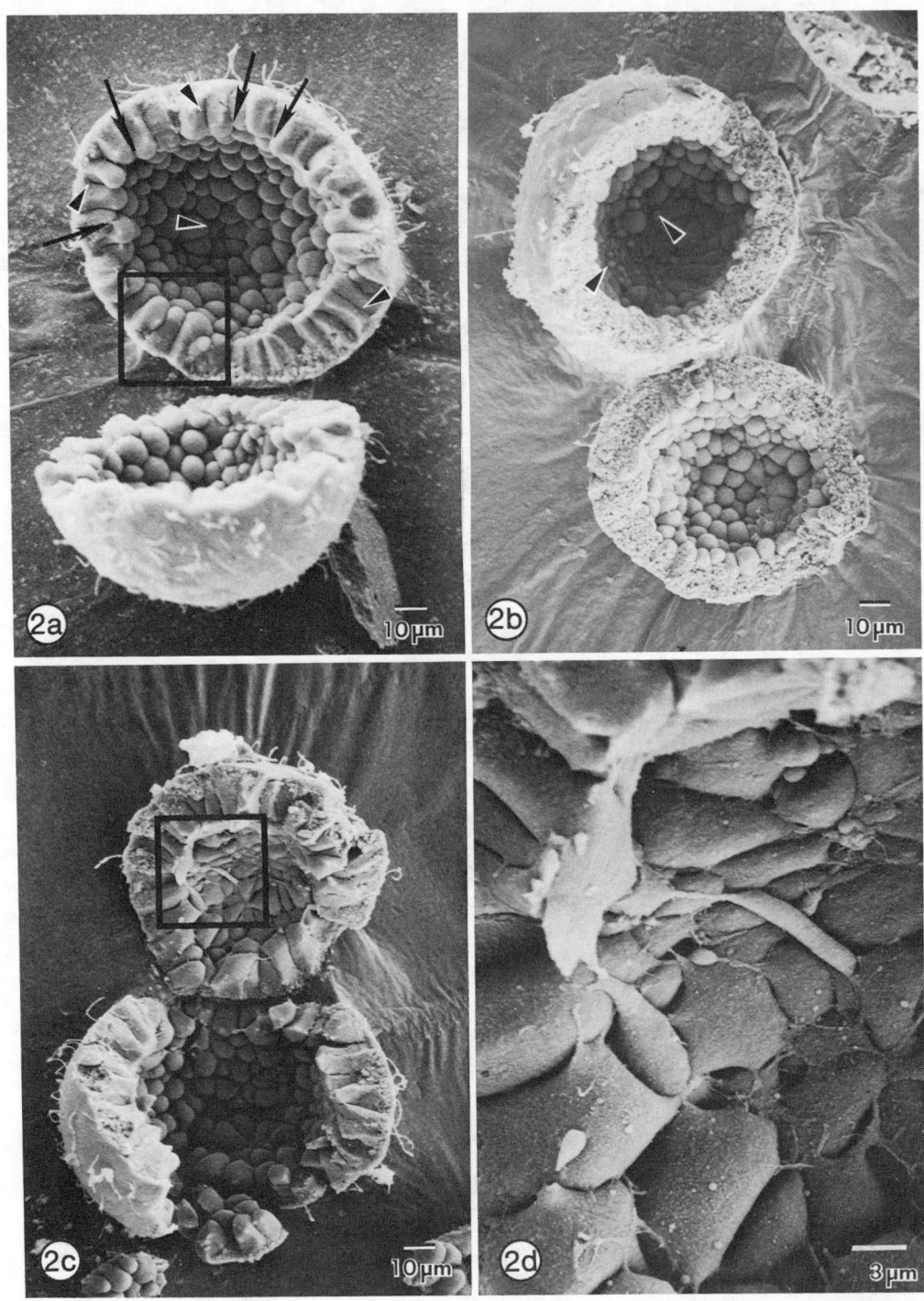

Fig. 2.   Fractured blastulae (7 h) one hour before hatching from the fertilization membrane. Embryos fixed in DG-RR (a & c) or DG-CPC (b). The blastomeres are rounded at their basal, blastocoelic ends. Short, spike-like attachment processes (black arrows) occur between neighboring cells. The boxed area in (a) shows the shapes of neighboring cells and how the angular cells are packed in the blastular wall. Spaces or clefts (arrowheads) between the cells in (a) may be shrinkage artifacts or spaces once occupied by the edges of adjoining cells. In a slightly older embryo (b) the cells of the blastular wall at the animal pole have begun to spread and converge toward an area where the wall has numerous vesicles (arrowheads) that range from less than 0.1 to 0.8 μm in diameter. Fine extracellular fibrils are common in this region. The spreading and flattening of the blastocoelic ends of the cells at the animal pole are more visible in (c). The lamellar spreading and filopodial activity of the cells in the square in (c) are shown at a higher magnification (2500X) in (d).

hemisphere. This results in a marked decrease in the diameter of the blastocoel in the vegetal hemisphere and the appearance of a pair of fan-shaped images composed of curved optical streaks in the living embryos. When Okazaki et al. (1962) first projected these fan-shaped figures on the developmental map of Horstadius (1939), the optical figures coincided with the An_2 and Veg_1 regions. The line that connected the apices of the two fans coincided with the edge of Veg_1 next to An_2. Following PMC ingression, a ring of PMCs developed at the level of the apices of the two fans, while the ventro-lateral branches of the PMCs extended toward the animal pole along the two lateral areas of the An_2 portion of the blastular wall.

In L. variegatus these optical fan-shaped patterns begin before hatching, persist through blastulation and then disappear with the flattening of the vegetal plate at the onset of gastrulation. We now show with SEM the cellular aspects of these optical patterns in the living embryo (Figs. 3 & 4). Henceforth we will refer to these cellular patterns as the Okazaki patterns.

SEMs of fractured premesenchyme blastulae (9 h) in Figures 3a and d reveal that the basal ends of the cells of the blastula elongate toward the vegetal pole and converge toward the bilaterally arranged apices of the Okazaki pattern. The inner ends of some cells of the blastular wall may extend toward the vegetal pole over as many as five cells when viewed in section (Fig. 3c). At the center of convergence, the apex of an Okazaki pattern, the basal ends of the cells are covered with a mat of extracellular fibers that may anchor the lamellopodial extensions that converge in this region. Figure 3d shows the discontinuity between the cellular elongation of the An_2 region and an edge of the apical plate region of the blastular wall. Figure 3a shows another view of this discontinuity in the shapes of the cells of the blastular wall near the animal pole. Thus, even

14 MORRILL & SANTOS

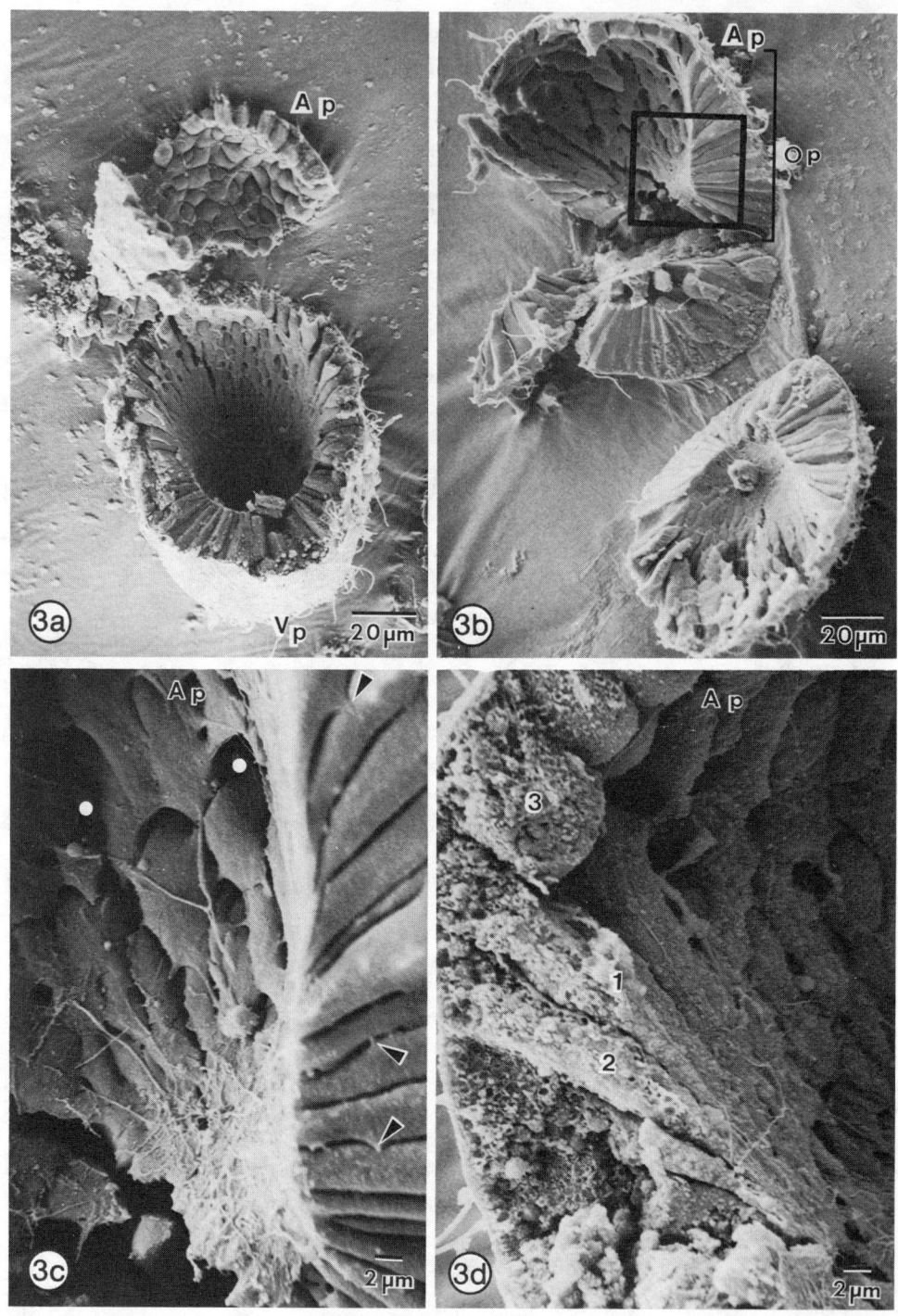

Fig. 3. Fractured, hatched, premesenchyme blastulae (9 h). In (a) and (b) are two different views of the pear- or ovoid-shaped blastulae. In (a) the horizontal fracture just below the animal pole (Ap) exposed the spreading animal pole cells. In contrast to the non-directional spreading of the cells at the animal pole, the blastocoelic ends of other cells in the animal hemisphere of the blastular wall exhibit a polarized elongation toward the vegetal pole. Cellular extensions of the lateral wall overlap one another like shingles on a roof, as shown in (b) where the embryo was fractured along the animal-vegetal axis. In (b) the distinct Okazaki pattern (Op bracket) and variation in thickness of the wall are evident. In this embryo the PMCs have just begun to elongate and extend into the blastocoel at the vegetal pole. At a higher magnification (2500X), (c) the boxed area in (b), shows a mat of fibrils and numerous extracellular granules at the center of convergence. Spaces (white dots) are present between the cellular extensions. Lateral margins of the cells have short projections (arrowheads) that attach to the adjoining cells. (d) is a fracture near the animal pole of an embryo fixed in DG-CPC at the same magnification as (c) fixed in DG-RR, and illustrates the differences in texture of the surfaces of the blastocoelic sides of the cells with the two fixatives. In (d) the elongated cells 1 and 2 are 35 to 45 µm long as compared to cell 3 that is 15 µm long.

before mesenchyme cell ingression, the future bilateral symmetry of the embryo is reflected in the regional patterns in cellular shapes in the blastular wall.

The PMCs begin to ingress between the 9th and 10th hour at the vegetal pole of the ovoid blastula (Fig. 4). At this stage, a thin, basal lamina has begun to line the inner wall of the blastula; coated fibers and granules are present on the inner surface of the basal lamina in the An_1 and An_2 regions of the blastular wall. In the vicinity of the ingressing PMCs, the basal lamina and the fibrous ECM is not apparent (Fig. 4d). In this region many extracellular vesicles appear on the basal surfaces of the presumptive endodermal and vegetal plate cells (Figs. 4e & f). The SEM micrographs of the four fractured embryos in Figure 4 (from the same batch of eggs) outline the morphological features of PMC ingression that occurs over a 30-min period in L. variegatus. It appears that in L. variegatus, PMC ingression differs markedly from that in Lytechinus pictus (Katow & Solursh, 1980) which has no distinct Okazaki pattern during blastulation. After approximately 30 PMCs have ingressed and accumulated at the vegetal end of the blastocoel, they begin to disperse and migrate toward the animal pole. The migratory behavior of individual PMCs is quite variable during this period that lasts about one hour. During this time the vegetal plate flattens and the blastocoel enlarges at the vegetal pole as the presumptive endodermal and vegetal plate cells (Veg_2 and Veg_1) shorten.

Gastrulation

As blastulation ends and gastrulation begins, SEMs of fractured embryos (Fig. 5a) show regional variations in the

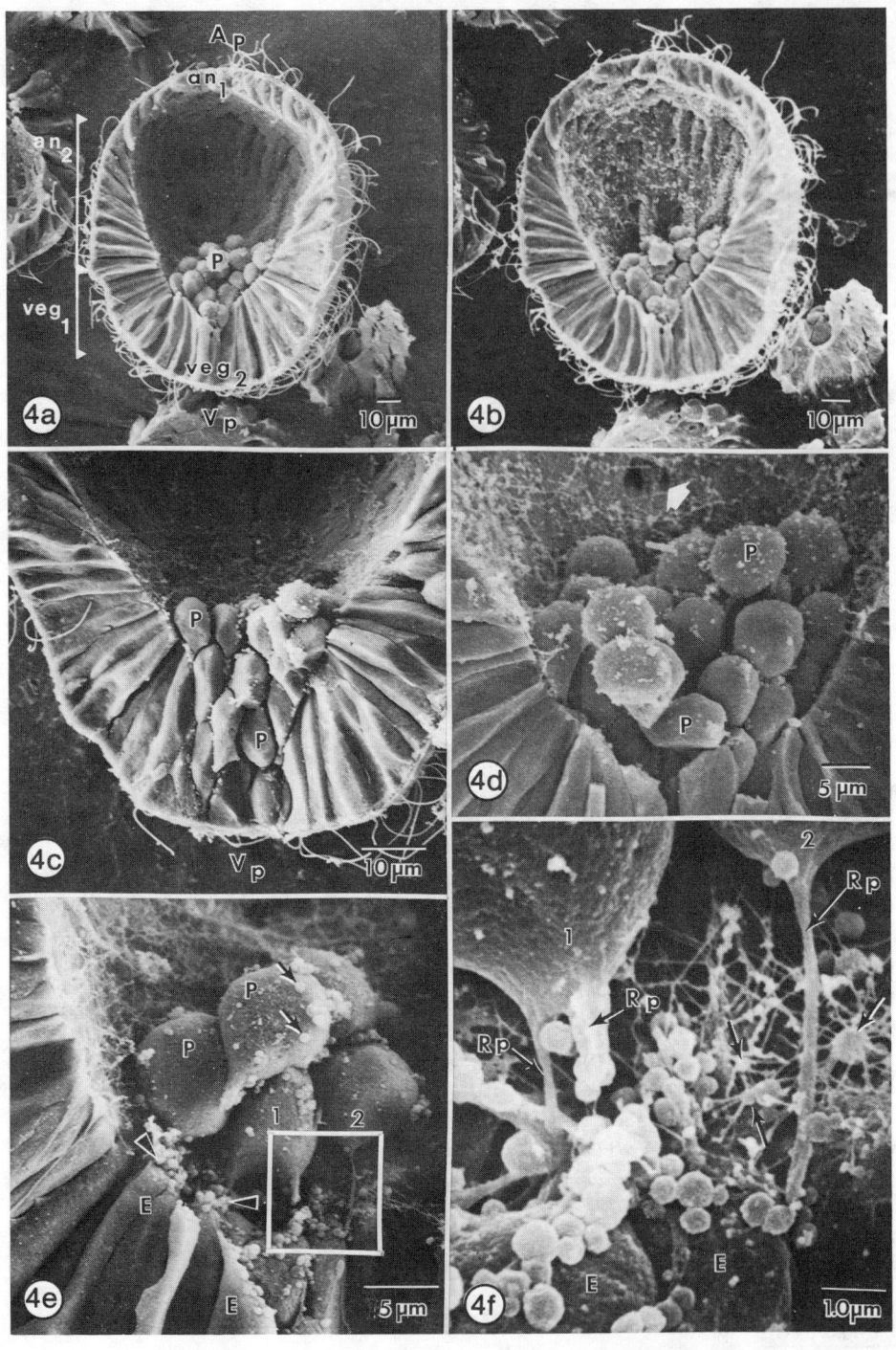

Fig. 4. Mesenchyme blastulae (10 h) fractured along the animal-vegetal axis. (a) and (b) show a whole embryo at 500X magnification to illustrate differences in a mono image when photographed at different working distances and with different secondary electron detector conditions. In (a) the working distance (w.d.) was 8 mm; only the chamber S.E. detector was on; gamma attenuation was on. In (b) the w.d. was 15 mm; both the chamber and column S.E. detectors were on; gamma attenuation was off. Many of the features of the blastocoelic wall in (a) are markedly different in (b). The flatter and wider image of the embryo in (b) is not apparent when the embryo is viewed in stereo.

In (a) and (b) nearly all the PMCs (P) have ingressed. The embryo is still ovoid and the fan-shaped Okazaki pattern of cells of the blastular wall is present. The developmental fate map of Horstadius (1939) is indicated in (a). In (c) a mat of coated fibrils is present on the blastular wall in the vicinity of the ingressing PMCs. The PMCs have extracellular vesicles on their surfaces. (d) shows an increase in the amount of coated fibers in the vicinity of the PMCs after PMC ingression is nearly completed. A basal lamina is now present and lines the blastocoel except where there are holes (arrow). Following ingression, the PMCs begin to migrate along the blastular wall (e). At this stage the PMCs are rounding up; some have elongated, retractile cellular processes at their posterior ends. Numerous vesicles (arrowheads) appear at and near the blastocoelic ends of presumptive endodermal cells (E) at the vegetal pole. Similar vesicles (arrows) occur on the PMCs. The square in (e) is enlarged to 10,000X in (f) to show the retraction processes (Rp) of PMCs 1 and 2, the extracellular vesicles and extracellular fibers at the tips of the endodermal cells. Fibers radiate from several vesicles (arrows).

shape of the blastular wall and an abundance of ECM in the blastocoel (on the animal pole side) of the ring of PMCs that become located at the interface between the An_2 and the Veg_1 regions of the embryo. At this stage a few PMCs have yet to associate with either the mesenchymal ring or its paired ventrolateral branches (Fig. 5a). Interestingly, the extent (in the animal pole direction) of these PMCs is a region where the dense meshwork of coated fibers along the blastular wall thins abruptly (Figs. 5a & c). These morphological patterns of the wandering PMCs and the coated extracellular fibers in L. variegatus are analogous to the FITC-ConA binding patterns in the early gastrula of L. pictus (Katow & Solursh, 1982).

At this stage the surfaces of the PMCs are covered with vesicles and granules of various sizes; extracellular fibers are attached to the PMCs as well as the basal lamina. The basal lamina is perforated with irregular holes (Fig. 5e) that are probably caused by the shrinkage of the ectodermal cells during preparation for SEM. While the basal lamina appears amorphous at magnifications up to 15,000X, its macromolecular complexity begins to emerge at 30,000X (Fig. 5f). At this magnification the basal lamina appears to be composed of chain-like arrays of globular subunits whose blastocoelic sides are coated with small granules and thin fibrils that differ morphologically from the coated fibers of the blastocoel proper. Figure 5f illustrates how the basal lamina proper may be contaminated by colloidal, blastocoelic

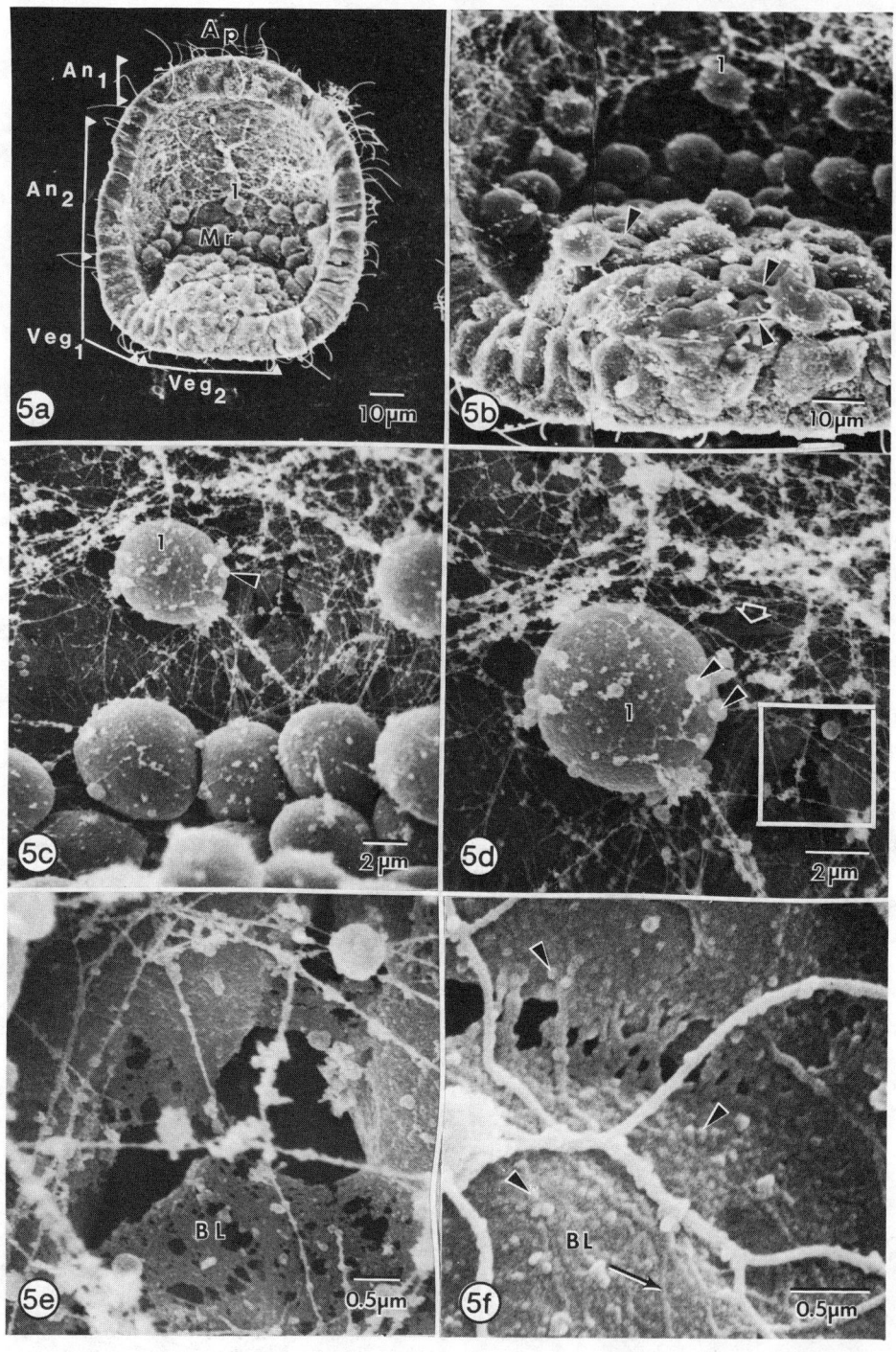

Fig. 5. Early gastrula (12 h) fractured along the A-V axis. At this stage a dense mat of coated fibers is attached to the basal lamina in the animal hemisphere (a). Near the equator the mat of fibers becomes thinner. Most of the wandering PMCs have begun to form the mesenchymal ring (Mr) near the sharp bend in the blastular wall near the vegetal pole. In (b) the blastocoelic side of the invaginating vegetal plate is composed of flattened cells with lamellopodia and filopodia (arrowheads). Numerous spaces occur between these cells; these cells are covered with vesicles 0.4 to 0.5 μm in diameter.

(c) and (d) show PMC 1 in (a) and (b) at 3000X and 5000X. Small granules and vesicles are on the cell's surface; extracellular fibers are attached to the surface of the cell; a filopodium (white outline arrow) of the cell extends along the basal lamina. The basal lamina (BL) and coated fibers in the square in (d) are shown at 15,000X in (e). However, the morphological features of the basal lamina only become apparent at 30,000X in (f) that shows the blastocoelic surface of the lamina to be covered with small granules (arrowheads) and fine fibrils (arrow) in addition to the larger coated fibers of the blastocoelic cavity.

substances such as coated fibers that collapse onto the basal lamina with preparatory procedures for SEM, TEM and biochemical analyses.

The basal lamina lines the blastocoel except at the vegetal pole where it appears to be absent in the vicinity of the inner end of the incipient archenteron wall. Here, there are motile cells with filopodia and lamellopodia (Fig. 5b). These cells may be derived from the smallest micromeres of the 32-cell stage and are the progenitors of the SMCs.

Between 12 and 13 h post-fertilization the endodermal plate proper begins to invaginate (Figs. 6a, b, & c). The invaginating endodermal cells as well as the adjoining ectodermal cells of the vegetal plate exhibit incremental changes in three dimensions related to their positions along the curvature of the blastular wall. Regretfully, SEM alone does not allow us to say whether the bending of the blastular wall is a totally passive response to the active pulsatory activity of the endodermal cells at the tip of the archenteron. Nevertheless, the SEM micrographs in Figure 6 indicate that planar models of archenteron formation in the sea urchin may be both simplistic and misleading. Modeling of this process in the future may require the use of analytical stereoimaging techniques.

Other features associated with this early phase of invagination in Figure 6 include the compression of the mesenchymal ring cells between the archenteron wall and blastular wall and the short cilia of the endodermal cells beneath the lip of the blastopore. The 'torn' nature of the hyaline layer lining the archenteron is associated with the contraction and retraction of the distal ends of the invaginating cells that have morphologically complex retractile processes at their distal ends (Figs. 6d & e). The lateral

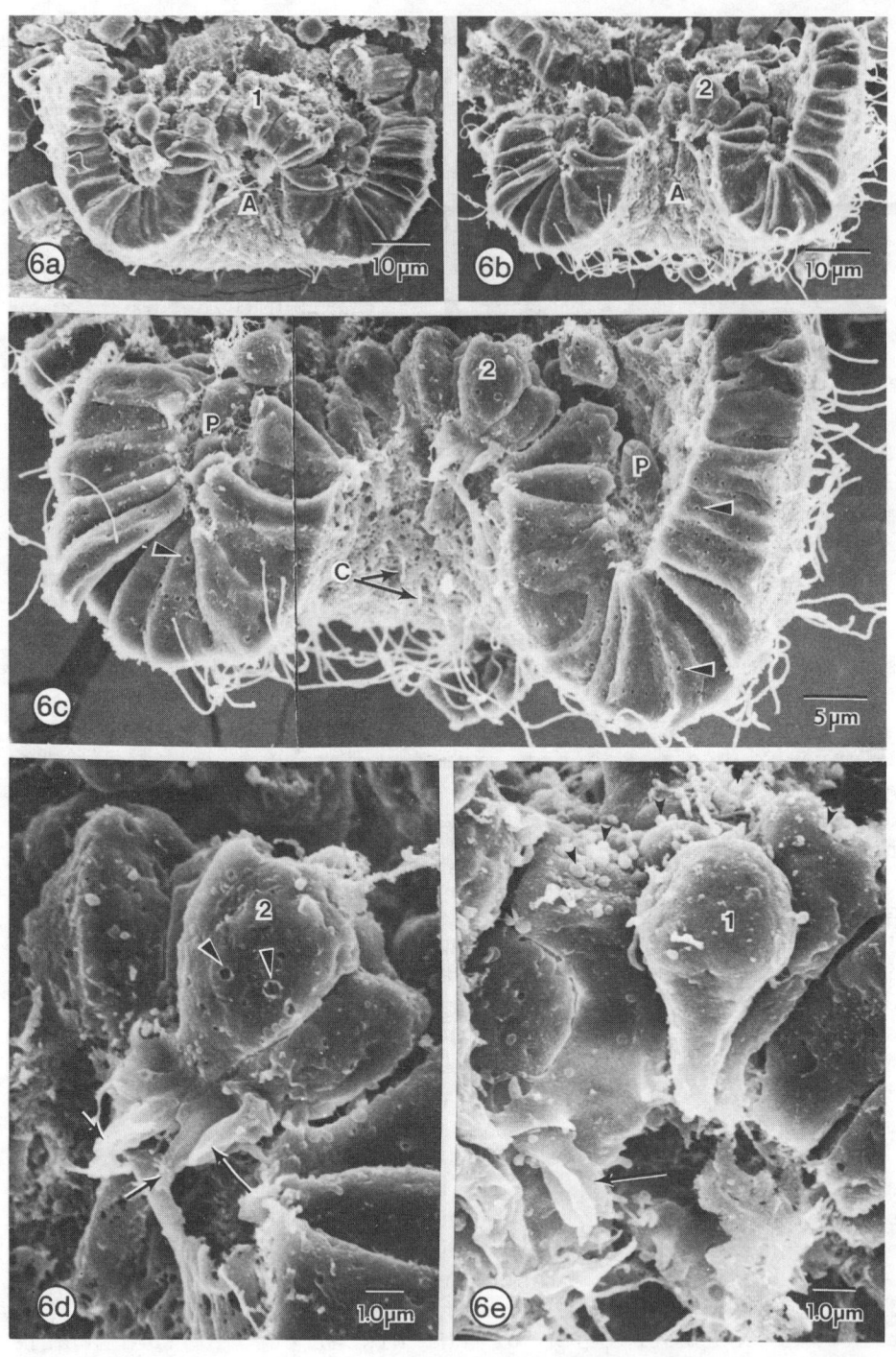

Fig. 6. An early gastrula (13 h) fixed in MW-fixative and fractured along the A-V axis. (a) and (b) show the arrangements and shapes of the cells of the invaginating wall of the archenteron (A) and the blastular wall near the vegetal pole in the two halves of the same embryo. At 2000X (c) the lateral sides of the angular cells of the blastular wall have numerous punctate holes or pores (arrowheads); the archenteron is lined with a perforated extracellular coating through which short cilia (C) extend. The motile cells at the inner end of the archenteron have a variety of shapes and retractile processes (arrows) shown in (d) and (e). In (e) numerous vesicles (black arrowheads) are present on blastocoelic ends of the cells at the tip of the archenteron.

borders of the cells of the blastular wall and the archenteron adhere closely except at the tip of the archenteron where clefts or spaces occur between the cells.

Another overt morphological feature we have observed only at this stage are punctate-like holes or pores (arrowheads, Figs. 6c & d) on the lateral sides of the cells of the wall of the archenteron and the vegetal region of the blastular wall. Since these pores appear to be rimmed by a raised extension of the cellular membrane, they may reflect some kind of exocrine, secretory activity. On the other hand these rimmed pores may be fractured holocrine, secretory vesicles similar to those shown by black arrowheads in Figure 6e.

For a brief period during the earliest phases of invagination of the archenteron there appears in the center of the blastopore a button-like morphological feature (Fig. 7a) that is probably the group of cells that will become the SMCs. The outer ends of these cells and the invaginating endodermal cells have short, stubby cilia. The epigenetic origins of these short cilia as well as the ciliary patterns of the post-gastrula larva have yet to be studied in detail.

Deciliation of the embryo is simple enough. A brief osmotic shock when the embryos are treated with 2x sea water results in the dehiscence of the cilia at the base of the ciliary shafts. Interestingly, this treatment may also remove the hyaline layer thereby exposing the outer surfaces of the cells of the blastular wall and the surface geography of the polygonal cellular patterns (Fig. 7b). Figure 7b illustrates the potential of SEM of whole embryos for supplementing light microscopical analyses (i.e., Honda et al., 1983) of surface patterns of embryos undergoing blastulation and gastrulation.

By the mid-gastrula (13 to 14 h) the tip of the conical archenteron extends nearly half way through the blastocoel (Figs. 7c & d). The SMCs at and near the tip of the archenteron are motile and exhibit lamellopodial and filopodial activity. Some filopodia extend and attach to the lateral

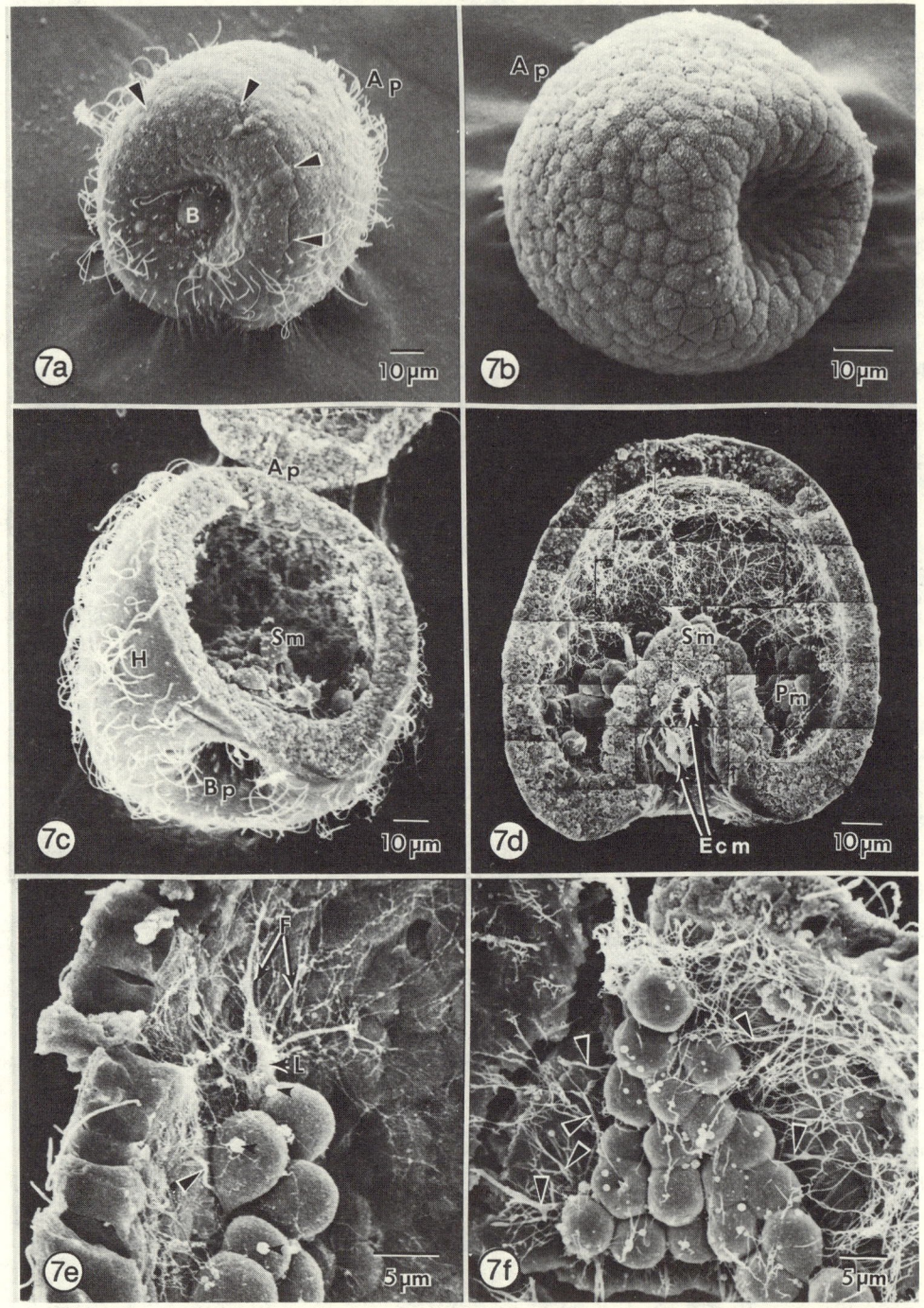

Fig. 7. External and internal views of early and mid-gastrulae (13 to 14 h). (a) external view of an early gastrula with a bulbous (B) grouping of the descendents of the smallest micromeres in the center of the invaginating endodermal cells. The cilia of the invaginating endodermal cells have shortened. A large portion of the hyaline layer has been rubbed off and the cilia broken on the upper side of the embryo to reveal a shrinkage artifact that delineates the lateral limits of the vegetal plate (arrowheads). (b) a mid-gastrula deciliated by treating the embryo for 20 sec in 2X sea water before fixing in MW-fixative.

(c) a mid-gastrula fixed in DG-CPC and fractured along the A-V axis. The hyaline layer (H) continues over the lips of the blastopore (Bp) and lines the wall of the archenteron. Secondary mesenchyme cells (Sm) at the inner end of the archenteron have filopodia and lamellopodia. Heavily coated extracellular fibers are confined primarily to the animal half of the blastocoel. These features in (c) and the ventrolateral branches of the PMCs (Pm) are more visible in (d) which is a reduction of a photomontage taken at 5000X. Large amounts of extracellular material (Ecm) are present in the lumen of the archenteron.

(e) and (f) are two views of a ventrolateral branch of PMCs. Only the lead cell of the branch in (e) has an exploratory lamellopodium (L) that branches into several filopodia (F) whose tips attach to the basal lamina. The other cells of the branch are attached to the basal lamina by thin cellular processes (arrowheads) and to each other and to the mesenchymal ring by a syncytial cable of cytoplasm (not shown). Typically, each of the 18 cells of each ventrolateral branch have a few prominent vesicles (black arrowhead) on their surfaces. The small triradiate spicule that is present near the base of the ventrolateral branch at this stage is not visible because it lies on the basal laminal side of the PMCs. (7b, courtesy of Matthew Wahl, unpubl.; 7c, courtesy of Shawn Murphy, unpubl.; 7e, courtesy of Deni Galileo, unpubl.).

walls. The relatively large, intercellular spaces that occur between the SMCs in this region may provide the means for reducing the hydrostatic pressure in the blastocoel - a pressure created by the osmotic nature of the colloidal extracellular substances in the blastocoel and the reduced permeability of the blastular wall associated with the development of septate junctions. At the same time mucous-like extracellular material accumulates in the lumen of the archenteron (Fig. 7d). This material coupled with the constriction of the blastopore may be causally involved in the inward elongation of the archenteron. Thus, following its initial invagination, the second phase of archenteron formation may involve not only the attachment and contraction of the filopodia of the SMCs but also a release of the hydrostatic pressure in the blastocoel and an increase in the hydrostatic pressure in the archenteron. The combined physical forces that result from these activities could be reflected in the elongation of the endodermal cells that become spindle-shaped in surface profile.

By mid-gastrula the mesenchymal ring is located near the vegetal pole and the ventrolateral branches of PMCs have extended to the equator of the embryo. Typically, each ventrolateral branch consists of 18 PMCs arranged as a

24 MORRILL & SANTOS

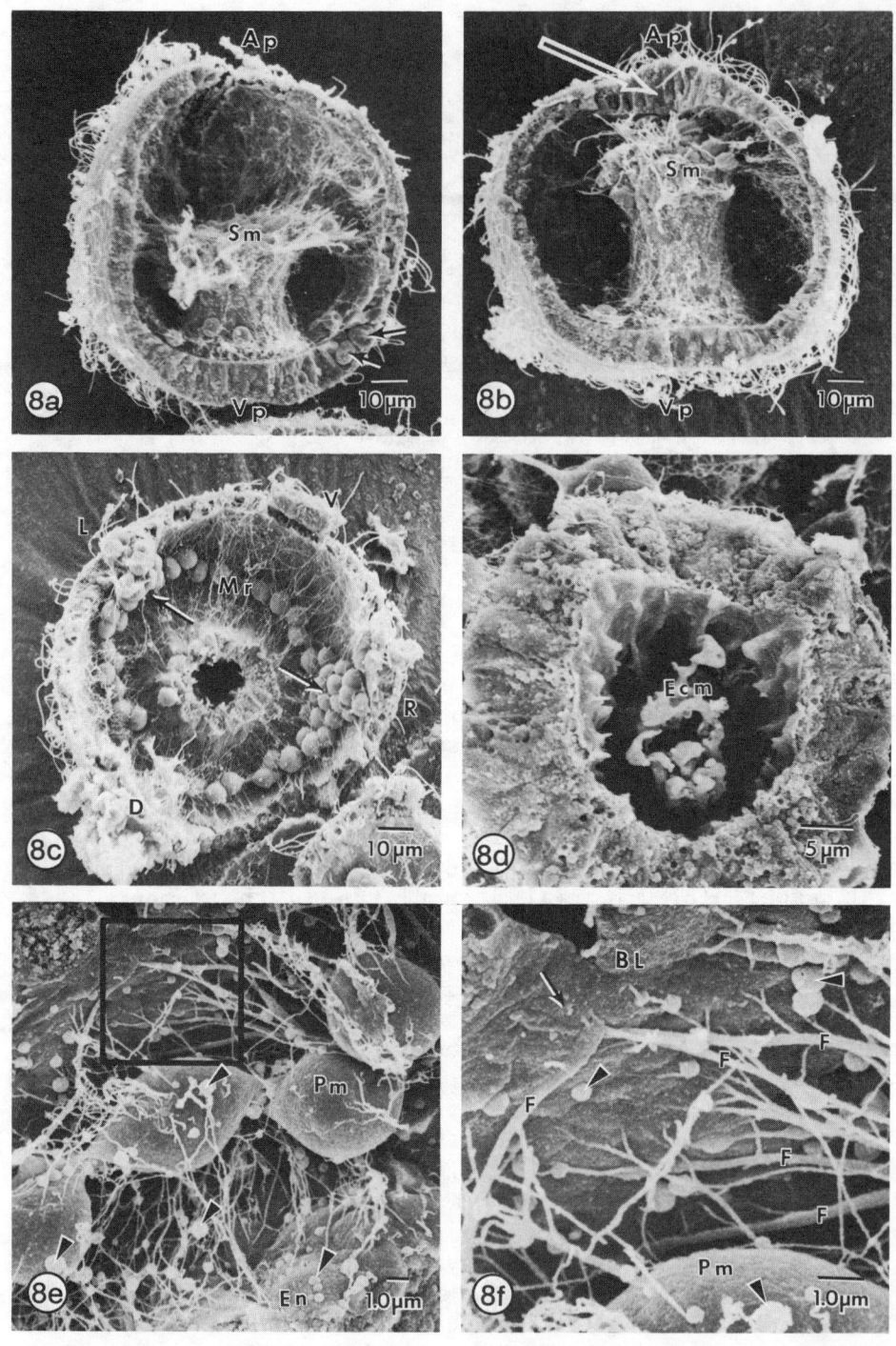

Fig. 8. Late gastrulae (15 h) fractured along the A-V axis (a & b) and transversely to show the vegetal hemisphere (c), a cross section of the archenteron (d) and details of the attachment of the ring of PMCs to the blastocoelic wall near the vegetal pole (e & f).

The gastrula in (a) is an intermediate phase between initial invagination of the endodermal plate in Figure 7d and the fully elongated archenteron in Figure 8b. In this gastrula the SMCs (Sm) at the tip of the archenteron extend filopodia and lamellopodia toward the lateral wall of the blastocoel. In the animal hemisphere above the tip of the archenteron there is still an abundance of blastocoelic fibers. When the cellular processes of the SMCs concentrate at the animal pole in (b) the cells of the blastular wall are elongated to form a thickened apical plate (white outline arrow). Near the right hand margin of the vegetal plate in (a) two cells (arrows) in the blastular wall are round; either these cells are dividing or they are pigment cells.

In the fractured gastrula in (c) the bilateral symmetry is evident in the shape of the embryo, the mesenchymal ring (Mr) and the position of the two ventrolateral branches (arrows). (V, ventral side; D, dorsal side; R, right side; L, left side). Even the archenteron wall and lumen (d) in cross section exhibits a lateral symmetry. The thick extracellular coat lining the inner wall of the archenteron is thrown into folds while the lumen has a dried coagulum of extracellular material (Ecm).

The mesenchymal ring is composed of PMCs connected to one another by a syncytium between the cells. The entire ring is anchored to the blastular wall by a complex array of unbranched and branched filopodia that extend from the individual PMCs into the basal lamina (BL) (e & f). The square in (e) shows the location of (f) (7000X). The arrow in (f) points to where three filopodia (F) extend into the basal lamina. In both (e) and (f) extracellular vesicles (arrowheads) 0.4 to 0.5 μm in diameter are common on the surfaces of the basal lamina (Bl), PMCs (Pm), and endodermal cells (En) and on the thin, blastocoelic fibers that extend from the wall of the archenteron to the PMCs and the basal lamina.

triangular array of cells that are anchored to the gastrular wall by filopodia (Figs. 7f & 9b). The lead or apical PMC of each ventrolateral branch has a large, arborescent lamellopodium (Fig. 7e). If the PMCs of the ventrolateral branches are joined to one another by a syncytial sheet of cytoplasm, then the extent (towards the animal pole) of the ventrolateral branches may depend in part on the extent to which these PMCs become anchored to the gastrular wall.

Shortly after the mid-gastrula phase shown in Figure 7d, the filopodial activities of the SMCs at the tip of the archenteron increase markedly. At first the filopodia attach to the lateral walls of the gastrula (Figs. 8a & 9a); then they appear to shift their attachments toward the animal pole near the end of gastrulation (15 h) when the thickened apical plate ectoderm becomes apparent (Fig. 8b). SEMs of fractured embryos during this last phase of gastrulation suggest that as the SMCs and archenteron advance toward the animal pole the ECM in the animal hemisphere decreases. At the same time, an ECM develops between the mesenchymal ring and the wall of the archenteron near the vegetal pole (Figs. 8c & e). The late gastrula has an overt bilateral and dorsoventral symmetry when viewed in cross section (Fig. 8c). Even the archenteron may appear bilaterally symmetrical (Fig.

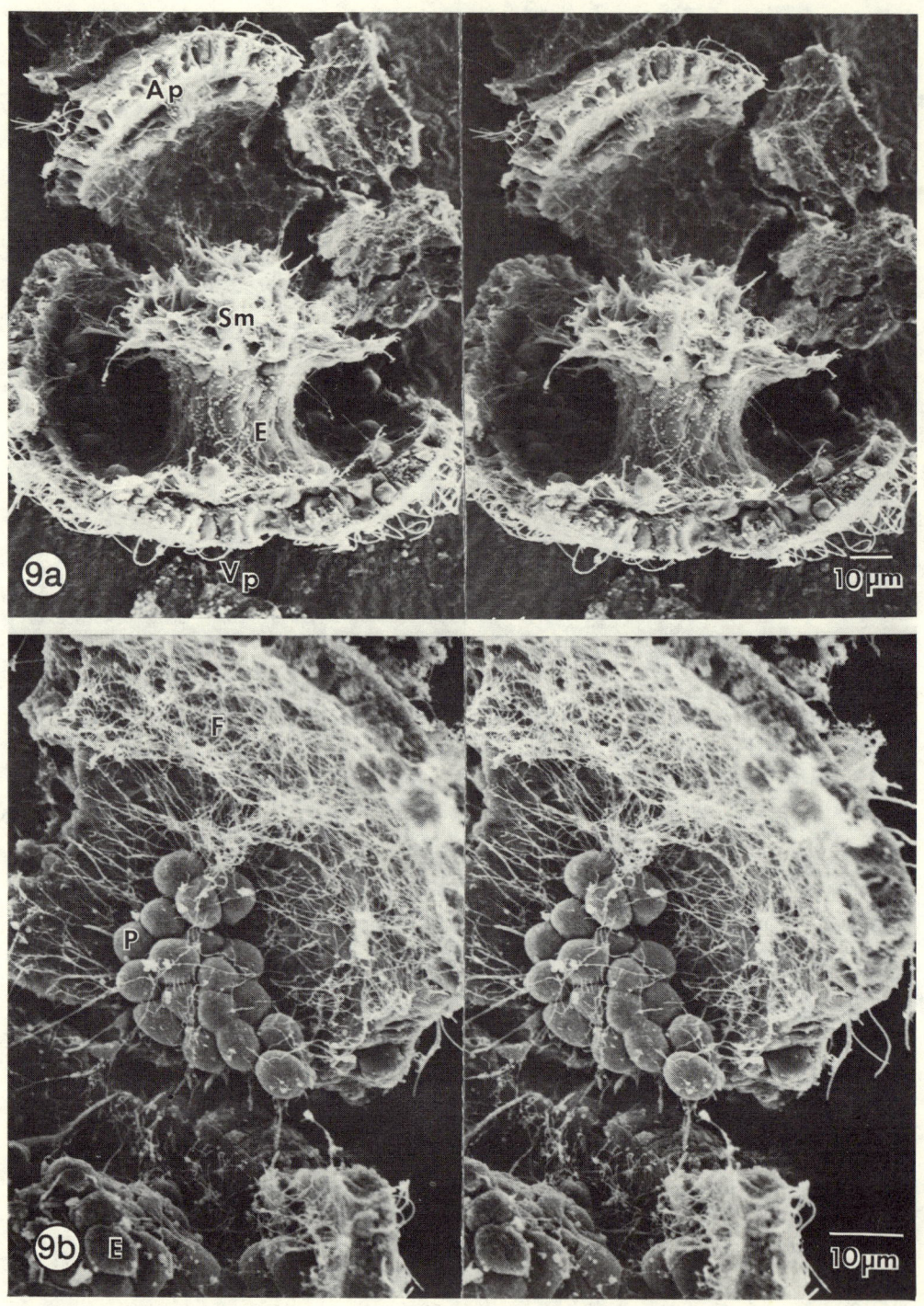

Fig. 9. Stereo pairs of (a) a fractured late gastrula at 500X and (b) a ventrolateral branch of PMCs and blastocoelic fibers attached to a piece of the lateral wall of a gastrula at 1000X. These two pairs of SEM micrographs illustrate the utility of stereo images in determining the true shapes and positions of structures relative to one another. In (a) the complexity of the cellular extensions of the SMCs (Sm) is particularly striking in the stereo image. In (b) the roundness of the PMCs, the curvature of the gastrula wall and the three dimensional nature of the meshwork of blastocoelic fibers is enhanced in the stereo image. Ap, animal pole; E, endodermal wall of archenteron; F, blastocoelic fibers; P, primary mesenchyme cells; Sm, secondary mesenchyme cells; Vp, vegetal pole.

8d). At this stage the mesenchymal ring appears anchored to the outer wall by numerous filopodia that extend from the PMCs to and actually beneath the basal lamina (Figs. 8e & f).

Lithium-Exogastrulae

L. variegatus eggs treated chronically or pulsed between the 16- and 32-cell stage develop into a variety of exogastrulae and abnormal larvae. In Figure 10 we show three views of one of the more common types of exogastrulae where the archenteron is everted and swollen. There is an overt vegetalization and radialization of these embryos that have more than two spicule-forming centers (not shown).

While the analysis of the development of these radialized exogastrulae is beyond the scope of this paper, radialization in these embryos may be related to the abolishment of a bilateral, oral field gradient of cytochrome oxidase in the ectoderm of blastulae and earlier stages described by Czihak (1971) and an alteration in the dorsoventral distribution of the Spec 1 and Spec 2 proteins described by Klein (this volume). Less clear is whether vegetalization results from the suppression of animal differentiation or the enhancement of the vegetal processes. However, in L. variegatus the onset of endodermal evagination and possibly vegetalization is preceded by the presence of an abnormally large quantity of non-fibrous ECM in the blastocoel before and during PMC ingression (Murphy, unpubl.).

Regardless of its origins, a 28-h exogastrula may have five distinct morphological regions (Fig. 10a): 1, an ectodermal region with long cilia and a smooth hyaline layer; 2, a transitional region where a constriction occurs between the ciliated ectoderm and Region 3, where the cells of the everted, endodermal wall bear short cilia that protrude through circular gaps in the stretched extracellular coat, presumably the hyaline layer; 4, a region near the tip of the everted, endodermal wall where the extracellular coat is less developed; and 5, an aggregation of small cells and extracellular mucous external to the tip of the everted endodermal

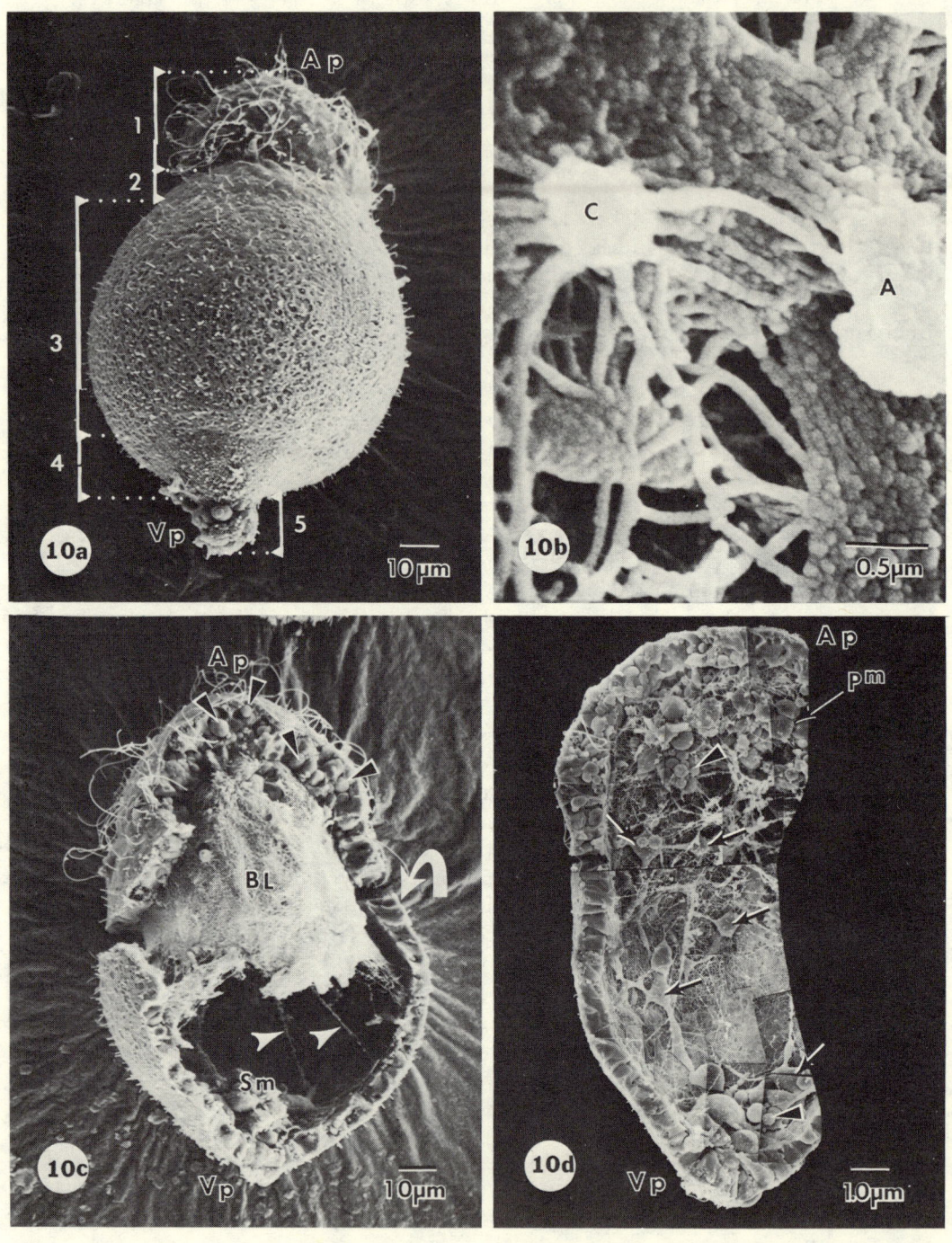

Fig. 10. Vegetalized, radialized exogastrulae (28 h) induced by chronic treatment with 26 mM LiCl. (a) Externally a typical exogastrula has five morphological regions delineated by brackets. In Region 1 the ciliated ectodermal cells are covered with the smooth hyaline layer; 2 is a transitional region where the cilia of the cells are shorter; 3 is the main, swollen portion of the wall of the evaginated endoderm whose cells have short cilia that extend through circular gaps in the stretched hyaline layer; in 4 the hyaline layer is less developed near the tip of the evaginated endoderm; 5 is a cluster of small cells, presumably SMCs, coated with a mucous-like secretion. At 30,000X (c) the stretched hyaline layer over the endodermal wall in (a) appears to be composed of strands of spherical subunits. Some strands converge on the bases of the ciliary shafts (C). Scattered on the outer surface of the stretched hyaline layer are aggregates (A) of material.

The fractured exogastrula in (c) has a thick, fibrous, basal lamina (BL) that has pulled away from the inner wall in Regions 1 and 2 and is thinner in Regions 3 and 4. The ectodermal wall near the animal pole has many small, spherical bodies (arrowheads) that may be cells. Beneath the tip of the evaginated endoderm is a cluster of cells that may be SMCs (Sm). Coated fibers (white arrowheads) span the blastocoelic cavity. The curved white arrow marks the position in the outer wall of the reversed Okazaki pattern of cells.

(d) This photomontage of a fractured exogastrula reduced from 5000X shows the morphologies and organization of the cells in the wall of the embryo and the cells and extracellular fibers in the blastocoelic cavity. In addition to the spherical PMCs (Pm) and smaller spherical cells (arrowheads) beneath the ectodermal wall, several flattened cells (arrows) have long, cytoplasmic processes that extend along the basal lamina lining the wall of the everted, endodermal cells in Regions 2 and 3.

wall. A high magnification (30,000X) SEM of the stretched hyaline layer over Region 3 (Fig. 10b) shows that this layer is composed of individual strands of spherical subunits. Beneath this layer is a fibrous network or apical lamina (Spiegel & Spiegel, 1979; Hall & Vacquier, 1982) that is especially visible in gaps in the hyaline layer.

When whole exogastrulae are fractured by plucking, the fracture may expose the outer surface of the basal lamina (Fig. 10c) or expose the entire inside of the embryo as a hemi-section (Fig. 10d). The SEMs of these fractured embryos show that the basal lamina is thickest toward the animal pole and becomes thinner toward the vegetal pole of the reversed Okazaki pattern (white arrow, Fig. 10c) at the margin of Region 2, the transitional region. Coated fibers form a fibrous mat on the inner surface of the basal lamina as well as forming extensive networks throughout the blastocoel. Round PMCs are located at both ends of the blastocoel along with smaller cells and spherical bodies that appear to originate in the ectodermal wall of Region 1. Some of these smaller cells may be echinophores that occur in these regions in live embryos. Finally, a variable number of flattened cells with long, branching, cellular processes occurs along the inner wall of the basal lamina of the everted endodermal wall of Region 3 (arrows, Fig. 10d). The processes of these cells appear to be oriented along the A-V axis of the wall as well as circumferentially. The overall impression from these SEMs is that in the radialized exogastrula the normal patterns of extracellular substances of the inner side of the blastular wall, the basal lamina and the blastocoel, are markedly altered. The result is that the normal morphogenetic patterns of PMCs and other cells are not established.

CONCLUDING REMARKS

For those intimately familiar with the historical and contemporary literature on the early embryogenesis of the sea urchin, the implications of our SEM overview of blastulation and gastrulation are self-evident. Less evident is the potential for incorporating SEM in the various types of microscopical and biochemical analyses of form and function during this period of development.

While high resolution SEM and stereoimagery can provide new views of old observations, descriptive secondary electron images need to be correlated with other descriptive and experimental methodologies. In particular, the elegant immunofluorescent monoclonal and polyclonal antibody studies (Spiegel et al., 1980; Spiegel & Burger, 1982; Spiegel et al., 1983; Wessel et al., 1984; McClay, this volume) and lectin-binding studies (Katow & Solursh, 1982; Spiegel & Burger, 1982) have set the stage for high resolution, ultrastructural, cytochemical marking experiments where colloidal gold is linked to cytochemical probes. The beauty of colloidal gold as a marker and tracer is that it may be visualized with both light and electron microscopy (reviewed by DeMey, 1984).

By combining gold particles of different sizes with other markers (i.e., antibodies and lectins), multiple marking of the same specimen is possible. Furthermore, three-dimensional SEM patterns of individual gold particles may be mapped quantitatively with either secondary or backscatter electron image modes. Thus, the spatial distribution and effects of microinjected compounds into the blastocoel (Spiegel & Burger, 1982) can be analyzed by SEM of fractured embryos. For example, native or fractionated blastocoelic fluid collected from embryos centrifuged at different stages (Pointer, 1978) may now be injected into the blastocoel and the macromolecular effects mapped with SEM.

Although SEM can further resolve the morphogenetic events causally related to blastulation and gastrulation, it does require that the embryos be killed, fixed and dried. Therefore, preparatory techniques and potential artifacts demand critical attention. Highly synchronous cultures of small numbers of embryos and short sampling intervals will be required to trace the details of regional morphogenetic patterns associated with changes in activities and shapes of cells and temporal and spatial changes in extracellular entities such as the ECM of the blastocoel, the apical and basal laminae, and the secretion of extracellular proteases and complex carbohydrases.

ACKNOWLEDGEMENTS

We are indebted to Deni Galileo, Shawn Murphy and Matthew Wahl for having developed the fixatives in this study, for sharing their SEM specimens and unpublished photomicrographs and for helpful discussions. We also thank R.H. Sawyer and R. Showman for their generous invitation to contribute to this volume. The study was supported by the Biology Research Fund, New College Foundation.

LITERATURE CITED

Czihak, G. 1971. Echinoids. pp. 363-506. In: Experimental Embryology of Marine and Freshwater Invertebrates. G. Reverberi (ed.). North-Holland Publ. Co., Amsterdam.
DeMey, J. 1984. Colloidal gold as marker and tracer in light and electron microscopy. EMSA Bull. 14: 54-66.
Guthrie, M.J. and H. Hibbard. 1919. Cleavage and mesenchyme formation in Toxopneustes variegatus. Biol. Bull. 37: 139-157.
Hall, H.G. and V.D. Vacquier. 1982. The apical lamina of the sea urchin embryo: Major glycoproteins associated with the hyaline layer. Dev. Biol. 89: 168-178.
Honda, H., M. Dan-Sohkawa, and K. Watanabe. 1983. Geometrical analysis of cells becoming organized into a tensile sheet, the blastular wall, in the starfish. Differentiation 25: 16-22.
Horstadius, S. 1939. The mechanics of sea urchin development, studied by operative methods. Biol. Rev. 14: 132-179.
Immers, J. 1961. Comparative study of the localization of incorporated ^{14}C-labeled amino acids and $^{35}SO_4$ in the sea urchin ovary, egg and embryo. Exp. Cell Res. 24: 356-378.

Just, E.E. 1939. Basic Methods for Experiments on Eggs of Marine Animals. P. Blakiston's Son & Co., Inc.

Katow, H. and M. Solursh. 1980. Ultrastructure of primary mesenchyme cell ingression in the sea urchin Lytechinus pictus. J. Exp. Zool. 213: 231-246.

Katow, H. and M. Solursh. 1982. In situ distribution of concanavalin A-binding sites in mesenchyme blastulae and early gastrulae of the sea urchin Lytechinus pictus. Exp. Cell Res. 139: 171-180.

Klein, W.H., C.D. Carpenter, L.E. Philpotts and B.P. Brandhorst. 1985. The sea urchin spec family of calcium-binding proteins: Characterization and consideration of possible role in larval development. In: The Cellular and Molecular Biology of Invertebrate Development. Roger H. Sawyer and Richard M. Showman (eds.). Belle W. Baruch Library in Marine Science, No. 15. University of South Carolina Press, Columbia.

McClay, D.R., V. Matanga and G. Wessel. 1985. Expression and appearance of germ layer-specific antigens on the surface of embryonic sea urchin cells. In: The Cellular and Molecular Biology of Invertebrate Development. Roger H. Sawyer and Richard M. Showman (eds.). Belle W. Baruch Library in Marine Science, No. 15. University of South Carolina Press, Columbia.

Mitsunaga, K., A. Fugiwara, T. Yoshimi, and I. Yasumasu. 1983. Stage specific effects on sea urchin embryogenesis of Zn, Li, several inhibitors of AMP-phosphodiesterase and inhibitors of protein synthesis. Develop. Growth and Different. 3: 249-260.

Motomura, I. 1960. Secretion of mucosubstance in the gastrula of the sea urchin embryo. Bull. Mar. Biol. Stat. Asamushi 10: 165-171.

Okazaki, K., T. Fukushi, and K. Dan. 1962. Cyto-embryological studies of sea urchins IV. Correlation between the shape of the ectodermal cells and the arrangement of the primary mesenchyme cells in sea urchin larvae. Acta Embryol. Morphol. Exp. 5: 17-31.

Okazaki, K. and L. Nijima. 1964. "Basement membrane" in sea urchin larvae. Embryologia 8: 89-100.

Pointer, D.A. 1978. Blastocoelic fluid of the sea urchin embryo: Its components and their significance in development. Ph.D. Dissertation, University of California, Berkeley. 139 pp.

Serafy, D.K. 1973. Variation in the polytypic sea urchin Lytechinus variegatus (Lamarck, 1816) in the Western Atlantic (Echinodermata: Echinoidea). Bull. Mar. Sci. 23: 525-534.

Slack, J.M.W. 1983. From Egg to Embryo. Cambridge University Press, p. 3.
Spiegel, E., M.M. Burger, and M. Spiegel. 1980. Fibronectin in the developing sea urchin embryo. J. Cell Biol. 87: 309-313.
Spiegel, E. and M. Spiegel. 1979. The hyaline layer is a collagen containing extracellular matrix in sea urchin embryos and reaggregating cells. Exp. Cell Res. 123: 434-441.
Spiegel, E., M.M. Burger, and M. Spiegel. 1983. Fibronectin and laminin in the extracellular matrix and basement membrane of sea urchin embryos. Exp. Cell Res. 144: 47-55.
Spiegel, E. and L. Howard. 1983. Development of cell junctions in sea urchin embryos. J. Cell Sci. 62: 27-48.
Spiegel, M. and M.M. Burger. 1982. Cell adhesion during gastrulation. Exp. Cell Res. 139: 377-382.
Wessel, G.M., R. Marchase, and D.R. McClay. 1984. Ontogeny of the basal lamina in the sea urchin embryo. Dev. Biol. 103: 235-245.
Young, R.S. 1958. Development of pigment in the larva of the sea urchin, Lytechinus variegatus. Biol. Bull. 114: 394-403.

EGG PLASMA MEMBRANE CHANGES AT FERTILIZATION
Frank J. Longo

ABSTRACT

Many changes characteristic of the fertilized ovum are related to or are a direct result of alterations in the structure and function of the egg plasma membrane. In this review morphological and chemical changes of the egg plasma membrane, as a result of gamete fusion and cortical granule dehiscence, are presented and discussed. Attention is directed mainly to investigations examining the fate of the sperm plasma and cortical granule membranes following their fusion with the egg plasmalemma, and to work that may lead to an understanding of fundamental mechanisms as they are related to metabolic changes of the activated ovum.

INTRODUCTION

As a result of the interaction and fusion of the gametes, processes are initiated which move the previously quiescent egg on to pathways leading to its cleavage, differentiation and eventual development into an adult organism. Some of the structural and metabolic changes that characterize activation in sea urchin eggs are listed in Table 1 with their time of occurrence following gamete binding and fusion. It is noteworthy that many of these processes are related to or are a direct result of alterations in the function of the egg plasma membrane. Analyses of these early aspects of development, particularly those related to the role of ions in egg activation, have been published (Gilkey et al., 1978; Epel et al., 1982; Whitaker & Steinhardt, 1982; Gilkey, 1983; Jaffe, 1983; Shen, 1983). In this review structural and chemical changes of the egg plasma membrane, as a result of gamete fusion and cortical granule dehiscence, are presented and discussed. Attention is directed mainly to investigations examining the fate of the sperm plasma and cortical granule membranes following their fusion with the egg plasmalemma, and to work that may lead to an understanding of fundamental mechanisms as they are related to metabolic changes of the activated ovum. Processes related to aspects presented here, including dynamic changes of the egg cortex and secretory functions of cortical granules in fertilization, have been published (Runnström, 1966; Schuel, 1978; Schroeder, 1979; Gulyas, 1980; Shapiro & Eddy, 1980; Shapiro et al., 1980; Schroeder, 1981; Vacquier, 1981).

Table 1. Timing of fertilization events in sea urchin eggs.

Membrane potential	
Ca-Na action potential	Before 13 s
Na activation potential	13-120 s
Increase in K conductance	500-3000 s
Intracellular calcium release	40-120 s
Cortical reaction	40-100 s
Activation of NAD kinase	40-120 s
Increases in reduced nicotinamide nucleotides	40-900 s
Acid efflux	1-5 min
Increases in intracellular pH	1-5 min
Increased oxygen consumption	1-3 min
Initiation of protein synthesis	5 min onwards
Activation of amino acid transport	15 min onwards
Initiation of DNA synthesis	20-40 min

Timing of these stages depends markedly upon species and temperature. Values given are for Lytechinus pictus at 16-18°C (taken from Whitaker & Steinhardt, 1982).

Cortical Granule Reaction

As previously indicated by Shapiro and Eddy (1980), although the cortical granule reaction has been studied in selected species of echinoderms, amphibians and mammals, and is considered a general process, many exceptions and modifications exist. Nicosia et al. (1977) have shown that 25% of the cortical granules in mouse eggs are released within 30 min of insemination, i.e., prior to sperm-egg fusion. The functional significance of such a premature release of cortical granules has not been determined. In the annelid, Sabellaria, cortical granule exocytosis occurs upon spawning of the egg into sea water and not at the time of fertilization (Pasteels, 1965c). In Urechis a subset of cortical granules dehisce at fertilization while the majority are released some time later (Paul, 1975). Eggs of the molluscs Mytilus, Spisula, Dentalium, and Barnea have organelles that are morphologically comparable to the cortical granules of sea urchin eggs; however, these structures do not change at gamete fusion and their role in fertilization, if any, has not been determined (Pasteels & deHarven, 1962; Rebhun, 1962; Pasteels, 1965a & b; Humphreys, 1967; Longo & Anderson, 1969, 1970a; Longo, 1976a & b; Dufresne-Dube et al., 1983). Despite the absence of a cortical granule reaction in Spisula there is a block to polyspermy (Ziomek & Epel, 1975; Longo, 1976a & b). The eggs of a number of different animals, such as ascidians, do not possess true cortical granules (Rosati et al., 1977; Guraya, 1982).

In sea urchins, sperm-egg binding/fusion is followed by the dehiscence of cortical granules that underlie the plasmalemma. Exocytosis spreads from the point of gamete contact in a wave to the opposite pole of the egg (Afzelius, 1956; Endo, 1961; Wolpert & Mercer, 1961; Anderson, 1968; Millonig, 1969). The mechanism by which the egg plasma and the cortical granule membranes fuse, and the nature of the intermediates in this process, are unclear. There is considerable evidence indicating that alterations in calcium metabolism are related to initiation of the cortical granule reaction and egg activation (Chambers et al., 1974; Steinhardt & Epel, 1974; Vacquier, 1975; Gilkey et al., 1978; Gilkey 1983; Sasaki, 1984; for reviews see Epel, 1978, 1980; Shapiro & Eddy, 1980; Epel et al., 1982; Whitaker & Steinhardt, 1982; Jaffe, 1983).

In sea urchins the cortical granules are tightly adherent to the plasma membrane. This association is sufficiently strong to survive the forces encountered during cortical isolation procedures (Detering et al., 1977). Based on this observation and the difficulty in dislodging cortical granules in centrifuged eggs (cf. Millonig, 1969; Anderson, 1970), Detering et al. (1977) have suggested that the cortical granule membrane may be confluent with the egg plasma membrane. This speculation is not supported by electron microscopic observations (Longo, 1981) nor by investigations demonstrating that cortical granules are dislodged by procaine and other amines (Millonig, 1969; Longo & Anderson, 1970b; Vacquier, 1975; Hylander & Summers, 1981; Decker & Kinsey, 1983). When the plasma membranes of Arbacia eggs are examined with a variety of ultrastructural techniques, sites associated with cortical granules are marked by a small, dome-shaped elevation free of intramembranous particles

(Longo, 1981). Clearings of intramembranous particles have also been reported by Chandler and Heuser (1979) at sites of cortical granule exocytosis in Strongylocentrotus. Similar appearing structures have been observed in the plasma membrane of starfish oocytes; however, they reportedly were not associated with cortical granules (Dolber, 1982). In Arbacia these modifications appeared to be formed as a result of the association of the cortical granule with the plasmalemma, i.e., they were absent in the plasma membrane of fertilized and immature eggs in which the cortical granules were either absent or not localized to the cortex (Longo, 1981). Furthermore, these structures disappeared when the mature egg was treated with amines (Longo, 1981).

Unique patterning of intramembranous particles possibly induced by structures subjacent to the plasma membrane have been described in numerous cell types demonstrating secretory activities (Satir et al., 1973; Friend & Fawcett, 1974; Beisson et al., 1976; Satir, 1976; Weiss et al., 1977a & b; Kinsey & Koehler, 1978). In addition, clearings of intramembranous particles have been observed in portions of the plasma membrane associated with secretory vesicles (Chi et al., 1975; Lawson et al., 1977; Orci et al., 1977; Amherdt et al., 1978; Orci & Perrelet, 1978; Swift & Murkherjee, 1978; Theodosis et al., 1978). The absence of intramembranous particles in such instances is generally held as evidence for the depletion of membrane proteins from the zone of membrane fusion.

Insertion of cortical granule membrane into the egg plasma membrane results in a dramatic structural reorganization of the oolemma. The resultant membrane of the fertilized egg has been referred to as a mosaic structure, indicating that it is derived from several sources, i.e., the egg plasma membrane, the cortical granule membrane and the sperm plasmalemma (Colwin & Colwin, 1967; Anderson, 1968; Epel, 1978; Schroeder, 1979). It is generally held that there is essentially a doubling of the surface area of the activated echinoid egg as a result of the cortical granule reaction, i.e., the sum total of membrane delimiting all of the cortical granules within the egg is equivalent to the surface area of the egg plasma membrane and both sources of membrane are believed to be completely incorporated with the cortical granule reaction (Epel, 1978; Schroeder, 1979; Vacquier, 1981). That the plasma membrane of the unfertilized egg and all of the membrane delimiting the cortical granules are incorporated into the plasmalemma of the activated ovum has not been established. In fact, observations by Anderson

(1968) and Millonig (1969) have indicated that this may not be the case. At the site of fusion between the egg plasma membrane and the cortical granule membrane, vesicles were noted, presumably derived from both the plasma and cortical granule membranes, which were released into the perivitelline space. Vesicles have also been seen in freeze fracture preparations of activated sea urchin eggs (Chandler & Heuser, 1979; F. J. Longo, pers. obs.). Whether these structures are real or artifacts of specimen preparation has not been determined. Nevertheless, it is possible that the surface area of the activated egg immediately following the cortical granule reaction is not the arithmetic sum of the membrane delimiting the entire complement of cortical granules and the original plasma membrane.

Formation of the mosaic plasma membrane and concomitant physiological changes in activity of the egg prompt a number of questions: (1) Is the formation of the mosaic membrane, specifically the insertion of cortical granule membrane into the egg plasma membrane, related to physiological and biochemical changes characteristic of the activated ovum? (2) What physical/chemical modifications occur in the egg plasma membrane as a result of the cortical granule reaction? and (3) Do identifiable domains exist in the plasma membrane of the fertilized egg that are derived from the cortical granule membrane or sperm plasma membrane?

As a result of egg activation and the cortical granule reaction, components of the cortical granules that formerly were sequestered into a well-defined compartment within the egg are externalized. Release of the cortical granule contents and changes associated with the plasma membrane dramatically alter the receptivity of the egg to sperm (Braden et al., 1954; Austin & Braden, 1956; Vacquier et al., 1973; Jaffe, 1976; Schuel, 1978; Gulyas, 1980; Bleil et al., 1981). Secretory materials from the cortical granules of sea urchin eggs contain a number of enzymatic activities that have profound effects on surface components of the ovum (Vacquier et al., 1973; Carroll & Epel, 1975). Carroll and Epel (1975) have demonstrated several protease activities which are released from the egg at the time of the cortical granule reaction. One cleaved the sperm attached to the developing fertilization membrane and another was involved in cleaving bonds attaching the vitelline layer to the plasma membrane, thereby permitting its elevation and development into a fertilization membrane. Peroxidase activity has also been found within the contents of the cortical granules and

has been shown to be involved in cross-linking components of the fertilization membrane (Foerder & Shapiro, 1977; Vernon et al., 1977; Shapiro et al., 1980; Hall, 1981).

It has been speculated that the contents of cortical granules serve other roles in the regulation of the egg's metabolism, e.g., they may elicit subtle perturbations of the plasma membrane that may be critical for activation (Schuel, 1978). Echinoid eggs can be parthenogenetically activated by treatment with trypsin and other proteases (Runnström, 1948, 1966; Moore, 1951; Hand, 1971). In this context, release of surface protein from the plasma membrane of sea urchin eggs at fertilization has been reported and it has been suggested that proteolytic processing of surface proteins may be an important aspect of activation (Mazia et al., 1975; Shapiro, 1975). The idea of a release of membrane-bound protein in initiating metabolic activation at fertilization is an attractive one, which needs to be supported by further evidence.

Although activation of transport systems for specific metabolites occur at activation, there is no evidence linking this change in activity with the insertion of cortical granule membrane into the plasma membrane (Epel et al., 1974; Epel, 1978). That transport systems develop in ammonia-activated eggs and when the cortical granule reaction is blocked, suggest that alterations in the properties induced by the cortical granule reaction may not be essential for the permeability changes characteristic of fertilization (Epel, 1978). Hence, the role of the cortical granule reaction in later developmental events is uncertain.

That many aspects of fertilization are membrane-mediated events leading to egg activation is consistent with the notion that a change in the state of the plasma membrane is an obligatory step in cellular activation (Pardee et al., 1974; Campisi & Scandella, 1978). Campisi and Scandella (1978, 1980a), using electron spin resonance spectroscopy, have indicated that there was an increase in the bulk membrane fluidity in sea urchin eggs after fertilization. Because the spin label (fatty acid) was equilibrated among all of the subcellular membrane fractions, they were unable to resolve whether (1) ovum activation is accompanied by a change in total cellular membranes to a more fluid state, or (2) more specialized membranes (such as the plasma membrane) entered a more fluid state and the probe was showing the average change experienced by altered and unaltered membranes. The structural changes involving membrane lipids accompanying activation are probably not a result of the cortical granule reaction as eggs partially activated by ammonia showed a similar effect. In experiments with cortical fractions, Campisi and Scandella (1980b) showed that the fluidity of the fertilized egg cortex is less than that of the unfertilized cortex. Addition of calcium to cortical fractions from unfertilized eggs resulted in a fluidity decrease in vitro. Campisi and Scandella (1980b) suggested that this change may represent a change in membrane structure rather than a direct interaction of calcium with phospholipid head groups.

Analysis of membrane lipid changes in sea urchin and mouse eggs by fluorescence photobleaching recovery suggested that fertilization is not accompanied by a change in bulk membrane viscosity but rather by alterations in the ensemble of lipid domains (Wolf et al., 1981a & b; Wolf & Ziomek, 1983). The different lipid analogs employed by Wolf et al. (1981a & b) probed different microenvironments and indicated the existence of lipid domains differing in composition or physical states from the average for the plasma membrane. Wolf et al. (1981a & b) explained their results in terms of the existence of gel and fluid lipid domains within the egg plasma membrane, the proportion and composition of which change upon fertilization. At fertilization there may be a reordering of lipid domains which release inactive proteins from gel regions of the plasma membrane into fluid regions, where they would become active. Changes in lipid compositions and extent of gel-fluid regions on fertilization could then act as a switch which would rapidly activate protein functions not requiring the synthesis or insertion of new materials into the membrane.

Fluorescence photobleaching recovery experiments have also been performed with mouse eggs using protein probes (Wolf & Ziomek, 1983) and have demonstrated that interactions with cytoskeletal components may regulate membrane protein diffusion (Tank et al., 1982; Wu et al., 1982). As in the case of membrane lipids, the proteins that were probed also demonstrated a heterogeneous distribution (Shimschick & McConnell, 1973; Klausner et al., 1980; Klausner & Wolf, 1980; Wolf et al., 1981a & b; Wolf & Ziomek, 1983). Moreover, although "new" membranes (i.e., cortical granule and sperm plasma membranes) were added to the egg plasmalemma at fertilization, there was no generalized effect on the diffusion of membrane protein in the mouse egg.

Binding studies with plant lectins have been utilized in an effort to demonstrate possible membrane changes between fertilized and unfertilized eggs. Investigations with mouse and hamster eggs have shown that concanavalin A-binding sites change quantitatively following fertilization (Yanagimachi & Nicolson, 1976; DeFelici & Siracusa, 1981). In ascidian eggs both the agglutinibility and number of concanavalin A receptors increased following activation (O'Dell et al., 1973). The qualitative changes in lectin-binding following fertilization may reflect modifications in the nature and/or structure of the sites themselves. Alterations in lectin-binding may also be influenced by membrane fluidity and functional states of the cytoskeleton (Karsenti et al., 1977; Marshall & Heiniger, 1979). In the sea urchin Strongylocentrotus two classes of concanavalin A-binding sites have been identified: a high-affinity site which is associated with the vitelline layer and a low-affinity site that is associated with the plasma membrane. The low-affinity sites doubled at fertilization, apparently as a result of the insertion of cortical granule membrane (Vernon & Shapiro, 1977). Although the increase in low-affinity binding sites may have been due to the appearance of cryptic sites, there was no doubling when eggs were activated with ammonia, supporting the notion that the increase in number of sites is a reflection of the addition of cortical granule membrane to the egg plasmalemma.

When examined with scanning electron microscopy, the surface of the activated sea urchin egg following cortical granule release consisted of relatively smooth areas, believed to be derived from cortical granule membrane, surrounded by microvillar regions, presumably derived from the original plasma membrane (Eddy & Shapiro, 1976; Shapiro & Eddy, 1980). This topographical mosaicism persisted for about one hour in Strongylocentrotus; in Arbacia it was of shorter duration (Longo, 1981; Carron & Longo, 1983). Although scanning electron microscopic investigations provided an overview of the topographical relations of the activated egg surface, they did not indicate whether or not (nor the degree to which) macromolecular components of both the plasmalemma and cortical granule membrane might be mixing. Several studies have examined this question by investigating intramembranous particle distribution and filipin "staining" of the egg plasma membrane following cortical granule discharge.

Examination of freeze fracture replicates of unfertilized sea urchin eggs demonstrated a significant difference in the number of intramembranous particles within the plasmalemma and cortical granule membrane. In Arbacia the number of intramembranous particles within the P-face of the cortical granule membrane was about 30% of that in the P-face of the egg plasma membrane (Table 2). In light of such a difference in the density of intramembranous particles, what happens to this dichotomy following cortical granule exocytosis? Are patches observed in the plasma membrane of the fertilized egg following the cortical granule reaction containing a density of intramembranous particles corresponding to the cortical granule membrane or is there a homogeneous distribution of particles within the plasma membrane of the activated egg, suggesting an intermixing of macromolecular components within the mosaic membrane (Fig. 1)? Studies by Pollack (1978), Chandler and Heuser (1979) and Longo (1981) in three different species of sea urchins demonstrated that the mosaic pattern of the fertilized egg plasmalemma, in terms of intramembranous particles, is temporary; recognizable differences between the original egg plasma membrane and cortical granule membrane are lost soon after activation.

Table 2. Intramembranous particle density of P- and E-faces of the plasmalemma and cortical granule membrane in fertilized and unfertilized Arbacia eggs (from Longo, 1981).

	Mean number of particles/μm^2 ± SD	
	Unfertilized	Fertilized
Plasmalemma		
P-face	1270 ± 570	1080 ± 280
E-face	160 ± 60	180 ± 110
Cortical granule membrane		
P-face	370 ± 220	1530 ± 280
E-face	250 ± 150	330 ± 50

Studies with Strongylocentrotus indicated that the membrane of exocytotic pockets displayed fewer intramembranous particles than the rest of the plasma membrane and corresponded to the cortical granule membrane (Chandler & Heuser, 1979). In specimens examined at slightly later intervals after cortical granule exocytosis, patches of decreased particle number were found in areas lacking microvilli and corresponding to the flat regions described by Eddy and Shapiro (1976) with scanning electron microscopy. Hence, collapse of the granule membrane took several seconds, while mixing of some cortical granule and plasma membrane components, as judged by intramembranous particle distribution, took longer.

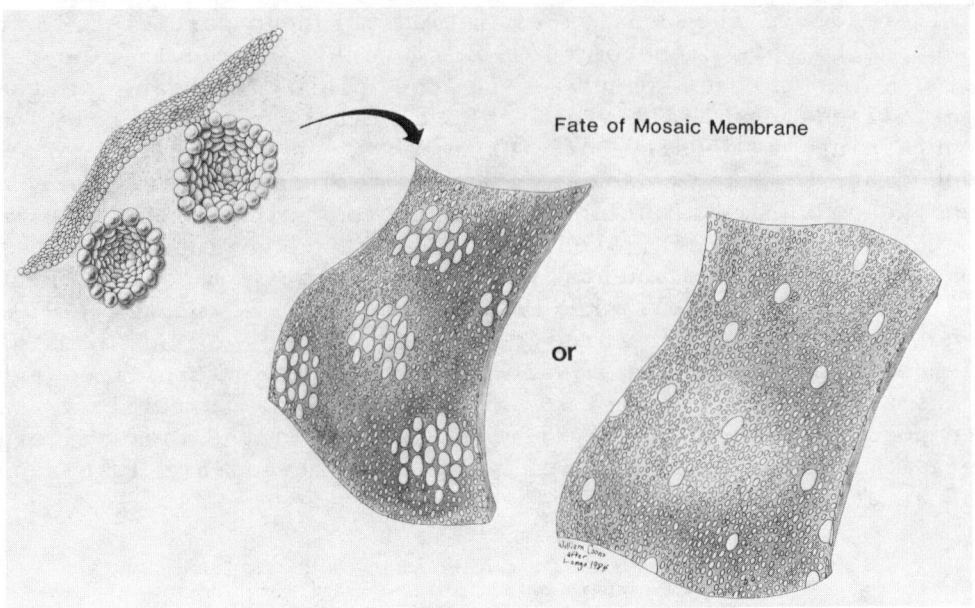

Fig. 1. Diagram depicting the possible fate of membrane components derived from cortical granules (large spheres). Membrane components may persist, in a time-dependent manner, as patches or disperse to varying extents among elements derived from the original egg plasma membrane (small spheres) (Williams Coons after Longo, 1984). Observations published thus far demonstrate a mixing of components derived from the egg plasmalemma within membrane domains originating from the cortical granule (Longo, 1981; Carron & Longo, 1983).

In Arbacia recognizable differences between cortical granule and plasma membranes were lost soon after cortical granule exocytosis (Table 2). Patches, containing a reduced number of intramembranous particles and corresponding to the cortical granule membrane, were not found, indicating a rapid alteration in the composition of cortical granule membrane following its fusion with the plasma membrane (Table 2). By four minutes postinsemination the density of intramembranous particles in the P-face of the plasma membrane of the fertilized egg was slightly reduced from that of the membrane of the unfertilized egg (1080 vs. 1270, respectively), suggesting a possible "flow" of intramembranous particles from the oolemma into membrane derived from the cortical granule. This suggestion is in keeping with the fluid character of membranes and consistent with schemes reported for other cells (Frye & Edidin, 1970; Singer & Nicolson, 1972; Singer 1976; Friend et al., 1977; Orci & Perrelet, 1977).

Changes in the density of intramembranous particles that occur within the P-face of the egg plasma membrane upon activation may not represent merely the integration and "lateral flow" of cortical granule membrane components into the oolemma but represent more complex interactions. Factors which may be relevant to apparent intermixing of membranes, in addition to the diffusion of components within the plane of the membrane, include rapid turnover of components, integration of units from cytoplasmic pools, and movements of components from one membrane site into the cytoplasm and their emergence at a new position (Frye & Edidin, 1970). Since the difference in intramembranous particles of the P-face of the plasma and cortical granule membranes was significant, and as a result of the cortical granule reaction the surface area of the plasma membrane is believed to double, a decrease in particle density within the P-face of the plasmalemma of the activated egg might be anticipated. Why such a decrease was not observed may be contingent upon a number of factors, which are highly speculative at this time.

Changes in the distribution of intramembranous particles have also been described in the plasma membrane of Spisula eggs which do not have a cortical granule reaction. The manner in which polyspermy is prevented in this organism has not been established. Changes in membrane structure/function have been demonstrated that may be related to insuring monospermic fertilization (Longo, 1976a & b; Ii & Rebhun, 1979; Finkel & Wolf, 1980). Following activation, there was an approximate two-fold increase in density of intramembranous particles within the plasma membrane of Spisula eggs (Longo, 1976a & b). The functional significance of this membrane change and whether or not it is related to the development of a block to polyspermy have not been determined.

Studies have also been performed in sea urchin (Arbacia) eggs treated with filipin in an effort to determine alterations in membrane sterols at activation. The plasma membranes of treated unfertilized eggs possessed numerous filipin-sterol complexes while fewer complexes were associated with membranes delimiting cortical granules, demonstrating that the plasma membrane is relatively rich in β-hydroxysterols (de Kruijff & Demel, 1974; Carron & Longo, 1983). This dichotomy which was not related to an impermeability to filipin of the plasma and cortical granule membranes may represent an inequality in sterol content. Other factors may be influential here as well, including masking of sterols and lack of deformation of structures (cytoskeletal) associated with the membrane (Norman et al., 1976; Elias et al., 1979; Tamm & Tamm, 1983).

Following fusion with the plasmalemma, the membrane formerly delimiting cortical granules underwent a dramatic alteration in sterol composition as indicated by a rapid increase in the number of filipin-sterol complexes (Carron & Longo, 1983). In contrast, portions of the fertilized egg plasma membrane, derived from the original plasma membrane of the unfertilized egg, displayed little or no change in filipin-sterol composition. Other than regions engaged in endocytosis, the plasma membrane of the zygote possessed a homogeneous distribution of filipin-sterol complexes and appeared structurally similar to that of the unfertilized ovum.

The absence of patches within the plasma membrane of the fertilized egg, relatively devoid of filipin-sterol complexes and corresponding morphologically to the membrane formerly delimiting intact cortical granules, indicates that the cortical granule membrane is significantly altered when it fuses with the plasmalemma. How the cortical granule membrane acquires an increase in filipin-sterol complexes has not been determined. Lateral displacement of sterols from membranous regions derived from the original egg plasma membrane may be involved (Friend, 1982). However, there is no evidence documenting such a process in activated eggs. Sterols have been shown to diffuse rapidly in bilayers (Träuble & Sackermann, 1972) which is consistent with these results. Rapid lateral displacement of sterols into membrane patches from the cortical granules would be difficult, if not impossible, to visualize by the sampling methods employed in studies by Carron and Longo (1983).

Recent investigations by Decker and Kinsey (1983), analyzing membrane lipids in Lytechinus eggs, indicated that the cortical granule membrane contains 2.3 times as much cholesterol as that of the plasmalemma, a result which is at odds with that obtained by filipin staining of Arbacia eggs (Carron & Longo, 1983). Based on these results, Decker and Kinsey (1983) speculated that the level of cholesterol in the plasma membrane of eggs would increase following the cortical granule reaction. However, preliminary studies have not confirmed this. Decker and Kinsey (1983) speculated that cholesterol may not remain with the cell surface or is recycled to the interior of the egg via endocytosis. Results of investigations in sea urchin zygotes and in somatic cells do not support this speculation (Pearse, 1976; Altstiel & Branton, 1983; Carron & Longo, 1983)

Integration of the Sperm and Egg Plasma Membranes

Relatively few studies have examined the incorporations of the sperm plasma membrane into the egg plasmalemma in terms of the integration of membrane domains, the fate of the sperm plasma membrane components and their possible role in development. Whether or not all of the sperm plasma membrane is incorporated into the egg plasmalemma is more-or-less assumed in many instances, although experimental evidence is required to verify this unequivocally. Electron microscopic studies of sperm incorporation in some invertebrates and mammals have demonstrated membranous elements at the site of gamete fusion that appear to have been derived from the fused sperm and/or egg plasmalemmae (Franklin 1965; Colwin & Colwin, 1967; Piko, 1969; Zamboni, 1971; Bedford, 1972; Bedford & Cooper, 1978). The quantity of membrane that is incorporated into the fertilizing egg in these cases and its fate have not been ascertained (Gundersen et al., 1982).

Investigations examining the integration of the sperm and egg plasma membranes at fertilization, where one of the gametes has been labelled, have been carried out in both invertebrates and mammals (Yanagimachi et al., 1973; Gabel et al., 1979a & b; Longo, 1982). Prior to sperm-egg fusion in hamsters the sperm plasma membrane of the post-acrosomal region did not bind colloidal iron hydroxide. Once gamete fusion had been initiated, however, the former sperm plasma membrane was able to bind this marker (Yanagimachi et al., 1973). The rapid increase in colloidal iron hydroxide-binding on the incorporating sperm head is believed to be a result of intermixing of sperm-egg membrane components analogous to the intermixing of antigenic determinants after fusion of somatic cells (Frye & Edidin, 1970; Edidin et al., 1976; Edidin & Wei, 1977; Schlessinger et al., 1977; Ziomek et al., 1980). The results of Yanagimachi et al. (1973) are not unexpected since glycoproteins and glycolipids intercalated in a fluid lipid bilayer may be diffusible in the plane of the membrane. These observations, however, do not exclude the possibility that colloidal iron hydroxide-binding anionic residues are enzymatically added to sperm plasma membrane oligosaccharides after fusion or that entirely new colloidal iron hydroxide-binding membrane components are inserted into the sperm plasma membrane shortly after fusion.

Similar experiments have been carried out with Spisula in which concanavalin A binding to the egg, but not the sperm plasma membrane, has been demonstrated by the horseradish peroxidase-diaminobenzidine reaction (HRP-DAB; Longo, 1979,

1982). Because of this dichotomy in lectin-binding, changes in the affinity of the sperm plasmalemma following its fusion and integration with components of the egg plasma membrane could be demonstrated.

The plasma membranes of fertilized Spisula eggs reacted with concanavalin A-HRP-DAB and examined within one minute postinsemination were associated uniformly with enzymatic precipitate except at sites of sperm incorporation. These portions of unstained plasma membrane were shown to be derived from the sperm and associated with the apex of the fertilization cone. By four minutes postinsemination no difference in staining of plasma membranes derived from the egg or the sperm was detected. These observations are consistent with the movement of concanavalin A-binding sites from the egg plasmalemma into the sperm plasma membrane.

Not all components of the sperm and egg plasma membrane appear to rapidly intermix following gamete fusion as described for labelled oligosaccharides (Yanagimachi et al., 1973; Longo, 1982). Sea urchin and mouse eggs fertilized with FITC- or ^{125}IFC-labeled sperm retained a topographically mosaic surface, as if the lateral mobility of sperm plasma membrane components was restricted and retained as a discreet patch (Gabel et al., 1979a & b). Gabel et al. (1979a & b) suggested that a sperm surface component is transferred to the egg plasma membrane and unequally distributed during cleavage, indicating that the egg surface rigidifies upon fertilization (Johnson & Edidin, 1978). The persistence of this patch through early development suggested the possibility that the sperm may function in morphogenesis, e.g., by determining developmental gradients.

More recent experiments with FITC-labelled sperm by Gundersen et al. (1982) indicated that labeled sperm components (surface and mitochondria) may be internalized after fertilization. These results are consistent with previous electron microscopic studies indicating the incorporation of portions of the sperm plasma membrane (Colwin & Colwin, 1967; Yanagimachi & Noda, 1970; Bedford, 1972; Bedford & Cooper, 1978).

Microvillar Elongation

Elongation of microvilli is generally viewed as one means of accommodating an increase in surface of the activated sea urchin egg (Eddy & Shapiro, 1976; Schroeder, 1979). Eddy and Shapiro (1976) indicated that there was no mixing of smooth and microvillar regions of the mosaic membrane before microvillar elongation and suggested that this may represent a restriction in mobility of topographical features. The basis for this limitation in mobility may be related to changes in the structural organization of the underlying cytoskeleton (Vacquier, 1981).

Schroeder (1979) has shown that the amount of surface area represented by the sum total of cortical granule membrane in Strongylocentrotus eggs is greater than that of the egg plasmalemma. Hence, if all of the cortical granule membrane was incorporated into the egg plasmalemma there would be at least a two-fold increase in surface area. However, by 16 min postinsemination the surface area of the activated egg was only slightly larger than that of the unactivated ovum, indicating a rapid accommodation in surface membrane. Elongation of microvilli was shown to be unable to compensate for all of the cortical granule membrane that might be incorporated and internalization of membrane was proposed as a mechanism to quantitatively modify the surface area of activated eggs (Schroeder, 1979). Based on observations of capacitance changes at fertilization, Nuccitelli (1980) also postulated a process of plasma membrane removal within one minute of insemination in medaka eggs.

Microvillar elongation occurs in phases that have been described by Chandler and Heuser (1981). Rapid elongation is claimed to occur primarily in areas occupied by the original plasma membrane. In this connection, microvillar elongation is believed to occur at sites on the egg surface only where cortical granules have exocytosed (Fisher et al., 1982; Carron & Longo, unpubl. obs.). By two minutes postinsemination upheavals of cytoplasm at the bases of the microvilli develop whereby mounds are formed which possess two to four microvilli projecting from their apices. Between the mounds were areas undergoing pinocytosis. By five minutes postinsemination the mounds were reduced in size and interconnected by ruffles of cytoplasm. Similar morphological changes along the bases of microvilli have also been described for Spisula eggs (Longo, 1976a,b). In sea urchins these changes have been shown to be due to a reorganization of the cortical cytoskeletal system (Vacquier, 1981).

When examined with ultrastructural techniques, few or no actin filaments are found associated with the small microvilli of the unfertilized egg (Begg & Rebhun, 1979; Carron & Longo, 1980, 1982; Begg et al., 1982). The microvilli of fertilized eggs, however, contain a core of actin microfilaments that may mediate their support (Burgess & Schroeder,

1977; Begg et al., 1982; Carron & Longo, 1982). Based on similarities to processes where elongation of microvillar-like structures occurred with microfilament formation, and implying that filament development provided the force for extension (Tilney & Cardell, 1970; Tilney, 1976, 1977; Tilney et al., 1978), it was initially hypothesized that actin within the egg cortex could be induced to polymerize as a result of the increase in intracellular pH at activation. Actin polymerization might then participate in microvillar elongation (Begg & Rebhun, 1979).

Investigation of this question, both in vivo and with isolated cortices exposed to different ionic conditions, demonstrated that microvillar elongation was stimulated by the calcium flux characteristic of egg activation (Begg et al., 1982; Carron & Longo, 1980; 1982). Microvillar elongation did not occur when eggs were incubated in media, such as ammonia, which induced an increase in intracellular pH (Carron & Longo, 1980, 1982; Begg et al., 1982). Actin-filament bundle formation was triggered by an increase in intracellular pH. Formation of actin-filament bundles was not necessary for microvillar elongation but was required to provide a rigid support of the microvilli. Hence, the events of activation prior to the increase in intracellular pH induced the formation of cortical microfilament networks and microvillar elongation. The microvillar network may provide the structural and/or contractile framework for support of the egg surface which is undergoing extensive rearrangement at this time. Organization of microfilaments within the microvilli (i.e., bundle formation) may then be a consequence of cytoplasmic alkalinization.

The actual mechanisms of force production for microvillar extension are unknown but may include cortical actin networks as described for phagocytosis in macrophages (Hartwig et al., 1980); an increase in hydrostatic pressure, in this case actin networks may determine the shape of the microvilli; and changes in the plasma membrane itself, thereby providing either the force and/or the directional information for microvillar elongation.

The actual mechanisms of cortical reorganization are not known but are likely to involve actin-binding proteins described in other systems (Schliwa, 1981; Craig & Pollard, 1982; Weeds, 1982). Aggregation of actin filaments and their association with bundling protein (e.g., fascin) may give rise to microfilament bundles in egg microvilli much in the same manner as filopodia formation in sea urchin coelomocytes

(Edds, 1977; Otto et al., 1979; De Rosier & Edds, 1980; Tilney & Jaffe, 1980). Actin may be prevented from polymerizing in the unfertilized egg by a profilin-like protein (Mabuchi, 1981; Hosoya et al., 1982). Although fascin is found in the unfertilized sea urchin egg and has been localized in microvilli of fertilized ova, its interaction with actin has not been demonstrated to be calcium- or pH-sensitive (Bryan & Kane, 1982). Hence, other actin-binding proteins may be instrumental in microvillar elongation; cytoplasmic alkalinization may give rise to microfilament bundle formation by promoting actin, actin-binding protein interactions.

Pinocytosis

Following the cortical granule reaction and concomitant with the elongation of microvilli is the development of pinocytotic pits and vesicles (Anderson, 1968; Chandler & Heuser, 1981; Donovan & Hart, 1982; Carron & Longo, 1983, 1984; Fisher & Rebhun, 1983). According to Fisher and Rebhun (1983), pinocytosis commenced as a burst three to five minutes postinsemination. All regions of the cortical granule membrane incorporated into the sea urchin egg plasma membrane disappeared and the egg surface was uniformly covered with microvilli. Fisher and Rebhun (1983) indicated that invaginating pits were concentrated in areas where microvillar and non-microvillar (presumably derived from cortical granules) regions of the surface abut one another. Examination of activated Arbacia eggs in which the plasma membrane was labelled indicated that endocytosis occurred in membrane domains derived from both the plasmalemma and the cortical granule (F.J. Longo, pers. obs.). Whether or not the cortical granule membrane, as a complete structure, is internalized, has not been unequivocally demonstrated. This would seem to be highly unlikely in light of previously mentioned studies demonstrating dramatic changes in the composition of the cortical granule membrane following its insertion into the egg plasmalemma (DeCamilli et al., 1976; Hausmann & Allen, 1976; Longo, 1981; Carron & Longo, 1983).

That the mosaic membrane does, in fact, undergo endocytosis has been demonstrated by studies employing fluid-phase and adsorptive tracers such as horseradish peroxidase, native ferritin and cationized ferritin, which become internalized within coated vesicles of activated sea urchin and zebra fish eggs (Donovan & Hart, 1982; Fisher & Rebhun, 1983; Carron &

Longo, 1984). Endocytosis in activated sea urchin eggs appeared to be independent of cytoplasmic alkalinization and its initiation is believed to be calcium-dependent (Fisher & Rebhun, 1983).

That endocytosis follows the cortical granule reaction suggests a mechanism for both surface area reduction and cell surface remodelling which may be relevant to physiological changes characteristic of fertilized eggs (Epel, 1978; Donovan & Hart, 1982; Fisher & Rebhun, 1983; Carron & Longo, 1984). Its presence is commensurate with previous studies of somatic cells which have established that after exocytosis in secretory cells, excess membrane may be removed from the cell surface in the form of endocytotic vesicles (Orci et al., 1973; Pelletier, 1973; Kalina & Robinovitch, 1975; Oliver & Hand, 1978).

The extent of membrane internalized by endocytosis beginning at fertilization appears to be extensive and persists up to the time of cleavage (Fisher & Rebhun, 1983). Whether or not pinocytosis remains constant over this period has not been established; however, Fisher and Rebhun (1983) have estimated that about 26,300 μm^2 of surface membrane per Strongylocentrotus egg is resorbed by endocytosis during the first four minutes of fertilization. This represents approximately 46% of the membrane presumably added to the egg surface by cortical granule exocytosis. The relationship between cortical granule exocytosis and endocytosis, in terms of the quantity of membrane in flux, is unclear since (1) the rate of membrane interiorization is unknown; (2) the actual amount of cortical granule membrane added to the zygote surface has not been firmly established and may be less than 100% (Millonig, 1969; Hausmann & Allen, 1976); and (3) mechanisms other than pinocytosis that may contribute to the reduction of surface area have not been eliminated (Steinman et al., 1976).

Using the antibiotic, filipin, Carron and Longo (1983) have demonstrated that the endocytotic pits and vesicles that form in fertilized Arbacia eggs are depleted in cholesterol, possibly for the attainment of fluidity presumably required for membrane internalization (Chapman, 1973; Jain, 1975; Demel & de Kruijff, 1976; Legault et al., 1979; Montesano et al., 1979; Davis, 1980). This is in agreement with biochemical data showing a lower cholesterol-to-phospholipid ratio in coated vesicles relative to their portions of the plasma membrane (Pearse, 1976; Altstiel & Branton, 1983).

Following the appearance of tracer in pinocytotic vesicles of fertilized Arbacia eggs, label was observed in organelles recognizable as lysosomes (Schellens et al., 1977; Steinman et al., 1983; Carron & Longo, 1984). This transition suggests that the tracer makes its way from one cellular compartment to another and involves a sorting of membrane and its components (Carron & Longo, 1983; Steinman et al., 1983). That tracer was localized to lysosomes of zygotes examined up to 60 min postinsemination suggests that surface membrane may be degraded or modified as well (Silverstein et al., 1977). Membrane components may then re-enter cytoplasmic precursor pools by traversing the lysosomal membrane to be utilized at later stages of embryogenesis (DeDuve & Wattiaux, 1966; Holtzman, 1976). Although experiments by Carron and Longo (1984) suggest that surface membrane recovered by endocytosis in fertilized sea urchin eggs may not be recycled through the Golgi apparatus, investigations with somatic cells show that cationized ferritin is taken up by pinocytosis, transported to Golgi cisternae and to newly formed secretion granules (Herzog & Farquhar, 1977; Ottosen et al., 1980; Farquhar, 1981; Steinman et al., 1983). Hence, in some cells, membrane internalized by endocytosis may be reutilized or recycled intact, and not in the form of dissociated membrane. In this context, additional surface membrane is required at cleavage (Bluemink & deLaat, 1977; Schroeder, 1981) and a possible source may be that derived from endocytosis.

The interiorization of zygote plasma membrane denotes the existence of a pathway from the cell surface to the cytoplasm and has significant biological implications. In a variety of cell types nutritional and regulatory molecules are selectively incorporated from the extracellular milieu (Silverstein et al., 1977; Goldstein et al., 1979). Receptor-mediated uptake and degradation systems may serve to modify hormones and/or surface receptors and, hence, represent a mechanism for alteration of the physiological response of cells to their environment (Goldstein et al., 1979). Whereas pinocytosis may have a role in zygote surface area reduction, the interiorization of the zygote plasma membrane may play an additional role in the modification of surface properties and may represent a mechanism for altering the number and pattern of developmentally significant cell-surface receptors.

ACKNOWLEDGMENTS

The assistance of JoAnn Barnes, Tena Perry and Frederick So is gratefully appreciated. The author's investigations described in this review were supported by grants from the NSF and NIH.

LITERATURE CITED

Afzelius, B.A. 1956. The ultrastructure of the cortical granules and their products in the sea urchin egg as studied with the electron microscope. Exp. Cell Res. 10: 257-285.

Altstiel, L. and D. Branton. 1983. Fusion of coated vesicles with lysosomes: Measurement with a fluorescence assay. Cell 32: 921-929.

Amherdt, M., M. Baggiolini, A. Perrelet, and L. Orci. 1978. Freeze fracture of membrane fusions in phagocytosing polymorphonuclear leukocytes. Lab. Invest. 39: 398-404.

Anderson, E. 1968. Oocyte differentiation in the sea urchin, Arbacia punctulata with particular reference to the origin of cortical granules and their participation in the cortical reaction. J. Cell Biol. 37: 514-539.

Anderson, E. 1970. A cytological study of centrifuged whole, half, and quarter eggs of the sea urchin Arbacia punctulata. J. Cell Biol. 47: 711-733.

Austin, C.R. and A.W.H. Braden. 1956. Early reactions of the rodent egg to spermatozoan penetrations. J. Exp. Biol. 33: 358-365.

Bedford, J.M. 1972. An electron microscopic study of sperm penetration into the rabbit egg after natural mating. Am. J. Anat. 133: 213-254.

Bedford, J.M. and G.W. Cooper. 1978. Membrane fusion events in the fertilization of vertebrate eggs. pp. 65-125. In: Cell Surface Reviews, Vol. 5. Membrane Fusion. G. Poste and G.L. Nicolson (eds.). Elsevier/North Holland Biomedical Press, The Netherlands.

Begg, D.A. and L.I. Rebhun. 1979. pH regulates the polymerization of actin in the sea urchin egg cortex. J. Cell Biol. 83: 241-248.

Begg, D.A., L.I. Rebhun, and H. Hyatt. 1982. Structural organization of actin in the sea urchin egg cortex: Microvillar elongation in the absence of actin filament bundle formation. J. Cell Biol. 93: 24-32.

Beisson, J., M. Lefort-Tran, M. Pouphile, M. Rossignal, and B. Satir. 1976. Genetic analysis of membrane differentiation in Paramecium. J. Cell Biol. 69: 126-143.

Bleil, J.D., C.F. Beall, and P.M. Wassarman. 1981. Mammalian sperm-egg interaction: Fertilization of mouse eggs triggers modification of the major zona pellucida glycoprotein, ZP2. Dev. Biol. 86: 189-197.

Bluemink, J.G. and S.W. deLaat. 1977. Plasma membrane assembly as related to cell division. pp. 402-461. In: The Synthesis, Assembly and Turnover of Cell Surface Components. G. Poste and G.L. Nicolson (eds.). Elsevier/North Holland Biomedical Press, The Netherlands.

Braden, A.W.H., C.R. Austin, and H.A. David. 1954. The reaction of the zona pellucida to sperm penetration. Aust. J. Biol. Sci. 7: 391-409.

Bryan, J. and R.E. Kane. 1982. Actin gelation in sea urchin egg extracts. Meth. Cell Biol. 25: 175-199.

Burgess, D.R. and T.E. Schroeder. 1977. Polarized bundles of actin filaments within microvilli of fertilized sea urchin eggs. J. Cell Biol. 74: 1032-1037.

Campisi, J. and C.J. Scandella. 1978. Fertilization-induced changes in membrane fluidity of sea urchin eggs. Science 199: 1336-1337.

Campisi, J. and C.J. Scandella. 1980a. Bulk membrane fluidity increases after fertilization or partial activation of sea urchin eggs. J. Biol. Chem. 255: 5411-5419.

Campisi, J. and C.J. Scandella. 1980b. Calcium-induced decrease in membrane fluidity of sea urchin egg cortex after fertilization. Nature 286: 185-186.

Carroll, E.J. and D. Epel. 1975. Isolation and biological activity of the proteases released by sea urchin eggs following fertilization. Dev. Biol. 44: 22-32.

Carron, C.P. and F.J. Longo. 1980. Relationship of intracellular pH and pronuclear development in the sea urchin, *Arbacia punctulata*. Dev. Biol. 79: 478-487.

Carron, C.P. and F.J. Longo. 1982. Relation of cytoplasmic alkalinization to microvillar elongation and microfilament formation in the sea urchin egg. Dev. Biol. 89: 128-137.

Carron, C.P. and F.J. Longo. 1983. Filipin/sterol complexes in fertilized and unfertilized sea urchin egg membranes. Dev. Biol. 99: 482-488.

Carron, C.P. and F.J. Longo. 1984. Pinocytosis in fertilized sea urchin (*Arbacia punctulata*) eggs. J. Exp. Zool. 231: 413-422.

Chambers, E.L., B.C. Pressman, and B. Rose. 1974. The activation of sea urchin eggs by the divalent ionophores A-23187 and X-537A. Biochem. Biophys. Res. Commmun. 60: 126-132.

Chandler, D.E. and J. Heuser. 1979. Membrane fusion during secretion: Cortical granule exocytosis in sea urchin eggs as studied by quick-freezing and freeze-fracture. J. Cell Biol. 83: 91-108.

Chandler, D.E. and J. Heuser. 1981. Postfertilization growth of microvilli in the sea urchin egg: New views from eggs that have been quick-frozen, freeze-fractured and deeply etched. Dev. Biol. 92: 393-400.

Chapman, D. 1973. Some recent studies of lipids, lipid-cholesterol and membrane systems. pp. 91-144. In: Biological Membranes, Vol. 2. D. Chapman and D.F.H. Wallach (eds.). Academic Press, N.Y.

Chi, E.Y., D. Lagunoff, and J.K. Koehler. 1975. Electron microscopy of freeze-fractured rat peritoneal mast cells. J. Ultrastruct. Res. 57: 46-54.

Colwin, L.H. and A.L. Colwin. 1967. Membrane fusion in relation to sperm-egg association. pp. 295-367. In: Fertilization, Vol. 1. C. B. Metz and A. Monroy (eds.). Academic Press, N.Y.

Craig, S. W. and T.D. Pollard. 1982. Actin-binding proteins. Trends Bio-chem. Sci. 7: 55-58.

Davis, B.K. 1980. Interaction of lipids with the plasma membrane of sperm cells. I. The antifertilization action of cholesterol. Arch. Androl. 5: 249-254.

Decker, S.J. and W.H. Kinsey. 1983. Characterization of cortical secretory vesicles from the sea urchin egg. Dev. Biol. 96: 37-45.

DeCamilli, P., D. Peluchetti, and J. Meldolesi. 1976. Dynamic changes of the luminal plasmalemma in stimulated parotid acinar cells. A freeze-fracture study. J. Cell Biol. 70: 59-74.

DeDuve, C. and R. Wattiaux. 1966. Function of lysosomes. Ann. Rev. Phys. 28: 435-493.

DeFelici, M. and G. Siracusa. 1981. Fertilization-induced changes in concanavalin A binding to mouse eggs. Exp. Cell Res. 132: 41-45.

de Kruijff, B. and R.A. Demel. 1974. Polyene antibiotic-sterol interactions in membranes of Acholeplasma laidlawii cells and lecithin lysosomes. III. Molecular structure of the polyene antibiotic-cholesterol complexes. Biochim. Biophys. Acta 339: 57-70.

Demel, R.A. and B. de Kruijff. 1976. The function of sterols in membranes. Biochim. Biophys. Acta 457: 109-132.

DeRosier, D.J. and K.T. Edds. 1980. Evidence for fascin cross links between the actin filaments and coelomocyte filopodia. Exp. Cell Res. 126: 490-494.

Detering, N.K., G.L. Decker, E.D. Schmell, and W.J. Lennarz. 1977. Isolation and characterization of plasma membrane-associated cortical granules from sea urchin eggs. J. Cell Biol. 75: 899-914.

Dolber, P.C. 1982. Distribution and nature of oolemmal dense plaques and particle-free patches in starfish oocytes. J. Cell Biol. 95: 175a.

Donovan, M. and N.H. Hart. 1982. Uptake of ferritin by the mosaic egg surface of Brachydanio. J. Exp. Zool. 223: 229-304.

Dufresne-Dube, L., B. Picheral, and P. Guerrier. 1983. An ultrastructural analysis of Dentalium vulgare (Mollusca, Scaphopoda) gametes with special reference to early events at fertilization. J. Ultrastruct. Res. 83: 242-257.

Edds, K.T. 1977. Microfilament bundles. I. Formation with uniform polarity. Exp. Cell Res. 108: 452-482.

Eddy, E.M. and B.M. Shapiro. 1976. Changes in the topography of the sea urchin egg after fertilization. J. Cell Biol. 71: 35-48.

Edidin, M., Y. Zagyansky, and T. J. Lardner. 1976. Measurement of membrane protein lateral diffusion in single cells. Science 191: 466-468.

Edidin, M. and T. Wei. 1977. Diffusion rates of cell surface antigens of mouse-human heterokaryons. I. Analysis of the population. J. Cell Biol. 75: 475-482.

Elias, P.M., D.S. Friend, and J. Goerke. 1979. Membrane sterol heterogeneity. Freeze-fracture detection with saponins and filipin. J. Histochem. Cytochem. 27: 1247-1260.

Endo, Y. 1961. Changes in the cortical layer of sea urchin eggs at fertilization as studied with the electron microscope. I. Clypeaster japonicus. Exp. Cell Res. 25: 383-397.

Epel, D. 1978. Mechanisms of activation of sperm and egg during fertilization of sea urchin gametes. Curr. Top. Dev. Biol. 12: 185-246.

Epel, D. 1980. Experimental analysis of the role of intracellular calcium in the activation of the sea urchin egg at fertilization. pp. 169-185. In: The Cell Surface: Mediator of Developmental Processes. S. Subtelny and N. K. Wessells (eds.). Academic Press, N.Y.

Epel, D., G. Perry, and T. Schmidt. 1982. Intracellular calcium and fertilization: Role of the cation and regulation of intracellular calcium levels. pp. 171-183. In: Membrane in Growth and Development. Prog. Clin. Biol. Res. Vol. 91. J.F. Hoffman and G.H. Giebisch (eds.). A.R. Liss.

Epel, D., R. Steinhardt, T. Humphreys, and D. Mazia. 1974. An analysis of the partial metabolic derepression of sea urchin eggs by ammonia: The existence of independent pathways. Dev. Biol. 40: 245-255.

Farquhar, M.G. 1981. Recovery of surface membrane in anterior pituitary cells. Variations in traffic detected with anions and cationized ferritin. J. Cell Biol. 77: R35-R42.

Finkel, T. and D.P. Wolf. 1980. Membrane potential, pH and the activation of surf clam oocytes. Gamete Res. 3: 299-304.

Fisher, G.W. and L.I. Rebhun. 1983. Sea urchin egg cortical granule exocytosis is followed by a burst of membrane retrieval via uptake into coated vesicles. Dev. Biol. 99: 456-472.

Fisher, G.W., R.G. Summers, and L.I. Rebhun. 1982. Cortical transformation in fertilized sea urchin eggs in the absence of cortical granule exocytosis. J. Cell Biol. 95: 164a.

Foerder, C.A. and B.M. Shapiro. 1977. Release of ovoperoxidase from sea urchin eggs hardens the fertilization membrane with tyrosine crosslinks. Proc. Nat. Acad. Sci. USA 74: 4212-4218.

Franklin, L.E. 1965. Morphology of gamete membrane fusion and of sperm entry into oocytes of the sea urchin. J. Cell Biol. 25: 81-100.

Friend, D.S. 1982. Plasma-membrane diversity in a highly polarized cell. J. Cell Biol. 93: 243-249.

Friend, D.S. and D.W. Fawcett. 1974. Membrane differentiations in freeze-fractured mammalian sperm. J. Cell Biol. 63: 641-664.

Friend, D.S., L. Orci, A. Perrelet, and R. Yanagimachi. 1977. Membrane particle changes attending the acrosome reaction in guinea pig spermatozoa. J. Cell Biol. 74: 561-577.

Frye, L.D. and M. Edidin. 1970. The rapid intermixing of cell surface antigens after formation of mouse-human heterokaryons. J. Cell Sci. 7: 319-335.

Gabel, C.A., E.M. Eddy, and B.M. Shapiro. 1979a. After fertilization, sperm surface components remain as a patch in sea urchin and mouse embryos. Cell 18: 207-215.

Gabel, C.A., E.M. Eddy, and B.M. Shapiro. 1979b. Persistence of sperm surface components in the early embryo. pp. 119-229. In: The Spermatozoan. D.W. Fawcett and J.M. Bedford (eds.). Urban and Schwarzenberg, Inc., Baltimore.

Gilkey, J.C. 1983. Roles of calcium and pH in activation of eggs of the medaka fish, Oryzias latipes. J. Cell Biol. 97: 669-678.

Gilkey, J.C., L.F. Jaffe, E.G. Ridgeway, and G.T. Reynolds. 1978. A free calcium wave traverses the activating eggs of the medaka, Oryzias latipes. J. Cell Biol. 76: 448-466.

Goldstein, J.L., R.G.W. Anderson, and M.S. Broan. 1979. Coated pits, coated vesicles, and receptor mediated endocytosis. Nature 276: 679-685.

Gulyas, B.J. 1980. Cortical granules of mammalian eggs. Intl. Rev. Cytol. 63: 357-392.

Gundersen, G.G., C.A. Gabel, and B.M. Shapiro. 1982. An intermediate state of fertilization involved in internalization of sperm components. Dev. Biol. 93: 59-72.

Guraya, S.S. 1982. Recent progress in the structure, origin, compository and function of cortical granules in animal eggs. Intl. Rev. Cytol. 78: 257-360.

Hall, H.G. 1981. Hardening of the sea urchin fertilization envelope by peroxidase-catalyzed phenolic coupling of tryosine. Cell 15: 343-355.

Hand, G.S. 1971. Stimulation of protein synthesis in unfertilized sea urchin and sand dollar eggs treated with trypsin. Exp. Cell Res. 64: 204-208.

Hartwig, J.W., J. Tyler, and T.P. Stossel. 1980. Actin-binding protein promotes the bipolar and perpendicular branching of actin filaments. J. Cell Biol. 87: 841-848.

Hausmann, K. and R.D. Allen. 1976. Membrane behavior of exocytotic vesicles. II. Fate of trichocyst membranes in Paramecium after induced trichocyst discharge. J. Cell Biol. 69: 313-326.

Herzog, V. and M.G. Farquhar. 1977. Luminal membrane retrieved after exocytosis reaches most Golgi cisternae in secretory cells. Proc. Nat. Acad. Sci., USA 74: 5073-5077.

Holtzman, E. 1976. Lysosomes: A Survey. Springer-Verlag, N.Y.

Hosaya, H., I. Mabuchi, and H. Sakai. 1982. Actin modulating proteins in the sea urchin egg. I. Analysis of G-actin-binding proteins by DNase I-affinity chromatography and purification of a 17,000 molecular weight component. J. Biochem. 92: 1853-1862.

Humphreys, W.J. 1967. The fine structure of cortical granules in eggs and gastrulae of Mytilus edulis. J. Ultrastruct. 17: 314-326.

Hylander, B.L. and R.G. Summers. 1981. The effect of local anesthetics and ammonia on cortical granule-plasma membrane attachment in the sea urchin egg. Dev. Biol. 86: 1-11.

Ii, I. and L.I. Rebhun. 1979. Acid release following activation of surf clam (<u>Spisula solidissima</u>) eggs. Dev. Biol. 72: 195-200.

Jaffe, L.A. 1976. Fast block to polyspermy in sea urchin eggs is electrically mediated. Nature 261: 68-71.

Jaffe, L.F. 1983. Sources of calcium in egg activation: A review and hypothesis. Dev. Biol. 99: 265-276.

Jain, M.K. 1975. Role of cholesterol in biomembranes and related systems. Curr. Top. Membr. Transp. 6: 1-57.

Johnson, M. and M. Edidin. 1978. Lateral diffusion in plasma membrane of mouse egg is restricted after fertilization. Nature 272: 448-450.

Kalina, M. and R. Robinovitch. 1975. Exocytosis couples to endocytosis of ferritin in parotid acinar cells from isoprenalin stimulated rats. Cell Tiss. Res. 163: 373-382.

Karsenti, E., M. Bornens, and S. Avrameas. 1977. Control of density and microredistribution of concanavalin-A receptors in rat thymocytes at 4°C. Eur. J. Biochem. 75: 251-256.

Kinsey, W.H. and J.K. Koehler. 1978. Cell surface changes associated with <u>in vitro</u> capacitation of hamster sperm. J. Ultrastruct. Res. 64: 1-13.

Klausner, R.D., A.M. Kleinfeld, R.L. Hoover, and M.J. Karnovsky. 1980. Lipid domains in membranes: Evidence derived from structural perturbations induced by free fatty acids and lifetime heterogeneity analysis. J. Biol. Chem. 255: 1286-1295.

Klausner, R.D. and D.E. Wolf. 1980. Selectivity of fluorescent lipid analogs for lipid domains. Biochemistry 19: 6199-6203.

Lawson, D., M.R. Raff, B. Gomperts, C. Fewtrell, and N.B. Gilula. 1977. Molecular events during membrane fusion: A study of exocytosis in rat peritoneal mast cells. J. Cell Biol. 72: 242-259.

Legault, Y., G. Bleua, A. Chapdelaine, and K. D. Roberts. 1979. The binding of sterol sulfates to hamster spermatozoa. Steroids 34: 89-99.

Longo, F.J. 1976a. Cortical changes in <u>Spisula</u> eggs upon insemination. J. Ultrastruct. Res. 56: 226-232.

Longo, F.J. 1976b. Ultrastructural aspects of fertilization in spiralian eggs. Amer. Zool. 16: 375-394.

Longo, F.J. 1979. Surface alterations of fertilized surf clam (Spisula solidissima) eggs induced by concanavalin A. Dev. Biol. 68: 422-439.
Longo, F.J. 1981. Morphological features of the surface of the sea urchin (Arbacia punctulata) egg: Oolemma-cortical granule association. Dev. Biol. 84: 173-182.
Longo, F.J. 1982. Integration of sperm and egg plasma membrane components at fertilization. Dev. Biol. 89: 409-416.
Longo, F.J. and E. Anderson. 1969. Cytological aspects of fertilization in the lamellibranch, Mytilus edulis. I. Polar body formation and development of the female pronucleus. J. Exp. Zool. 172: 69-96.
Longo, F.J. and E. Anderson. 1970a. An ultrastructural analysis of fertilization in the surf clam Spisula solidissima. I. Polar body formation and development of the female pronucleus. J. Ultrastruct. Res. 33: 495-574.
Longo, F.J. and E. Anderson. 1970b. A cytological study of the relation of the cortical reaction to subsequent events of fertilization in urethane-treated eggs of the sea urchin, Arbacia punctulata. J. Cell Biol. 47: 646-665.
Mabuchi, I. 1981. Purification from starfish eggs of a protein that depolymerizes actin. J. Biochem. 89: 1341-1344.
Marshall, J.D. and H.J. Heiniger. 1979. High affinity concanavalin A binding to sterol-depleted cells. J. Cell Physiol. 100: 539-550.
Mazia, D., G. Schatten, and R. Steinhardt. 1975. Turning on of activities in unfertilized sea urchin eggs: Correlation with changes of the surface. Proc. Nat. Acad. Sci., USA 72: 4469-4473.
Millonig, G. 1969. Fine structure analysis of the cortical reaction in the sea urchin egg: After normal fertilization and after electric induction. J. Submicr. Cytol. 1: 69-84.
Montesano, R., A. Perrelet, P. Vassalbi, and L. Orci. 1979. Absence of filipin-sterol complexes from large coated pits on the surface of cultured cells. Proc. Nat. Acad. Sci., USA 76: 6391-6395.
Moore, A.R. 1951. Action of trypsin on the eggs of Dendraster excentricus. Exp. Cell Res. 2: 284-287.
Nicosia, S.V., D.P. Wolf, and M. Inoue. 1977. Cortical granule distribution and cell surface characterization in mouse eggs. Dev. Biol. 57: 56-74.

Norman, A.W., A. M. Spelvogel, and R.G. Wong. 1976. Polyene antibiotic-sterol interaction. Adv. Lipid Res. 14: 127-170.

Nuccitelli. R. 1980. The electrical changes accompanying fertilization and cortical vesicle secretion in the medaka egg. Dev. Biol. 76: 483-498.

O'Dell, D.S., G. Ortolani, and A. Monroy. 1973. Increased binding of radioactive con A during maturation of ascidian eggs. Exp. Cell Res. 83: 408-411.

Oliver, C. and A.R. Hand. 1978. Uptake and fate of luminally administered horseradish peroxidase in resting and isoproterenol-stimulated rat parotid acinar cells. J. Cell Biol. 76: 207-220.

Orci, L., F. Malaisse-Lage, M. Ravazzola, M. Amherdt, and A. E. Reynold. 1973. Exocytosis-endocytosis coupling in pancreatic beta cell. Science 181: 561-562.

Orci, L. and A. Perrelet. 1977. Morphology of membrane systems in pancreatic islets. pp. 171-210. In: The Diabetic Pancreas. B. W. Volk and K.F. Wellman (eds.). Plenum Press, N.Y.

Orci, L. and A. Perrelet. 1978. Ultrastructural aspects of exocytotic membrane fusion. pp. 629-656. In: Membrane Fusion. G. Poste and G.N. Nicolson (eds.). Elsevier/North Holland, N.Y.

Orci, L., A. Perrelet, and D.S. Friend. 1977. Freeze-fracture of membrane fusions during exocytosis in pancreatic β-Cells. J. Cell Biol. 75: 23-30.

Otto, T.T., R.E. Kane, and J. Bryan. 1979. Formation of filopodia in coelomocytes localization of fascin, a 58,000 dalton actin cross linking protein. Cell 17: 285-293.

Ottosen, P.D., P.J. Courtoy, and M.G. Farquhar. 1980. Pathways followed by membrane recovered from the surface of plasma cells and myeloma cells. J. Exp. Med. 152: 1-19.

Pardee, A.B., J. De Asua, and E. Rozengurt. 1974. Functional membrane changes and cell growth significance and mechanism. pp. 547-561. In: Control of Proliferation in Animal Cells. B. Clarkson and R. Baserga (eds.) Cold Spring Harbor, N.Y.

Pasteels, J.J. and E. DeHarven. 1962. Etude au microscope eletronique du cortex de l'oeuf de Barnea candida (Mollusque Bivalve), et son evolution au moment de la ficondation, de la maturation, et de la segmentation. Arch. Biol. 73: 465-490.

Pasteels, J.J. 1965a. Aspects structuraux de la fécondation vus an microscope électronique. Arch. Biol. 76: 463-509.
Pasteels, J.J. 1965b. La fécondation étudiée an microscope électronique. Bull. Soc. Zool. Fr. 90: 195-224.
Pasteels, J.J. 1965c. Etude au microscope électronique de la réaction corteals. J. Embryol. Exp. Morph. 13: 327-339.
Paul, M. 1975. The polyspermy block in eggs of Urechis caupo. Exp. Cell Res. 90: 137-142.
Pearse, B.M.F. 1976. Clathrin: A unique protein associated with intracellular transfer membrane by coated vesicles. Proc. Nat. Acad. Sci., USA 73: 1257-1259.
Pelletier, G. 1973. Secretion and uptake of peroxidase by rat adenohypophyseal cells. J. Ultrastruct. Res. 43: 445-459.
Piko, L. 1969. Gamete structure and sperm entry in mammals. pp. 325-403. In: Fertilization. Vol. 2. C.B. Metz and A. Monroy (eds.). Academic Press, N.Y.
Pollack, E.G. 1978. Fine structural analysis of animal cell surfaces: Membranes and cell surface topography. Amer. Zool. 18: 25-69.
Rebhun, L.I. 1962. Electron microscope studies on the vitelline membrane of the surf clam, Spisula solidissima. J. Ultrastruct. Res. 6: 107-122.
Rosati, F., A. Monroy, and P. de Prisco. 1977. Fine structural study of fertilization in the ascidian, Ciona intestinalis. J. Ultrastruct. Res. 58: 261-270.
Runnström, J. 1948. On the action of trypsin and chymotrypsin on the fertilized sea urchin egg. A study concerning the mechanism of formation of the fertilization membrane. Archiv. Zool. 40: 1-16.
Runnström, J. 1966. The vitelline membrane and cortical particles in sea urchin eggs and their function in maturation and fertilization. Adv. Morphogenesis 5: 221-325.
Sasaki, H. 1984. Modulation of calcium sensitivity by a specific cortical protein during sea urchin egg cortical vesicle exocytosis. Dev. Biol. 101: 125-135.
Satir, B., C. Schooley, and P. Satir. 1973. Membrane fusion in a model system. J. Cell Biol. 56: 153-176.
Satir, B. 1976. Genetic control of membrane mosaicism. J. Supramol. Struct. 5: 381-389.
Schellens, J.P.M., W.T. Daems, J.J. Emeis, P. Brederoo, W.C. DeBruijn, and E. Wisse. 1977. Electron microscopical identification of lysosomes. pp. 147-208. In: Lyso-

somes: A Laboratory Handbook. J. T. Dingle (ed.). Elsevier/North Holland Biomedical Press, N.Y.

Schlessinger, J., L.S. Barak, G.G. Hammes, K.M. Yamada, I. Pastan, W.W. Webb, and E.L. Elson. 1977. Mobility and distribution of a cell surface glycoprotein and its interaction with other membrane components. Proc. Nat. Acad. Sci., USA 74: 2909-2913.

Schliwa, M. 1981. Proteins associated with cytoplasmic actin. Cell 25: 587-590.

Schroeder, T.E. 1979. Surface area change at fertilization: Resorption of the mosaic membrane. Dev. Biol. 70: 306-326.

Schroeder, T.E. 1981. Interrelations between the cell surface and the cytoskeleton in cleaving sea urchin eggs. pp. 170-216. In: Cytoskeletal Elements and Plasma Membrane Organization. G. Poste and G.L. Nicolson, (eds.). Elsevier/North Holland Biomedical Press, N.Y.

Schuel, H. 1978. Secretory functions of egg cortical granules in fertilization and development. Gamete Res. 1: 299-382.

Shapiro, B.M. 1975. Limited proteolysis of some egg surface components is an early event following fertilization of the sea urchin, Strongylocentrotus purpuratus. Dev. Biol. 46: 88-102.

Shapiro, B.M. and E.M. Eddy. 1980. When sperm meets egg: Biochemical mechanisms of gamete interaction. Intl. Rev. Cytol. 66: 257-302.

Shapiro, B.M., R.W. Schackmann, C.A. Gabel, C.A. Foerder, M.L. Farance, E.M. Eddy, and S.J. Klebanoff. 1980. Molecular alterations in gamete surfaces during fertilization and early development. pp. 257-302. In: The Cell Surface: Mediator of Developmental Process. S. Subtelny and N. K. Wessells (eds.). Academic Press, N.Y.

Shen, S.S. 1983. Membrane properties and intracellular ion activities of marine invertebrate eggs and their changes during activation. pp. 213-267. In: Mechanism and Control of Animal Fertilization. J. F. Hartmann (ed.). Academic Press, N.Y.

Shimschick, E.J. and N.H. McConnell. 1973. Lateral phase separation in phospholipid membranes. Biochemistry 12: 2351-2360.

Singer, S.J. and G.L. Nicolson. 1972. The fluid mosaic model of the structure of cell membranes. Science 178: 720-731.

Singer, S.J. 1976. The fluid mosaic model of membrane structure: Some applications to ligand-receptor and cell-cell interactions. pp. 1-24. In: Surface Membrane Receptors. R.A. Bradshaw, W.A. Frazier, R.C. Merel, D.I. Gottlieb, and R.A. Horigne-Angeletti (eds.). Plenum Press, N.Y.

Silverstein, S.C., R.M. Steinman, and Z.A. Cohn. 1977. Endocytosis. Ann. Rev. Biochem. 46: 669-722.

Steinhardt, R.A. and D. Epel. 1974. Activation of sea urchin eggs by calcium ionophore. Proc. Nat. Acad. Sci., USA 71: 1915-1919.

Steinman, R.M., S.E. Brodie, and Z.A. Cohn. 1976. Membrane flow during pinocytosis: A stereologic analysis. J. Cell Biol. 68: 665-687.

Steinman, R.M., I.S. Mellman, W.A. Muller, and Z.A. Cohn. 1983. Endocytosis and the recycling of plasma membrane. J. Cell Biol. 96: 1-27.

Swift, J.G. and T.M. Murkherjee. 1978. Membrane changes associated with mucus production in intestinal goblet cells. J. Cell Sci. 33: 301-316.

Tamm, S.L. and S. Tamm. 1983. Distribution of sterol-specific complexes in a continually shearing region of a plasma membrane and at procaryotic-eucaryotic cell junctions. J. Cell Biol. 97: 1098-1106.

Tank, D.W., E.S. Wu, and W.W. Webb. 1982. Enhanced molecular diffusibility in muscle membrane blebs: Release of lateral constraints. J. Cell Biol. 92: 207-212.

Theodosis, D.T., J.J. Dreifuss, J. Jacques, and L. Orci. 1978. A freeze fracture study of membrane events during neurophyophysis secretion. J. Cell Biol. 78: 542-553.

Tilney, L.G. 1976. The polymerization of actin. III. Aggregates of nonfilamentous actin and its associated proteins: A storage form of actin. J. Cell Biol. 69: 73-89.

Tilney, L. 1977. Actin: Its association with membranes and the regulation of its polymerization. pp. 388-402. In: International Cell Biology, 1976-1977. B.R. Brinkley and K.R. Porter (eds.). Rockefeller Univ. Press, N.Y.

Tilney, L. and R.R. Cardell. 1970. Factors controlling the reassembly of the microvillus border of the small intestine of the salamander. J. Cell Biol. 47: 408-422.

Tilney, L.G. and L.A. Jaffe. 1980. Actin, microvilli and the fertilization cone of sea urchin eggs. J. Cell Biol. 87: 771-782.

Tilney, L.G., P. Kiehart, C. Sardet, and M. Tilney. 1978. Polymerization of actin. IV. Role of Ca^{++} and H^+ in the assembly of actin and in membrane fusion in the acrosomal reaction of echinoderm sperm. J. Cell Biol. 77: 536-550.

Träuble, H. and E. Sackermann. 1972. Studies of the crystalline-liquid crystalline phase transition of lipid model membranes. III. Structure of a steroid-lecithin system below and above the lipid phase transition. J. Amer. Chem. Soc. 94: 4499-4510.

Vacquier, V.D., M.J. Tegner, and D. Epel. 1973. Protease released from sea urchin eggs at fertilization alters the vitelline layer and aids in preventing polyspermy. Exp. Cell Res. 80: 111-119.

Vacquier, V.D. 1975. The isolation of intact cortical granules from sea urchin eggs: Calcium ions trigger granule discharge. Dev. Biol. 43: 62-74.

Vacquier, V.D. 1981. Dynamic changes of the egg cortex. Dev. Biol. 84: 1-26.

Vernon, M., C. Foerder, E.M. Eddy, and B.M. Shapiro. 1977. Sequential biochemical and morphological events during assembly of the fertilization membrane of the sea urchin. Cell 10: 321-328.

Veron, M. and B.M. Shapiro. 1977. Binding of concanavalin A to the surface of sea urchin eggs and its alteration upon fertilization. J. Biol. Chem. 252: 1286-1292.

Weeds, A. 1982. Actin-binding proteins - regulators of cell architecture and motility. Nature 296: 811-816.

Weiss, R.L., D.A. Goodenough, and U.W. Goodenough. 1977a. Membrane differentiations at sites specialized for cell fusion. J. Cell Biol. 72: 144-160.

Weiss, R.L., D.A. Goodenough, and U.W. Goodenough. 1977b. Membrane particle arrays associated with the basal body and with contractile vacuole secretion in Chlamydomonas. J. Cell Biol. 72: 144-160.

Whitaker, M.J. and R.A. Steinhardt. 1982. Ionic regulation of egg activation. Quar. Rev. Biophys. 15: 593-666.

Wolf, D.E., M. Edidin, and A.H. Handyside. 1981a. Changes in the organization of the mouse egg plasma membrane upon fertilization and first cleavage. Indications from the lateral diffusion rates of fluorescent lipid analogs. Dev. Biol. 85: 195-198.

Wolf, D.E., W. Kinsey, W. Lennarz, and M. Edidin. 1981b. Changes in the organization of the sea urchin egg plasma membrane upon fertilization: Indications from the lateral diffusion rates of lipid-soluble fluorescent dyes. Dev. Biol. 81: 133-138.

Wolf, D.E. and C.A. Ziomek. 1983. Regionalization and lateral diffusion of membrane proteins in unfertilized and fertilized mouse eggs. J. Cell Biol. 96: 1786-1790.

Wolpert, L. and E.H. Mercer. 1961. An electron microscope study of fertilization of the sea urchin egg Psammechinus milliaris. Exp. Cell Res. 22: 45-55.

Wu, E.S., D.W. Tank, and W.W. Webb. 1982. Unconstrained lateral diffusion of concanavalin A receptors of bulbous lymphocytes. Proc. Nat. Acad. Sci., USA 79: 4962-4966.

Yanagimachi, R. and Y.D. Noda. 1970. Electron microscope studies of sperm incorporation into the golden hamster egg. Am. J. Anat. 128: 429-462.

Yanagimachi, R., G.L. Nicolson, Y.D. Noda, and M. Fujimoto. 1973. Electron microscopic observations of the distribution of acidic anionic residues on hamster spermatozoa and eggs before and during fertilization. J. Ultrastruct. Res. 43: 344-353.

Yanagimachi, R. and G.L. Nicolson. 1976. Lectin-binding properties of hamster egg zona pellucida and plasma membrane during maturation and preimplantation development. Exp. Cell Res. 100: 249-257.

Zamboni, L. 1971. Fine Morphology of Mammalian Fertilization. Harper and Row, N.Y.

Ziomek, C.A. and D. Epel. 1975. Polyspermy block of Spisula egg is prevented by cytochalasin B. Science 189: 139-141.

Ziomek, C.A., S. Schulman, and M. Edidin. 1980. Redistribution of membrane proteins in isolated mouse intestinal epithelial cells. J. Cell Biol. 86: 849-857.

PHYSICAL INTERACTIONS BETWEEN ASTERS AND THE CORTEX IN ECHINODERM EGGS
Thomas E. Schroeder

ABSTRACT

   Microscopic observations of living eggs of sea urchins and other echinoderms suggest that several kinds of physical interactions, previously unrecognized or little-known, occur between asters and the egg cortex at specific times in development. Asters of various sorts (pre-meiotic asters, sperm asters, meiotic asters, mitotic asters, and cytasters) interact with the cortex by simply attaching, by moving toward the cortex, or by exerting "pulling forces" and deforming the cortex and contour of the egg. Examples are drawn from a wide variety of species, indicating the generality of the phenomena. The evidence is presently at a circumstantial and descriptive level; the underlying mechanisms of physical interaction are unknown, nevertheless the data suggest that such interactions may be physiologically relevant to the overall behavior of asters and the cortex in echinoderm eggs.

INTRODUCTION

   Two prominent morphological elements of eggs are the asters and the cortex. In this article I will present evidence suggesting some unusual physical interactions between these elements. According to my interpretations, the findings indicate that astral rays can "attach" to the cortex and can exert "pulling" forces upon it. Such ideas are somewhat unorthodox and, indeed, have not been proven. Asters are conventionally considered to be involved in positioning a mitotic apparatus within an egg and to be rather separate from the cortex, except when conveying "instructions" to the cortex, as during cleavage stimulation (Rappaport, 1971). The circumstantial evidence cited here raises the possibility that asters can mediate physical actions upon the cortex that have not generally been recognized.

The cortex of an egg underlies the plasma membrane. It is operationally defined as the outermost few microns of cytoplasm and is one of the most dynamic domains of an egg (Schroeder, 1981a; Vacquier, 1981). The cortical matrix is the mechanically significant component, but within the cortex also resides a number of granules and vacuoles. The cortical matrix has been extensively studied and is known to be rich in actin and actin-associating proteins that serve both cytoskeletal and contractile functions. In general, the cortical matrix mediates the establishment and maintenance of cell shape and it also stabilizes the cell surface.

Asters, in contrast, reside in the egg's endoplasm, usually at some distance from the cortex. They are typically found during meiosis and mitosis as structures organized around the two centrosomes that also serve as poles for the division spindles. The complex classical and modern literature concerned with the central regions of asters (i.e., centrioles, centrosomes and centrospheres) is extensive, and it would not be fruitful to review it here. An aster is comprised in large part of radial fascicles of microtubles (astral rays). Endoplasmic vesicles may also contribute to the structure of an aster, but their role is not well understood. When astral rays extend into the yolky cytoplasm, their presence can be detected with the simplest microscope as radiating patterns of aligned cytoplasmic granules.

Diversity of Astral Configurations

The eggs and embryos of sea urchins are much more frequently studied than the eggs and embryos from members of the other echinoderm classes (starfish, brittle-stars, crinoids, and sea cucumbers). In some ways, sea urchins are simply more convenient to use; they are widely distributed and their eggs are easily obtained and readily fertilizable. Nevertheless, a sea urchin egg is biologically quite different from nearly all other eggs in the animal kingdom, because it completes its meiotic maturation divisions in the ovary long before ovulation and is fully fertilizable at that time. Ovarian sea urchin eggs are haploid cells arrested at the pronucleus stage, whereas oocytes of the other echinoderms

ASTERS AND THE CORTEX   71

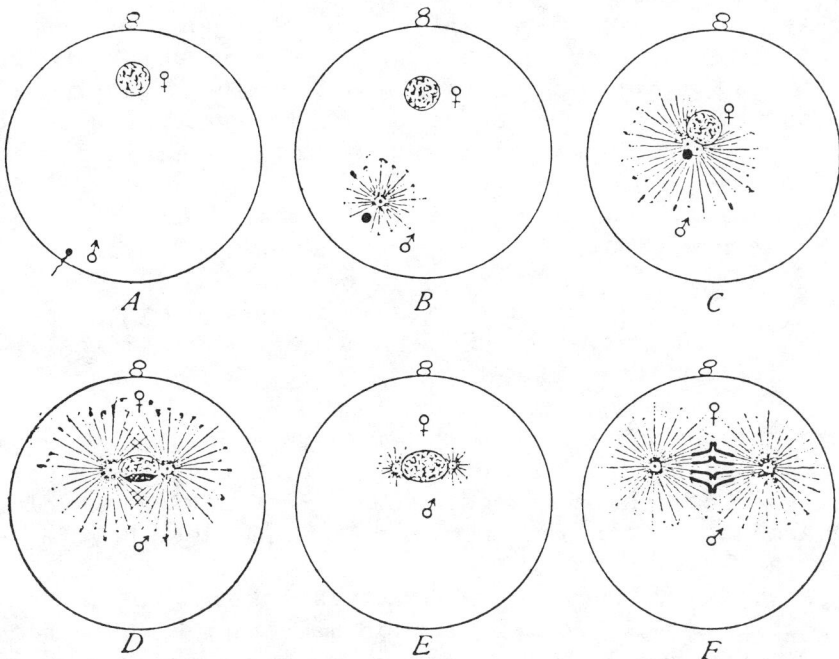

Fig. 1.   There are two cycles of aster formation between fertilization and first division during normal development of a sea urchin embryo, here shown in drawings by Wilson (1928, Fig. 186). A sperm aster (B) forms soon after fertilization and is the device that propels the male pronucleus to the center of the egg and also attracts the female pronucleus (B & C). Originally, the astral rays of the sperm aster appear to emanate from a single centrosome, but this soon divides into two (C). As pronuclei merge and fuse, the two centrosomes assume diametrically opposite positions adjacent to the zygote nucleus (D). The sperm aster develops into the diaster and its rays extend far out into the cytoplasm during the "pause" before nuclear envelope breakdown. Immediately prior to nuclear envelope breakdown, the astral rays of the diaster disappear (E). This completes the first aster cycle. The mitotic spindle sets up synchronously with nuclear envelope breakdown and mitotic asters form during the second cycle of aster assembly. During metaphase (F), astral rays of mitotic asters gradually elongate into the yolky cytoplasm.

are arrested at the germinal vesicle stage of meiotic prophase. Also, oocytes contain both microtubles and centrotubules or centrosomes prior to fertilization.

Early development in echinoderms is characterized by cycles of aster assembly and disassembly, nevertheless some of the basic facts of these cycles are still poorly known and there is substantial disagreement over morphological terminology. My interpretations of early events in sea urchin eggs agree with those of Wilson (1895, 1928) and Fry (1936) that there is one (and only one!) aster cycle preceding the formation of the spindles and asters involved in the first mitotic division (Fig. 1). Thus, soon after fertilization a sperm aster forms (Figs. 1a-c); it subsequently develops directly (in my opinion) into a diaster as the centrosome splits (Fig. 1c) and the astral rays elongate virtually to fill the egg cytoplasm (Figs. 1d & 6). The diaster survives until shortly before nuclear envelope breakdown and the onset of mitosis. The enlarged sperm aster is sometimes referred to as a "monaster" and the early diaster as the "streak" because of its peculiar morphology in some cases.

Some investigators interpret these early events differently (Harris et al., 1980; Bestor & Schatten, 1981). They think that the sperm aster fully disassembles and is then replaced by a newly assembled diaster. These investigators further believe that the diaster represents a unique "interphase aster" and that therefore an extra aster cycle exists between the sperm aster and the mitotic apparatus. By either interpretation, it is agreed that astral structures do exist during the interphase preceding the first division cycle; the disagreement concerns the number of unique aster cycles up to this stage.

Regardless, the diaster exists during the protracted pre-mitotic period from fusion of male and female pronuclei until breakdown of the zygote nucleus, i.e., the "pause", according to the terminology of Wilson (1895). At the height of the diaster configuration, astral rays extend throughout the egg cytoplasm, apparently as far as the cortex. During this period the zygote nucleus is deformed into a prolate spheroid (Figs. 1d-e), as if it experiences a pulling force along an axis between the centrosomes. A few minutes before any further change is seen in the nucleus, the astral rays of the diaster suddenly disappear (Fig. 1e). The nucleus deforms by indentations at the poles (Fig. 6), the envelope begins to break down at these points, and the bona fide mitotic spindle forms. The mitotic apparatus is not a direct descendant of the diaster but represents a separate cycle of assembly. During the early portions of mitotic metaphase, astral rays usually radiate only a short distance into the yolky cytoplasm, and later they gradually elongate.

Extreme elongation of astral rays of the mitotic asters is experienced during anaphase and, at least in some species, these persist through telophase. Consequently, in those species, long-lasting remnants of the mitotic asters persist into the early stages of interphase after cytokinesis is complete. Some investigators attribute such astral rays to yet another uniquely assembled "interphase aster" (Harris et al., 1980). In any case, these astral rays eventually disassemble later in interphase before the next mitotic apparatus begins to be set up.

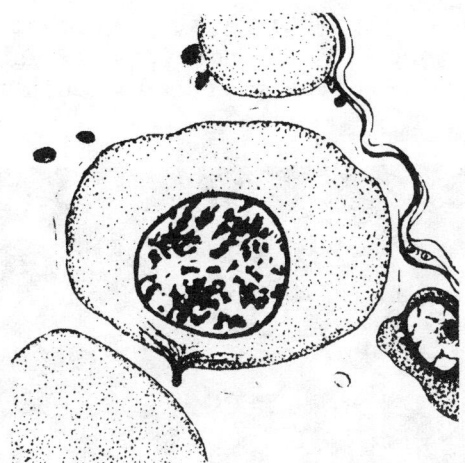

Fig. 2. A pre-meiotic aster exists near the animal pole of starfish oocytes. (a) The centrosome and astral ray microtubules of a starfish (Pisaster ochraceus) immature oocyte are stained by the indirect immunofluorescence technique with antitubulin antibody. The centrosome is located adjacent to the germinal vesicle (GV). In (b) a drawing of an oocyte of Asterias forbesi, reproduced from Wilson and Mathews (1895), similarly shows two centrosomes next to the germinal vesicle. Astral rays are indicated by long wavy lines. In such oocytes, astral rays may interact with the cortex to maintain the germinal vesicle in its eccentric position beneath the animal pole. Bar = 10 µm.

In echinoderms other than sea urchins, fertilization occurs after breakdown of the germinal vesicle but before complete egg maturation and so additional aster cycles can be distinguished. Of course, asters are associated with the meiotic apparatuses of the maturation divisions. But there is yet another astral configuration -- the pre-meiotic aster -- that exists in the narrow cytoplasmic space between the germinal vesicle and the animal pole of echinoderm oocytes at the germinal vesicle stage, probably including sea urchin oocytes. Such pre-meiotic asters were originally described in echinoderm oocytes by Wilson and Mathews (1895), as reproduced in Figure 2b, and they have recently been redescribed using immunocytochemical techniques for microtubules (Otto & Schroeder, 1984; Schroeder & Otto, 1984). A moderate number of long microtubules radiate from one or two centrosomes into or very near the cortex and adjacent to the germinal vesicle (Fig. 2a). The aster is probably attached somehow to the cortex near the animal pole and also to the germinal vesicle. Unlike sea urchin eggs, therefore, echinoderm oocytes have centrosomes and organized patterns of microtubules.

As illustrated in Figure 3, the cell surface of some echinoderm oocytes protrudes at the animal pole. Such animal

74 SCHROEDER

pole protrusions are typical of sea cucumber oocytes and have also been described in the oocytes of some sea urchins. Fibrillar material is often pictured within these protrusions

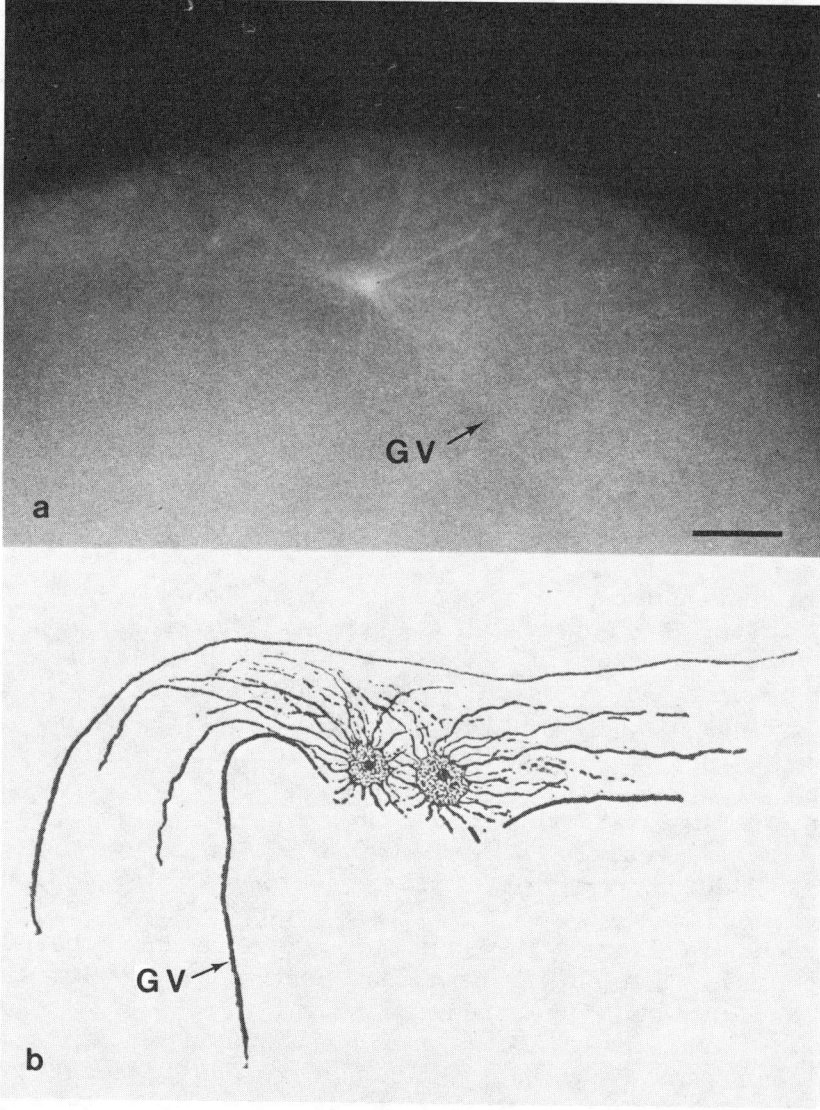

Fig. 3. Immature oocytes of certain echinoderms exhibit a distinctive protrusion at the animal pole, here seen in a <u>Paracentrotus lividus</u> oocyte, as reproduced from Lindahl (1932). Fibrous structures within the protrusion likely represent microtubules constituting a "pre-meiotic aster" (see Fig. 2). The existence of the protrusion may indicate an interaction between the pre-meiotic aster and cortex whereby the cortex is "attracted" or otherwise "pulled in" toward the aster.

(Boveri, 1901; Lindahl, 1932; Monne, 1947) and it has recently been identified as composed of microtubules (Smiley, 1984). Although the analysis of these structures is incomplete, the circumstantial evidence appears strong that entire pre-meiotic asters (probably including centrosomes) exist within these protrusions. If this is correct, then the specific existence of a protrusion is conceivably explained by the ability of pre-meiotic asters to attract or "gather" the cortex around them, an unusual phenomenon for which additional evidence will be presented below.

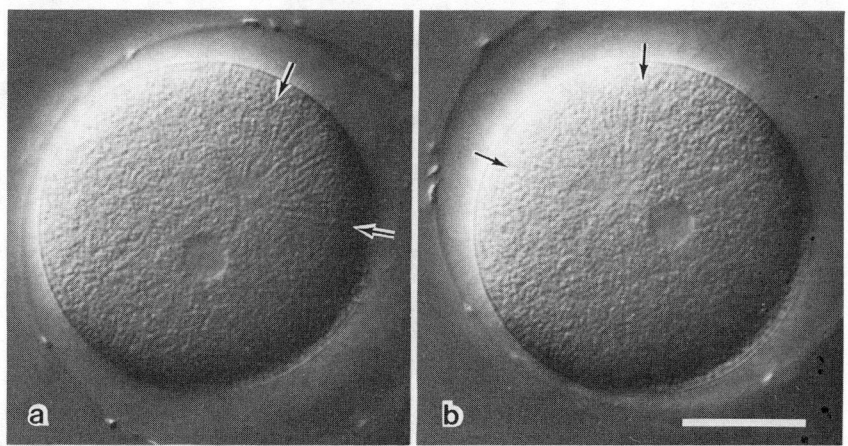

Fig. 4. A spherically asymmetric sperm aster is shown in two Nomarski views (rotated 90°) of a Lytechinus pictus egg 9 min after fertilization. Astral rays are concentrated in a narrow sector (between arrows). The clustered astral rays are evident in certain orientations (a) because of their orientation in the biased optical system, but they become virtually invisible when the specimen is rotated (b). The fact that numerous additional astral rays are not visible outside of the sector indicates that the clustering is not an optical artifact. One explanation of the cluster of astral rays is that they become embedded in the cortex near the site of sperm entry at an early stage, as postulated diagrammatically in Figure 5. Bar = 50 µm.

Asymmetry of the Sperm Aster

Observations. The sperm aster is an array of radiating astral rays containing microtubules. The growing sperm aster propels the sperm head (or male pronucleus) and centrosome toward the egg's center by elongating its astral rays and "fending off" against the cortex. It also attracts the female pronucleus and mediates its motion toward the male pronucleus for eventual fusion. The sperm aster is usually thought of as being spherically symmetrical.

In certain circumstances, the sperm aster is obviously not spherically symmetrical. Rather, astral rays can be much more concentrated in the sector facing the cortical region near the site of sperm entry, i.e., away from the direction of sperm aster migration. Figure 4 illustrates such a case in a fertilized egg of Lytechinus pictus. A cluster of astral rays occupies one sector of the egg. Since the Nomarski optics used for Figure 4 is a "biased" illumination system, two views of the same egg rotated 90° are shown, in order to demonstrate that the asymmetry is not an optical artifact. The cluster of astral rays is visible in one orientation (Fig. 4a) and disappears in the other (Fig. 4b), yet in neither image is there evidence for the existence of additional clustered astral rays in adjacent sectors.

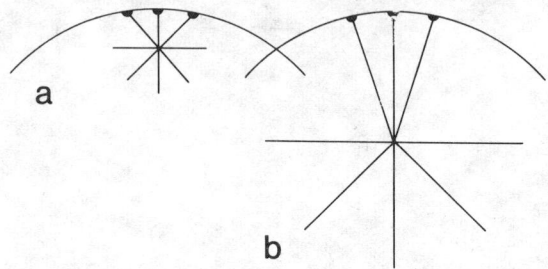

Fig. 5. An interpretation of the mechanism whereby astral rays, initially spherically symmetrical as in (a), become clustered near the site of sperm entry at a later stage (b). At an early stage, short astral rays become embedded in the cortex. Attachments between astral rays and the cortex are depicted by densities at the surface. As astral rays elongate, the sperm aster moves toward the egg center (b) and astral rays previously embedded in the cortex subtend a progressively smaller angle, i.e., they become clustered in a sector.

Discussion. Asymmetry of the sperm aster has been noted previously by Hamaguchi and Hiramoto (1980, see Figs. 4, 5 & 7) and Schatten (1981) in living eggs and is evident in the immunofluorescence studies of Harris et al. (1980, see Figs. 3 & 25) and Bestor and Schatten (1981, see Fig. 4). What circumstances can explain this asymmetry?

Conceivably, the cluster of astral rays could arise by the preferential proliferation of microtubules in the direction of the site of sperm entry; however, I find it difficult to image a precise mechanism for this. Alternatively, I favor the possibility that the distal ends of astral rays become embedded in the adjacent cortex during an early stage, while the sperm aster is still symmetrical (diagrammed in Fig. 5). As the sperm aster moves, elongating astral rays with embedded ends are confined to a sector of decreasing subtended angle.

In the present context, the essential feature of this

simple explanation of asymmetry in the sperm aster is physical attachment of astral rays to the cortex, as symbolized in Figure 5 by dense plaques. The postulated existence of such novel attachments is reinforced by additional suggestions of aster-cortex attachments described below.

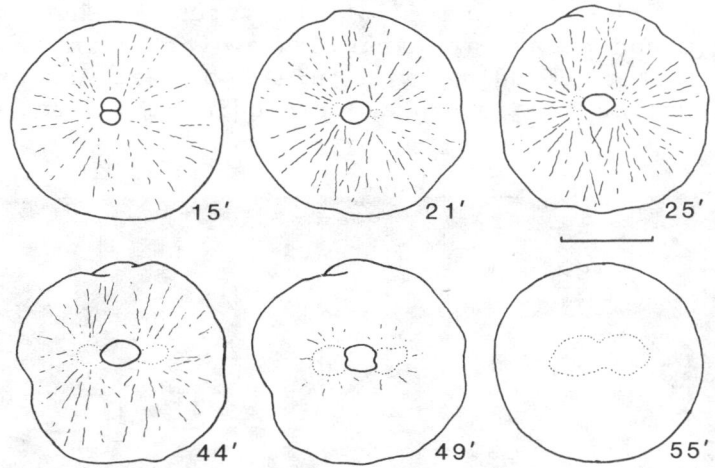

Fig. 6. "Crenation" in a sand dollar egg. This phenomenon involves a wrinkling of the surface during the pre-mitotic "pause" when the diaster maximally fills the cytoplasm with radiating astral rays. These drawings were traced from Nomarski photographs of a single Clypeaster japonicus egg (25°C). Times after fertilization are indicated in minutes. Dotted lines indicate yolk-free clear zones around the centrosomes. Crenation begins at 21 min, just as pronuclei fuse; it progresses during most of the subsequent 30 min "pause" and begins to subside simultaneously with disappearance of the astral rays of the diaster. Indentations in the egg surface are thought to be caused by astral rays "pulling in" the cortex. Bar = 50 μm.

"Crenation" During the "Pause"

Observations. In certain batches of fertilized sea urchin eggs of some species, the egg contour is markedly wrinkled or "crenated" (Fig. 6) during the "pause" (between pronuclear fusion and nuclear envelope breakdown). A striking temporal correlation exists between crenation and the condition of the sperm aster/diaster: (1) The egg contour is smooth soon after fertilization when the sperm aster is small. (2) Crenation begins when the astral rays of the early diaster noticeably fill the cytoplasm, i.e., when they first make contact with the cortex. (3) Irregular wrinkles and bulges become more pronounced and persist for 20-30 min while the diaster is large. (4) Crenations abruptly disappear and a smooth egg contour is reestablished when the astral rays of the diaster disappear a few minutes prior to nuclear envelope breakdown.

In Figure 6 a complete episode of crenation is illustrated in tracings from photomicrographs of a Clypeaster japonicus egg. Eggs of this species are exceptionally transparent (Fig. 7), allowing simultaneous observations of egg contour, nuclear condition, and astral rays within the cytoplasm. Crenation is characteristic of Clypeaster japonicus eggs in about one-third of the batches I have observed. The phenomenon occurs in natural seawater or in calcium-free seawater after denuding with urea, but it has only been observed during the interphase preceding the first division.

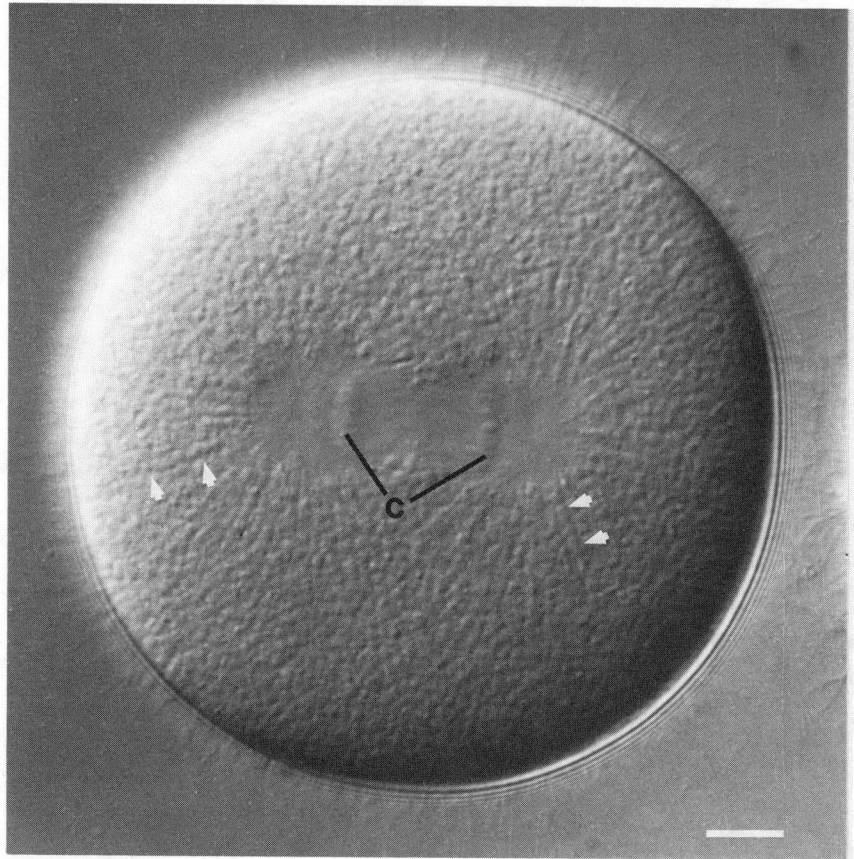

Fig. 7. Eggs and blastomeres of the irregular sea urchin Clypeaster japonicus are transparent, allowing detailed identification of cytological features in living cells. This Nomarski photomicrograph shows a single isolated blastomere from a two-cell embryo that has been denuded and raised in calcium-free seawater. The two sets of chromosomes (C) are widely separated, indicating the anaphase/telophase transition. At this stage astral rays of the mitotic apparatus (arrows) extend far out into the yolky cytoplasm and appear to have penetrated the cortex at polar regions (left and right), but they fail to contact the cortex at the equator. Bar = 10 μm.

Discussion. I have noticed prominent crenation during the "pause" in various species of sea urchin besides Clypeaster japonicus, including Paracentrotus lividus and Eucidaris tribuloides. In the literature, crenation has also been described for Arbacia punctulata (Harvey, 1956, see Plate III, Figs. 5-8 and p. 96; Heilbrunn, 1920, see p. 222), for Echinus esculentus (Harvey, 1935), for Paracentrotus lividus (Runnstrom, 1975, see p. 646), for Pseudocentrotus depressus (Iida, 1943, see p. 144; Dan, 1943, see p. 335), for Sphaerechinus granularis (Harvey, 1935), for Clypeaster japonicus (Dan, 1943), for Psammechinus miliaris (Lorch, 1952), and for Hemicentrotus pulcherrimus (Yoneda et al., 1978). It has also been demonstrated in brittle-star eggs (Olsen, 1942). In all cases, crenation is restricted to the pre-mitotic "pause".

To date there has been no satisfactory explanation of the crenation phenomenon. Harvey (1935) ruled out anoxia as a cause (but see Runnstrom, 1975, p. 646), and others have felt that crenation is not related to hypertonicity, even though this may enhance it.

In my opinion, the precise temporal correlation between the enlarged diaster and crenation is persuasive circumstantial evidence that astral rays are mechanistically involved. Specifically, I suggest that crenation occurs because the astral rays impinge upon, attach to, and somehow "pull in" the egg cortex. The actual force acting upon the cortex could be generated by a retraction of the elongated astral rays or by some independent action that draws the cortex (at least at some points) along the astral rays toward the centrosomes of the diaster. The ability of the cortex to respond by wrinkling may partially depend upon its intrinsic mechanical condition; for example, if it were too stiff, wrinkling would be resisted and so not occur. In actuality, it is known that cortical stiffness during the "pause" is quite low (Hiramoto, 1970) and begins to rise at the time of nuclear envelope breakdown (Schroeder, 1981b; Yoneda & Schroeder, 1984).

"Oblation" Prior to Furrow Formation

Observations. At about the beginning of anaphase when chromosomes initially move apart, the astral rays of the mitotic asters begin to extend progressively farther into the yolky cytoplasm. At some time shortly before the first appearance of the cleavage furrow, the astral rays apparently extend all the way to the cortex, at least at the polar regions (Fig. 7). They may never reach the equatorial cortex in appreciable numbers (Asnes & Schroeder, 1979). In several species of sea urchin eggs, just before the cleavage furrow pinches in, the equator bulges outward as the spherical egg is transiently deformed to an oblate spheroid, i.e., "M&M" shape (Fig. 8). The shortened axis of this "oblation" phenomenon corresponds to the axis of the mitotic apparatus. One or two minutes later, a cleavage furrow forms at the equator. Figure 8 illustrates oblation in an egg of Eucidaris tribuloides photographed over a 5-min interval. The extent of oblation is about 6%, that is, the expanded equatorial plane has a diameter about 6% greater than the shortened axial dimension.

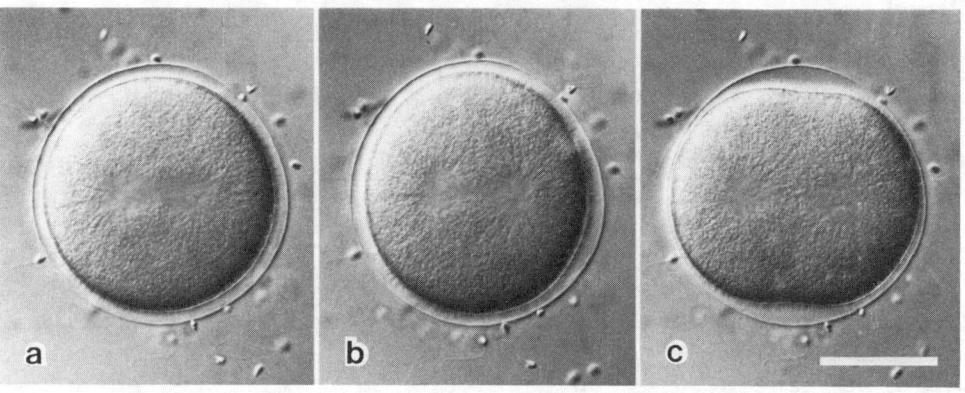

Fig. 8. "Oblation" as seen in a series of Nomarski photomicrographs of an egg of the sea urchin Eucidaris tribuloides at (a) 55 min, (b) 57 min, and (c) 60 min after fertilization (26°C). In (a) the egg is spherical in shape, but in (b) it has assumed the shape of an oblate spheroid; soon afterward, a cleavage furrow is underway in (c). "Oblation" is the consequence of shortening along the axis of the mitotic aster and expansion at the equator. This transient phenomenon is thought to be due to an unidentified axial contraction in the endoplasm that is transmitted by telophase astral rays to the cortex at the poles to which they are physically attached. Bar = 50 μm.

I have also recorded oblation phenomena by time-lapse cinemicrography in Clypeaster japonicus, Strongylocentrotus purpuratus, and Dendraster excentricus eggs under various conditions of culture. The phenomenon occurs in undenuded eggs in natural seawater or in denuded eggs in calcium-free seawater at both first or second divisions. In Figure 9 the course of an oblation event is plotted from a film of a denuded Clypeaster egg at first division; the transparency of this egg yields precise temporal relationships between the times of chromosome separation (anaphase onset at 71.5 min after fertilization), oblation (from 74.5 to 77.5 min), the time when astral rays appear to make first contact with the polar cortex (75 min), and the onset of furrowing (77.5 min). Considering that the moment when astral rays actually make contact with the polar cortex may be slightly earlier than it can actually be perceived, it seems fair to infer that oblation occurs simultaneously with aster-cortex interaction.

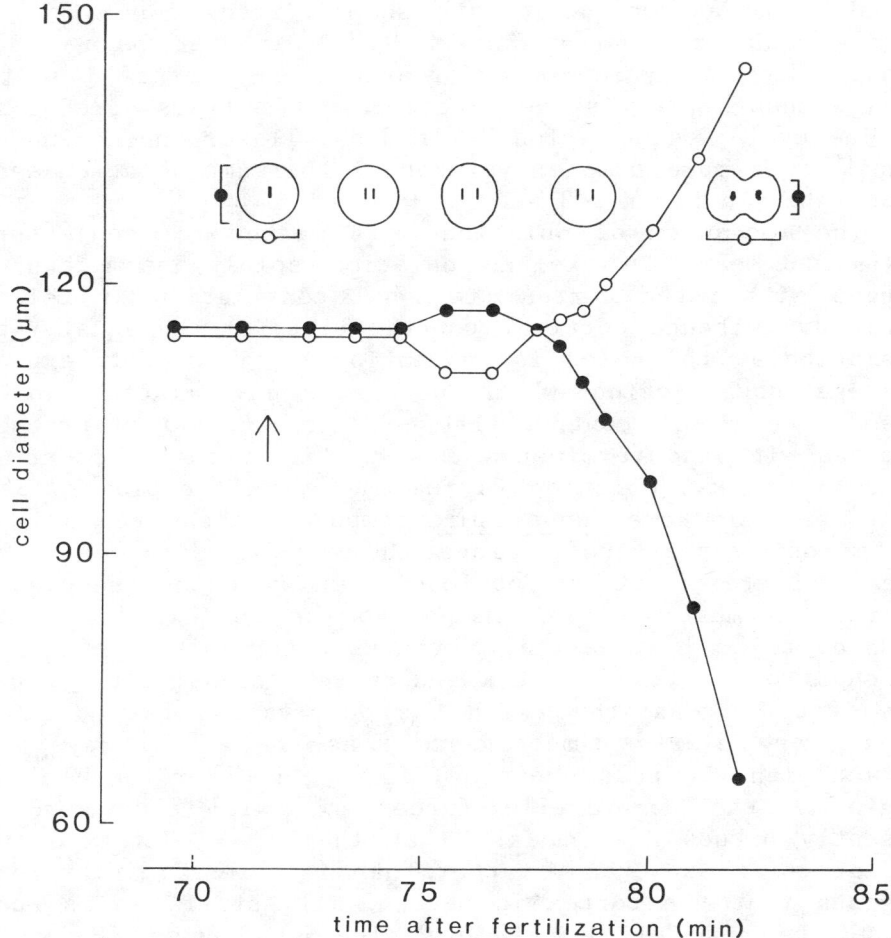

Fig. 9. Graphic illustration of the temporal relation of an episode of "oblation" in an egg of <u>Clypeaster japonicus</u> (23°C) before and during the early stages of first division; the data are taken from a time-lapse film using Nomarski optics. The egg diameters along two axes (open circles along the mitotic axis and filled circles for the equatorial diameter) are plotted against time after fertilization. The egg becomes an oblate spheroid from 74.5 min to 77.5 min and thereafter immediately progresses into cleavage furrow formation. The initial separation of chromosome sets (the onset of anaphase) is indicated by the arrow (71.5 min). "Oblation" commences at the time that astral rays first detectably make contact with the cortex at the poles. This temporal correlation suggests that physical attachment between astral rays and the cortex is necessary for this shape-change, although the force-generating mechanism is unknown.

Discussion. Oblation is often a fairly subtle event. Among the rare references to it, Usui and Yoneda (1982, see Fig. 1b) describe oblation in Hemicentrotus pulcherrimus and Mamaguchi and Hiramoto (1978) mention its occurrence in Cylpeaster. In addition, Yoneda and Dan (1972, see p. 578) recount a colleague's observations of a related phenomenon in eggs of Astriclypeus manni: "...4 min before cleavage...the egg abruptly assumes an oblate shape...then elongates to a prolate, the maximum change in the diameter being about ±3-5%. Such a cycle of deformation is repeated twice or more, although progressively decaying. It takes some 45 sec for one cycle of pulsation." This oscillatory phenomenon is presumably a more complex version of the same oblation event illustrated in Figures 8 and 9.

The mechanism of oblation is a matter for conjecture. It is not easy to explain oblation solely by mechanical changes within the cortex (regional contraction or relaxation), even though cortical contractility at the equator and relaxation at the poles is conventionally invoked to explain cleavage constriction which occurs a few minutes later. Indeed, cortical contractility at the equator probably competes with and terminates the oblation event. Rather, it seems to me that oblation is reasonably well explained as a contraction of some endoplasmic component along the axis of the mitotic apparatus. Since the egg is deformed to an oblate spheroid, it is obvious that such an endoplasmic contraction must be transmitted to the cortex; for accomplishing this, I suggest that the astral rays are physically attached to the polar cortex and thereby transmit the force. This idea is consistent with the observation that oblation occurs very nearly simultaneously as the astral rays make contact with the polar cortex (Fig. 9); however the ultimate origin of the contractile force for axial shortening is presently unknown. The fact that the egg is deformed as a regular oblate spheroid, rather than in a less regular form, suggests that the cortex is mechanically stiff, as is known to be the case shortly before cleavage begins (Hiramoto, 1970).

Oblation occurs at about the same time that the cleavage stimulus is being imparted by the mitotic asters to the cortex, although there is still considerable uncertainty regarding the nature and locus of aster-mediated cortical influence that initiates cleavage (Rappaport, 1971; Schroeder, 1981b). Nevertheless, according to the "global contraction-polar relaxation" hypothesis of cleavage initiation (Schroeder, 1981b), asters stimulate relaxation of the polar cortex as a step in cleavage initiation. Is such a phenomenon consistent with aster-mediated "pulling in" of the polar cortex, as postulated here for oblation? At present, this issue cannot be clearly settled, but it seems worth bearing in mind that other important interactions between asters and the cortex may be occurring simultaneously with oblation.

Aster-mediated Deformation in Cytochalasin-inhibited Eggs

<u>Observations</u>. The formation and progress of a cleavage furrow is inhibited by cytochalasin B. In addition, peculiar cell morphologies sometimes occur when such inhibited eggs enter the interphase after the first mitosis (Fig. 10). At about the time of nuclear envelope reformation, the eggs become oblate in shape (Figs. 10a & b). In these eggs the mitotic asters persist for protracted periods. The shape-change becomes more extreme as the equator gradually bulges outward and pseudo-furrows are created nearby (Fig. 10c); the polar and subpolar regions are at first drawn tightly around the perimeter of the persisting asters and then pseudo-furrows appear. At later stages the cell outline becomes irregularly indented, especially at the poles (Fig. 10d), and the pseudo-furrows deepen.

<u>Discussion</u>. In these inhibited eggs, cell deformation becomes progressively more extreme as the mitotic asters persist for abnormally long times. Without attempting to explain all of these phenomena, it appears that at later stages the polar and subpolar cortex becomes attached to the aster perimeter, whereas regions beyond the influence of astral rays, such as the equatorial cortex, are free to bulge outward. Pseudo-furrows indicate transitions between these regions.

My interpretation of such odd images is that they represent variations of some of the same aster-mediated events already discussed. That is, wherever astral rays make significant contact with the cortex that cortex is physically "pulled in". Cortical regions beyond the influence of astral rays are unaffected and bulge outward (by internal pressure). Clearly, in Figure 10 the early stages resemble "oblation", as previously described during normal cell division (Figs. 8 & 9). The later deformations, including prominently the indentations in Figure 10d, are reminiscent of the "crenation" phenomenon in the interphase preceding mitosis in certain normal eggs (Fig. 6).

The differences between early and later deformations may reflect changes in the response of the cortex to a constant

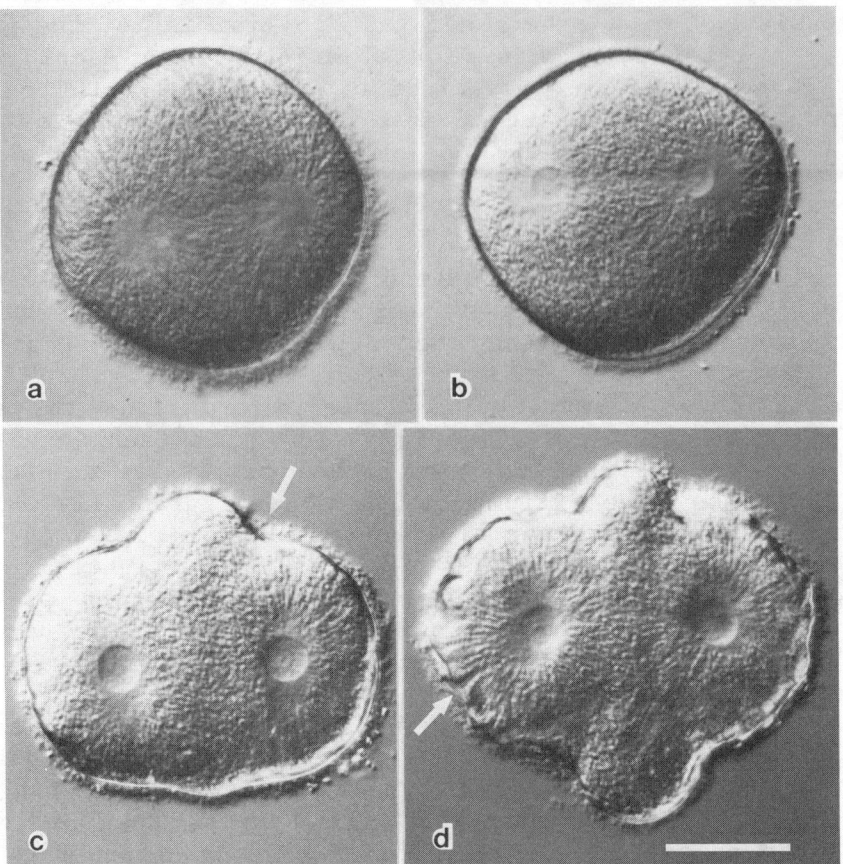

Fig. 10. Denuded eggs of the sand dollar <u>Dendraster</u> <u>excentricus</u> (16°C) exhibiting shapes that suggest a physical interaction between astral rays and the polar cortex. These eggs were treated with 10 μg/ml cytochalasin B before the onset of mitosis. Untreated control eggs began to furrow at 70 min after fertilization. Times after fertilization: (a) 82 min, (b) 83 min, (c) 99 min, and (d) 101 min. Cytochalasin B inhibits cleavage and results in persistent asters. Eggs first become deformed as oblate spheroids (a & b), progressively show pseudo-furrows (c, arrow) and outward bulges around the equator, and finally exhibit indentations at the poles (d, arrow). The contours of these eggs give the impression that the asters physically "attract" or somehow draw in the cortex wherever astral rays make significant physical contact, namely at the poles and subpolar regions. Since astral rays do not make contact with the equatorial cortex, these regions are free to bulge outward. Bar = 50 μm.

aster-mediated activity. With the advance of time, the mechanical stiffness of the cortex could change. Thus, at early stages a fairly high cortical stiffness may lead to oblation, whereas at later stages cortical stiffness may decline, allowing some local regions of the cortex to be indented as they are "pulled in" along (or between) particular astral rays.

GENERAL DISCUSSION

The above examples present circumstantial evidence that has been interpreted to signify that astral rays can attach to and exert (or transmit) inward-directed pulling forces upon the cortex. Accordingly, asters appear to interact physically with the cortex in ways that presently seem unfamiliar. Although it is impossible at this stage of investigation to identify the origin of the dynamic forces, their effect upon the cortex can be related to both geometrical aspects of the interaction and to the mechanical properties of the cortex.

During sperm aster growth (Fig. 4), astral rays may simply attach to the cortex without exerting any observable inward force. Other kinds of asters may significantly "pull upon" the cortex, e.g., the pre-meiotic asters in some immature oocytes (Fig. 2) to form animal pole protrusions, the interphase diaster to cause "crenation" (Fig. 6), and mitotic asters to cause "oblation" prior to cleavage (Fig. 8). Aberrant versions of these same interactions may be responsible for deformations of inhibited eggs (Fig. 10).

All of the present examples have been taken from echinoderm eggs and oocytes, however one can encounter related events in other cell types (e.g., Schrader, 1953, see Figs. 11 & 12; Foe & Alberts, 1983, see Figs. 4 & 14). Thus, aster-mediated physical effects upon the cortex may be widespread and physiologically significant.

As suggested here, aster-cortex interactions can often result in the cortex being deformed, especially if the cortex is more able to move than the aster. If the reverse occurs, and an aster is more free to move, then the same interactions and forces could conceivably cause an aster to be drawn toward the cortex and perhaps even become pressed tightly against it. Does this kind of interaction explain the migration of free cytasters toward the cortex in artificially polyspermic eggs, as shown in Figure 11? A similar phenomenon may explain the persistent migration of a meiotic apparatus toward the animal pole (Inoue & Dan, 1980; Hamaguchi et al., 1983) and of other mitotic apparatuses toward the polar cortex in general (Kronebusch & Borisy, 1981) or to selected eccentric positions for asymmetrical cell divisions (Rebhun, 1959; Dan, 1979). It seems possible that there are common elements in all of these examples and that novel, yet fundamental, physical interactions between these asters and the cortex are important features in some of the cytological phenomena of cell division and early development.

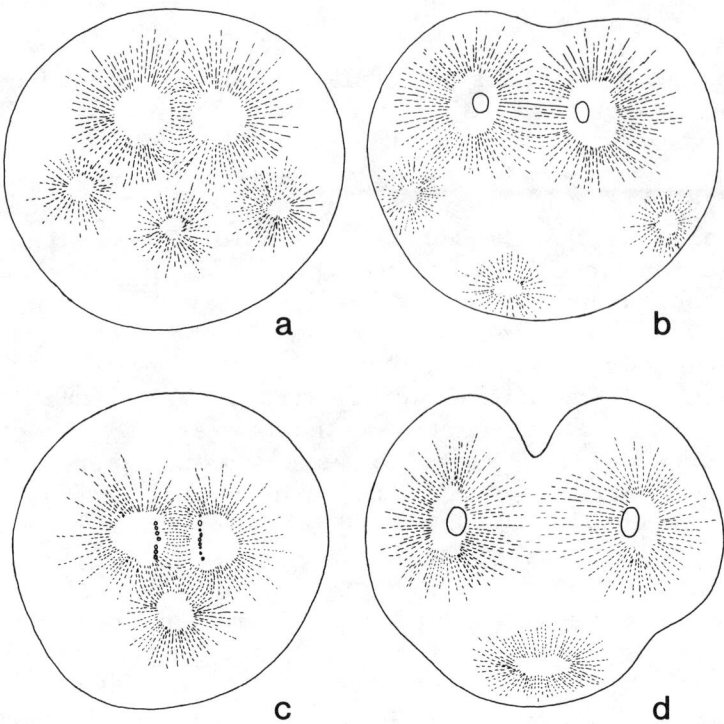

Fig. 11. Figures redrawn from Wilson (1901, see Figs. 5 & 6) showing progressive stages of two polyspermic sea urchin eggs just before cell division (a-b and c-d). In addition to the normal mitotic apparatus, extra cytasters have formed around the centrosomes of supernumerary sperm. At anaphase/telophase (b & d), the cytasters migrate toward the cortex, suggesting that they are either physically attracted "from a distance" or are pulled by some unvisualized astral rays that actually make contact with the cortex.

ACKNOWLEDGMENTS

This work is based upon observations made in many laboratories, including Tokyo Institute of Technology, Japan (thanks to help from Drs. Y. Hiramoto, Y. Hamaguchi and Y. Shoji) and Discovery Bay Laboratory, Jamaica (thanks to Dr. J.D. Woodley). It was supported in part by grant PCM-8201866 from the National Science Foundation.

LITERATURE CITED

Asnes, C.F. and T.E. Schroeder. 1979. Cell cleavage: Ultrastructural evidence against equatorial stimulation by aster microtubules. Exp. Cell Res. 122: 327-338.

Bestor, T.H. and G. Schatten. 1981. Anti-tubulin immunofluorescence microscopy of microtubules during the pronuclear movements of sea urchin fertilization. Dev. Biol. 88: 80-91.

Boveri, T. 1901. Die Polaritat von Ovocyte, Ei und Larve des Strongylocentrotus lividus. Zool. Jahrb. 14: 630-653.

Dan, K. 1943. Behavior of the cell surface during cleavage. VI. On the mechanism of cell division. J. Tokyo Imp. Univ. Fac. Sci. (Sect. 4) 6: 323-368.

Dan, K. 1979. Studies on unequal cleavage in sea urchins. I. Migration of the nuclei to the vegetal pole. Develop. Growth Differ. 21: 527-535.

Foe, V.E. and B.M. Alberts. 1983. Studies on nuclear and cytoplasmic behavior during the five mitotic cycles that precede gastrulation in Drosophila embryogenesis. J. Cell Sci. 61: 31-70.

Fry, H.J. 1936. Studies of the mitotic figure. V. The time schedule of mitotic changes in developing Arbacia eggs. Biol. Bull. 70: 89-99.

Hamaguchi, M.S. and Y. Hiramoto. 1978. Protoplasmic movement during polar-body formation in starfish oocytes. Exp. Cell Res. 112: 55-62.

Hamaguchi, M.S. and Y. Hiramoto. 1980. Fertilization process in the heart-urchin, Clypeaster japonicus observed with a differential interference microscope. Develop. Growth Differ. 22: 517-530.

Hamaguchi, Y., D.A. Lutz, and S. Inoue. 1983. Cortical differentiation, asymmetric positioning and attachment of the meiotic spindle in Chaetopterus pergamentaceus oocytes. J. Cell Biol. 97: 254a.

Harris, P., M. Osborn, and K. Weber. 1980. Distribution of tubulin-containing structures in the egg of the sea urchin Strongylocentrotus purpuratus from fertilization through first cleavage. J. Cell Biol. 84: 668-679.

Harvey, E.B. 1935. Some surface phenomena in the fertilized sea urchin eggs influenced by centrifugal force. Biol. Bull. 69: 298-304.

Harvey, E.B. 1956. The American Arbacia and Other Sea Urchins. Princeton University, Princeton.

Heilbrunn, L.V. 1920. An experimental study of cell-division. J. Exp. Zool. 30: 211-237.

Hiramoto, Y. 1970. Rheological properties of sea urchin eggs. Biorheology 6: 201-234.

Iida, T.T. 1943. Changes of electric capacitance following fertilization in sea urchin eggs. J. Tokyo Imp. Univ. Fac. Sci. (Sect. 4) 6: 141-151.

Inoue, S. and K. Dan. 1980. Mitotic spindle behavior in unequal cleavage of Spisula solidissima. Biol. Bull. 159: 443-444.

Kronebusch, P.J. and G.G. Borisy. 1981. Anaphase pole movement after central spindle disruption in PtK_1 cell. J. Cell Biol. 91: 319a.

Lindahl, P. 1932. Zur Kenntnis des Ovarialeis bei dem Seeigel. Arch. Entwicklungsmech. Organ 126: 373-390.

Lorch, I.J. 1952. Enucleation of sea-urchin blastomeres with and without removal of asters. Quart. J. Microscop. Sci. 93: 475-486.

Monne, L. 1947. Some observations on the polar and dorsoventral organization of the sea-urchin egg. Ark. Zool. 38A: 1-13.

Olsen, H. 1942. The development of the brittle-star Ohiopholis aculeata (O.F. Muller), with a short report on the outer hyaline layer. Bergens Mus. Arbok. 6: 1-107.

Otto, J.J. and T.E. Schroeder. 1984. Microtubule arrays in the cortex and near the germinal vesicle of immature starfish oocytes. Dev. Biol. 101: 274-281.

Rappaport, R. 1971. Cytokinesis in animal cells. Int. Rev. Cytol. 31: 169-313.

Rebhun, L.I. 1959. Studies on early cleavage in the surf clam, Spisula solidissima, using methylene blue and toluidine blue as vital stains. Biol. Bull. 117: 518-545.

Runnstrom, J. 1975. Integrating factors. pp. 646-670. In: The Sea Urchin Embryo. G. Czihak (ed.). Springer, Berlin.

Schatten, G. 1981. The movements and fusion of the pronuclei at fertilization of the sea urchin Lytechinus variegatus: time-lapse microscopy. J. Morphol. 167: 231-247.

Schrader, F. 1953. Mitosis. Second Edition. Columbia University, New York.

Schroeder, T.E. 1981a. Interrelations between the cell surface and the cytoskeleton in cleaving sea urchin eggs. pp. 169-216. In: Cytoskeletal Elements and Plasma Membrane Organization. G. Poste and G.L. Nicolson (eds.). Elsevier/North Holland Press, Amsterdam.

Schroeder, T.E. 1981b. The origin of cleavage forces in dividing eggs. Exp. Cell Res. 134: 231-240.

Schroeder, T.E. and J.J. Otto. 1984. Cyclic assembly-disassembly of cortical microtubules during maturation and early development of starfish oocytes. Dev. Biol. 103: 493-503.

Smiley, S.T. 1984. A description and analysis of the structure and dynamics of the ovary, of ovulation, and of oocyte maturation in the sea cucumber Stichopus californicus. M.S. Thesis, University of Washington, Seattle.

Usui, N. and M. Yoneda. 1982. Ultrastructural basis of the tension increase in sea-urchin eggs prior to cytokinesis. Develop. Growth Differ. 24: 453-465.

Vacquier, V.D. 1981. Dynamic changes of the eggs cortex. Dev. Biol. 84: 1-26.

Wilson, E.B. 1895. An Atlas of Fertilization and Karyokinesis of the Ovum. Macmillan, N.Y.

Wilson, E.B. and A.P. Mathews. 1895. Maturation, fertilization and polarity in the echinoderm egg. J. Morphol. 10: 319-342.

Wilson, E.B. 1901. Experimental studies in cytology. I. A cytological study of artificial parthenogenesis in sea urchin eggs. Arch. Entwicklungsmech. Organ. 12: 529-596.

Wilson, E.B. 1928. The Cell in Development and Heredity. Third Edition, Macmillan, N.Y.

Yoneda, M. and K. Dan. 1972. Tension at the surface of the dividing sea-urchin egg. J. Exp. Biol. 57: 575-587.

Yoneda, M., M. Ikeda, and S. Washitani. 1978. Periodic change in the tension at the surface of activated non-nucleated fragments of sea urchin eggs. Develop. Growth Differ. 20: 329-336.

Yoneda, M. and T.E. Schroeder. 1984. Cell cycle timing in colchicine-treated sea urchin eggs: Persistent coordination between the nuclear cycles and the rhythm of cortical stiffness. J. Exp. Zool. 231: 367-378.

TRANSLATIONAL CONTROL IN ECHINOID EGGS AND EARLY EMBRYOS

M.B. Hille, M.V. Danilchik,
A.M. Colin, and R.T. Moon

ABSTRACT

The historical aspects of translational control are reviewed, including translational rates, the effects of Ca^{2+} and alkalinization on translation after fertilization, and the masked messenger hypothesis. Evidence suggesting that mRNA is not the rate-limiting component in unfertilized eggs or zygotes is given: that message-containing ribonucleoprotein particles are efficient templates for in vitro protein synthesis, and that mRNAs injected into eggs and zygotes compete with maternal mRNAs for a limited component of the translational machinery. Thus, we conclude that some component such as ribosomes or a discriminatory factor must be rate limiting. The activities of egg monoribosomes and active embryonic ribosomes are compared. Egg ribosomes are less active, contain tightly bound inhibitory molecules, and are activated by alkalinization of the cytoplasm. Finally, the regulation of translational rates by initiation factors and by compartmentalization is reviewed. The complexity of the activation of protein synthesis in echinoid eggs and embryos suggests that translation is controlled at several levels.

INTRODUCTION

Sea urchin eggs provide an interesting example of translational regulation. The unfertilized egg of the sea urchin is a metabolically quiescent cell that actively uses only about 1% of its stored ribosomes and 1-5% of its mRNA. Fertilization triggers a cascade of changes in both the plasma membrane and the cytoplasm. By the first cleavage, or about 2 h after fertilization, 20-25% of the ribosomes are functioning in polyribosomes (Infante & Nemer, 1967; Humphreys, 1971; Goustin & Wilt, 1981), and the total rate of protein synthesis has increased some 50-fold (reviewed in Regier & Kafatos, 1977). Part of this increase is due to a 2-fold increase in elongation rate, i.e., the rate at which new amino acids are added to a growing peptide chain (Brandis & Raff, 1978; Hille & Albers, 1979). The remaining 20- to 25-fold increase is due to the recruitment of more mRNA and ribosomes into functioning polyribosomes. The early rise in the rate of protein synthesis occurs even after an egg is enucleated by physical (Brachet et al., 1963; Denny & Tyler, 1964) or chemical methods (Gross & Cousineau, 1963, 1964; Gross et al., 1964), so it does not require new mRNA or ribosome synthesis. Apparently, all of the components required for the post-fertilization rise in protein synthesis coexist in the unfertilized egg cytoplasm, but they are not utilized. This chapter explores how such changes in protein synthesis might occur. We show that the classical masked message hypothesis mechanism is not supported by recent experiments, and hence the recruitment of polyribosomes is not limited by the presence of repressor molecules on mRNAs, and we argue that a large portion of regulation may occur at the level of ribosomes or discriminatory factors.

Ionic Events at Fertilization

The activation of protein synthesis in echinoid eggs is temporally linked to two ionic events that occur within minutes after fertilization. The first is a transient intracellular release of Ca^{2+} (Steinhardt et al., 1977) and the second is a decrease in cytoplasmic hydrogen ion concentration (Johnson et al., 1976; Shen & Steinhardt, 1978). The relationship between ionic fluxes and protein synthesis has been studied by treating eggs with the calcium ionophore, A23187, which induces changes in intracellular Ca^{2+} concentrations (Grainger et al., 1979; Winkler et al., 1980); and by treating eggs with weak bases, such as ammonia, which raise the pH of the cytoplasm (Shen & Steinhardt, 1978). The alkalinization of the cytoplasm with ammonia in the absence of a calcium release stimulates the recruitment of mRNAs and ribosomes into polyribosomes (Chambers, 1975; Johnson et al., 1976; Paul & Johnston, 1978; Zucker et al., 1978). Such artificial alkalinization of the cytoplasm of eggs, however, does not increase peptide elongation rates (Brandis & Raff, 1979; Winkler et al., 1980). The Ca^{2+} release, when coupled with alkalinization, stimulates the additional 2-fold increase in elongation rate at fertilization (Winkler et al., 1980).

Regulation by Masking of Maternal Messenger RNAs

Much of the study of the activation of protein synthesis in these eggs has focused on the role of maternally derived mRNA, because mRNAs exist not as naked RNAs but as ribonucleoprotein particles, abbreviated mRNPs (Spirin, 1966). Spirin (1966) and others suggested that the mRNA-associated proteins could qualitatively and quantitatively modulate the rates of protein synthesis in eggs or even prevent the mRNAs from being translated -- the masked message hypothesis (reviewed by Raff, 1980; Raff & Showman, In press). Two groups have reported that populations of sea urchin mRNAs isolated with their associated proteins (mRNPs) are inactive, while naked mRNAs purified from the same population are active (Ilan & Ilan, 1978; Jenkins et al., 1978). In addition, as much as 70% of the poly A RNAs of both sea urchin eggs (Costantini et al., 1980) and Xenopus oocytes (Anderson et al., 1982) have short, interspersed repetitive sequences that appear to render them nontranslatable (Richter et al., 1984), while fewer transcripts from embryos of these animals contain poly A RNAs with interspersed repeats. Whether the interspersed sequences are significant for translational control in these oocytes and eggs has not been established. Nevertheless, there are probably sufficient stored mRNAs in sea urchin eggs that lack interspersed repetitive sequences to supply all of the mRNAs recruited after fertilization (Richter et al., 1984).

In contrast to studies on the translatability of mRNPs (Ilan & Ilan, 1978; Jenkins et al., 1978), we and others have found that egg mRNP complexes are efficient templates for in vitro protein synthesis in wheat germ (Moon et al., 1982) and rabbit reticulocyte (Murray & Sosnowski, 1980; Moon, 1983) lysates. We tested whether the proteins specifically associated with the free mRNPs in eggs repressed translation by comparing the translational efficiency of isolated mRNPs with the translational efficiency of deproteinized mRNAs. Using mRNA-dependent lysates from wheat germ, we found that the mRNP particles were at least 50% as efficient as their deproteinized mRNAs and generally as efficient (Fig. 1).

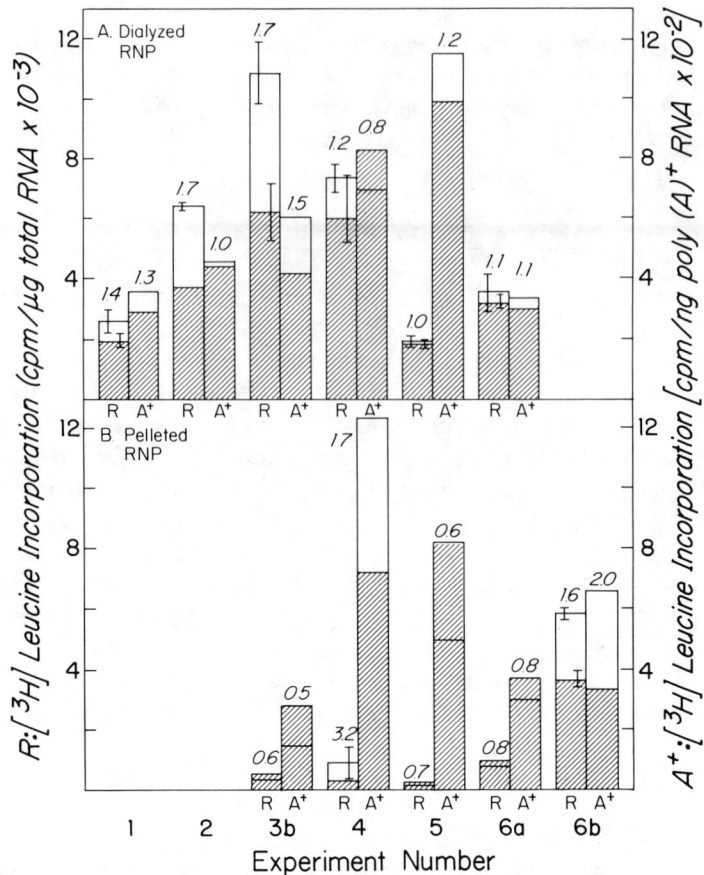

Fig. 1. Relative translational activities of crude egg mRNPs and mRNP-RNAs in wheat germ lysates. mRNPs were concentrated (A) by negative-pressure dialysis or (B) by pelleting. Details for the isolation of mRNPs and mRNP-RNAs in experiments 1-6 are given below. The translational activity of each sample was normalized both to RNA content (R), as determined by the absorbance at 260 nm (40 μg/ml $A_{260\ nm}$), and to poly A content (A^+), as determined by hybridization to ^{3}H-poly U. The number above each column is the ratio of the translational activities of mRNP-RNAs (open bar) divided by the translational activities of the intact mRNPs (hatched bar). If the translational activities of mRNP-RNAs are less than those of intact mRNPs, then the mRNP-RNA activity is represented by a horizontal line. The range of translational activities in each experiment is indicated by error bars for the samples normalized by RNA content. Standard reactions in wheat germ contained 110 mM K^+, 2.5 mM Mg^{2+}, pH 7.2 (Exp. 1 & 2) or pH 7.5 (Exp. 3b - 6b) and several subsaturating concentrations (4-20 μg RNA/50-μl assay) of RNA. Exp. 1 and 2: mRNPs were isolated from postmitochondrial supernatants prepared in PIB (220 mM K^+, 40 mM Na^+, 5 mM Mg^{2+}, 80 mM Cl^-, 185 mM OAc^-, 67 mM PO_4^{2-}, 1 mM dithiothreitol, pH 6.8, 0°C) and centrifuged on 10-30% preparative sucrose gradients. The fractions containing mRNPs were concentrated by negative-pressure dialysis against IB pH 7.0 (220 mM K^+, 5 mM Mg^{2+}, 80 mM Cl^-, 145 mM OAc^-, 20 mM Pipes, 1 mM dithiothreitol). RNA was purified by phenol-chloroform extraction. Exp. 3b: mRNPs were isolated from postmitochondrial supernatants prepared in IB pH 7.2. mRNPs from the sucrose gradients were concentrated by pelleting or by negative-pressure dialysis. RNA was purified either by phenol-chloroform extraction or by CsCl

centrifugation. Exp. 4: mRNPs were isolated from postmitochondrial supernatants prepared in HSBK-Mg^2, pH 7.4 (350 mM KCl, 5 mM Mg^{2+}, 50 mM triethanolamine) containing 0.25 mg/ml heparin. mRNPs were concentrated as in Exp. 3b. RNA was prepared by phenol-chloroform extraction. Exp. 5: mRNPs were isolated from postmitochondrial supernatants prepared in HSBK-Mg^{2+}, pH 7.4 supplemented with 1 mg/ml soybean trypsin inhibitor and 1 mM EGTA and centrifuged in 5-20% sucrose gradients. mRNP fractions were concentrated by negative-pressure dialysis or by pelleting at 130,000g$_{ave}$ for 17 h (1°C). The RNA was phenol-chloroform extracted. Exp. 6a: mRNPs were isolated essentially as in Exp. 5 except that those mRNPs concentrated by pelleting were centrifuged at 102,000g$_{ave}$ for 7 h (1°C). The flocculent and compact pellets (see Moon et al., 1982, for description) were dissolved together by shaking for 12 h (0°C). Exp 6b: mRNPs were prepared and pelleted in parallel with those in Exp. 6a except that the flocculent pellet was removed and dissolved separately from the compact pellet (Moon et al., 1982).

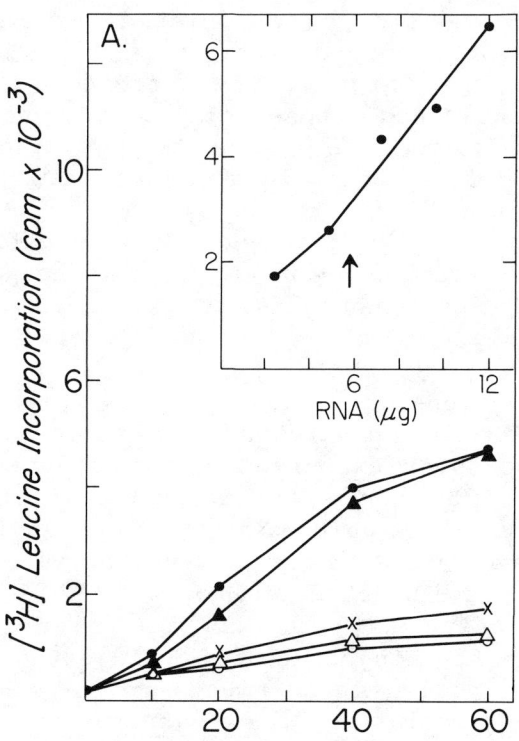

Fig. 2. Kinetics of mRNPs and mRNP-RNAs in a wheat germ lysate and inhibition by pactamycin. A crude preparation of dialyzed mRNPs prepared as in Exp. 4, Figure 1, which contained 5.8 µg of total RNA and 42 ng of poly A RNA, was translated in 50 µl at 110 mM K$^+$ and 2.5 mM Mg^{2+} in the absence of inhibitors (▲) or in the presence of 14 µM pactamycin (Δ), an inhibitor of peptide initiation. Samples were withdrawn at the times indicated. mRNA extracted from these mRNPs and containing 6 µg of total RNA and 63 µg of poly A-containing mRNA was translated under identical conditions in the absence (●) or in the presence (o) of 14 µM pactamycin. Endogenous incorporation is shown (x). Inset: Several concentrations of mRNP-RNA were translated for 60 min, and aliquots were then measured for incorporation above endogenous levels. The arrow denotes the concentration of RNAs used for the kinetic analysis (Moon et al., 1982).

These tests included isolation and extraction of the mRNPs under many different conditions as well as optimization of the translational systems. Some of our isolation procedures duplicated those of previously published experiments which indicated that egg mRNPs were "masked" (Jenkins et al., 1978). The possibility that mRNPs were being activated by our heterologous system was tested by following the kinetics of the translation of the isolated mRNPs. No lags in translation were observed compared to deproteinized mRNAs (Fig. 2), although this does not rule out very rapid activation of mRNPs in vitro. None of our in vitro translation experiments support masking of mRNPs by associated proteins as a rate-limiting process in echinoid eggs.

Because of the unresolved differences between our in vitro measurements and those of others, we have tested whether the in vivo rate of protein synthesis could be augmented by injecting functional mRNAs into echinoid eggs. An increase in the total translation mRNAs in mRNA-injected eggs would imply that there are normally insufficient amounts of active mRNAs available in eggs. No change would imply that something other than mRNA is rate limiting. Thus, we can test directly, as has been done with Xenopus oocytes (Laskey et al., 1977), whether these eggs have the capacity to translate additional messages.

Eggs of Strongylocentrotus droebachiensis, Lytechinus pictus, and Dendraster excentricus were injected with buffer or with poly A RNA from rabbit reticulocytes, and then incubated in ^{3}H-amino acid mixtures and analyzed for the newly synthesized proteins by SDS-polyacrylamide gel electrophoresis. Eggs injected with poly A RNA synthesized a protein co-migrating on polyacrylamide gels with rabbit globin, while buffer-injected and uninjected eggs did not (Fig. 3). Quantitative measurements showed that the synthesis of putative globin in D. excentricus unfertilized eggs was 10% (not shown) or 29% (Fig. 4) of the total peptide synthesis when the amount of injected mRNA was about 50% or 400%, respectively, of the endogenous mRNA in the sand dollar eggs (estimated at 0.4 ng/egg). Significantly, this synthesis of globin was at the expense of the translation of endogenous mRNAs in unfertilized eggs since total protein synthesis did not increase. It appears that the injected mRNAs compete with endogenous mRNAs for translation in unfertilized eggs. Hence we conclude that some component other than mRNA limits the translational capacity in unfertilized eggs.

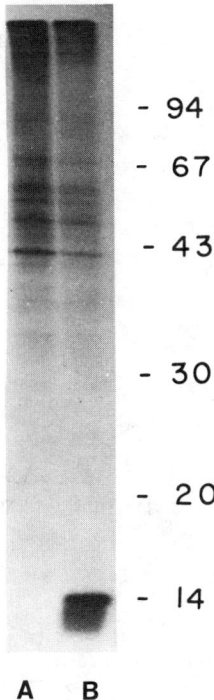

Fig. 3. Fluorograph of a 10% SDS-polyacrylamide gel of proteins from unfertilized eggs of D. excentricus. Lane (A) contains proteins from 30 buffer-injected eggs, and lane (B), proteins from 30 RNA-injected eggs. The eggs were dejellied in pH 5.0 sea water prior to injection. Messenger RNAs from rabbit reticulocytes (Gilbert & Anderson, 1970; Crystal et al., 1974; Danilchik & Hille, 1981) were purified with phenol-chloroform (Palmiter, 1974), then isolated on oligo(dT)-cellulose (Aviv & Leder, 1972), and dissolved in buffer containing 0.3 M glycine (pH 6.8), 0.01 M NaCl, 0.01 M $MgCl_2$, 0.001 M EGTA, and 0.01 M glycylglycine. Eggs were placed in plastic culture wells pretreated with 1% protamine sulfate (Steinhardt et al., 1971), and kept on ice under a dissecting microscope during the injection period. The 5 μ oil-filled micropipettes used for the injection were attached to a syringe and micrometer that was driven by a reversible motor that delivered a continuous flow of solution at 1 nl/sec. The eggs were injected with about 0.2 nl of solution containing 1.4 ng of mRNA (15-20% of the volume of the eggs). After injection, eggs were labeled for 4 h with 12.5 μCi/ml of tritiated amino acid mixture (New England Nuclear) at 12°C. At the end of the incubation, the eggs were iced, and those that appeared normal and had no fertilization envelopes were collected with a braking pipette and stored in absolute ethanol at -20°C. Samples for gel electrophoresis were centrifuged and dried, then dissolved in gel sample buffer and electrophoresed (Laemmli, 1970). The gel was treated with Enhance (New England Nuclear), dried, and exposed to preflashed (Laskey & Mills, 1975) Kodak XRP-1 X-ray film at -80°C. Three species are preferred for these studies: the sea urchins Strongylocentrotus droebachiensis (diameter = 180 μm) and Lytechinus pictus (d = 120 μm), and the sand dollar Dendraster excentricus (d = 120 μm). The commonly studied S. purpuratus eggs (d = 80 μm) are difficult to inject since the eggs have only 1/10 the volume of S. droebachiensis eggs.

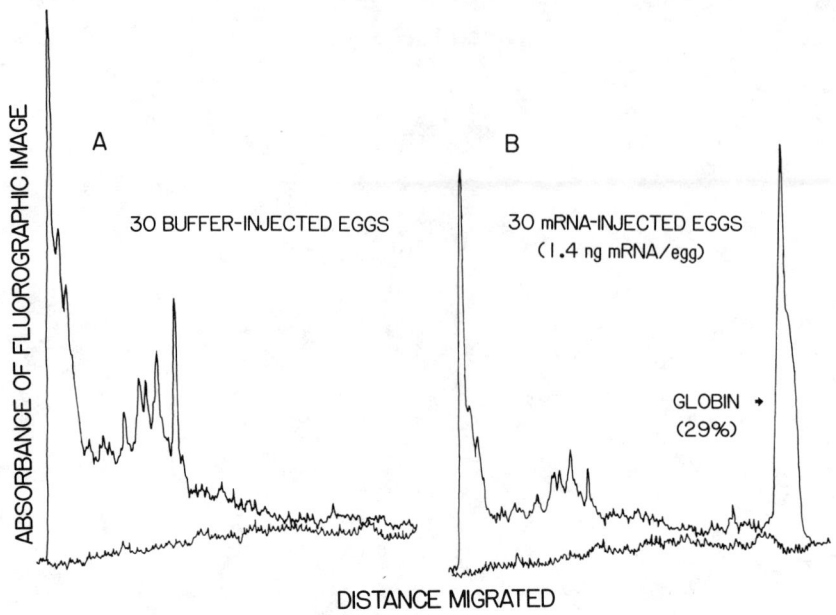

Fig. 4. Scans of the fluorograph in Figure 3 of proteins from unfertilized eggs of D. excentricus injected with buffer (A) or with 1.4 ng of poly A RNA from rabbit reticulocytes (B). The eggs were labeled with a mixture of ^{3}H-amino acids for 2 h at 12°C. The top of the gel containing the large peptides is to the left. The area under the peaks was determined with an electronic graphics calculator (Numonics). The baseline was determined by scanning an adjacent blank gel lane. Total incorporation into buffer-injected and mRNA-injected eggs was similar. In the mRNA-injected eggs, 29% of the total incorporation was in the putative globin band.

To determine whether, after fertilization, the rate-limiting component in translation is mRNA or another component, we injected S. droebachiensis zygotes with amounts of globin mRNA equal to about half the amount of endogenous mRNA (estimated as 1.0 ng/egg). Total protein synthesis did not increase although globin synthesis contributed about 25% to the total (Fig. 5). Thus, the continued recruitment of mRNAs up to the first cleavage must result from the activation of a component other than the mRNA molecule.

In their mathematical model, Raff et al. (1981) elegantly relate the exponential increase in protein synthesis to the elongation rate and to a gradual increase in a limiting component. They show that the activation of protein synthesis is an early response to fertilization that commences within minutes but continues gradually as observed by others (Hultin, 1964; Epel, 1967; Humphreys, 1971). Although they assumed that mRNA was the rate-limiting component, any

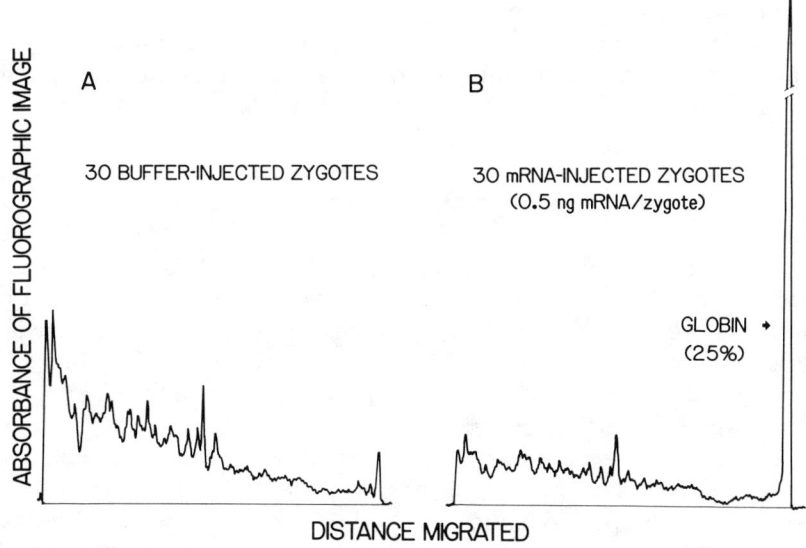

Fig. 5.   Scans at 540 nm of a fluorograph of a 10% polyacrylamide gel of proteins from zygotes of S. droebachiensis that were injected with buffer (10 mM Tris, pH 6.8, 50 µM EGTA), (A), or with 0.5 ng of poly A RNA from rabbit reticulocytes in the same buffer (B). Zygotes were prepared for microinjection by fertilizing the eggs in sea water containing 1.0 mM 3-amino 1,2,4 triazole to prevent hardening of the fertilization envelope (Showman & Foerder, 1979). The fertilized eggs were dejellied, washed, adhered to dishes, and injected as described in Figure 3 with about 0.5 nl of solution between 30 and 60 min after fertilization. The zygotes were then labeled for 1 h in sea water containing 40 µCi/ml tritiated amino acid mixture. The total incorporation into mRNA-injected zygotes was 20% less than that into buffer-injected zygotes. In the mRNA-injected zygotes, 25% of the total protein was in putative globin.

component that limits the initiation step would give the same theoretical fit to the experimental data.

Regulation of Translation by pH Changes

The changes in the intracellular pH of sea urchin eggs after fertilization have been variously reported as an increase from pH 6.8 to 7.4 and 6.85 to 7.27 for L. pictus (Shen & Steinhardt, 1978; Johnson & Epel, 1981, respectively), pH 7.08 to 7.43 for Strongylocentrotus purpuratus (Johnson & Epel, 1981), and pH 7.38 to 7.64 for Paracentrotus lividus (Payan et al., 1983). Alkalinization of the cytoplasm is essential for the stimulation and maintenance of high levels of protein synthesis in embryos of L. pictus (Grainger et al., 1979), and for the stimulation of protein synthesis in S. purpuratus (Johnson & Epel, 1981). These observations raise major questions. Does alkalinization of the cytoplasm by itself optimize the translational rates in these embryos? Additionally, or alternately, does it induce a permanent modification of molecular components of the protein synthesis apparatus and/or does it induce a permanent reorganization of materials in compartments in the cytoplasm? No doubt some of the stimulation of protein synthesis at fertilization can be strictly attributed to the electrostatic influence of pH on the translational optimum (1) since in vitro peptide synthesis is known to have an optimum between pH 7.3 and 7.6 (Palmiter, 1973) and (2) since Winkler and Steinhardt (1981) observed about an 8-fold stimulation of peptide synthesis with lysates from both egg and embryos of L. pictus when the pH was increased from 6.9 to 7.4. About half of this increase is due to a change in elongation rates. The stimulation of protein synthesis that they observed, however, did not overcome an additional 7-fold difference in the rates of protein synthesis between the egg and embryo lysates at both pH values. Thus, it is likely that the in vivo alkalinization also leads to a chemical modification and/or solubilization of some components of the protein synthesis apparatus.

Echinoid eggs seem to differ from asteroid eggs and amphibian oocytes in their dependence on an increase in the intracellular pH for the stimulation of protein synthesis. In the starfish, Pisaster ochraceus, the increase in pH after hormonally induced maturation or after fertilization is about 0.05 units above the basal level of 7.5, and is not essential for germinal vesicle breakdown (GVBD) and has little effect on protein synthesis rates (Johnson & Epel, 1982). Similarly the increase in pH of 0.18 units from a basal level of 7.4-7.8 after progesterone stimulation of Xenopus oocytes is not required for meiotic maturation (Lee & Steinhardt, 1981). Calcium seems to be the central ionic trigger of maturation for these animals. As suggested by Johnson and Epel (1982), the greater post-fertilization increase in protein synthesis in echinoids may be due to stimulation by alkalinization, which in turn is possible because of the lower basal pH and the lower resting rate of protein synthesis in echinoid eggs compared to asteroid and amphibian oocytes.

Current Prospects

Recapitulating, isolated mRNP particles from unfertilized sea urchin eggs are functional in cell-free lysates; globin mRNAs injected into echinoid eggs do not increase the total amount of protein synthesis, but rather compete with endogenous mRNAs for translational machinery. Therefore, the masking of mRNAs by associated proteins may not be the primary rate-limiting feature of protein synthesis in these eggs. In the absence of definitive evidence for the masking of mRNAs in sea urchin eggs it is important to consider alternative mechanisms. Alternatives include: (1) inhibitory factors in the cytoplasm or attached to egg monoribosomes that prevent the association of aminoacyl-tRNA, initiation factors and/or mRNA to ribosomes; (2) a deficiency in eggs of message recruitment factors that are activated after fertilization and function by binding to mRNAs before initiation occurs; and (3) the storage of ribosomes in cellular compartments which disperse after fertilization, e.g., ribosomes could be attached to membranes or encompassed by them. Currently, not enough evidence exists to support one of these possibilities more than the others. However, considerable information is available on activities of ribosomes and associated inhibitors, which we will describe. We will then describe studies in other systems that illustrate the merits of other alternatives.

Regulation by Ribosomes

Monroy et al. (1965) suggested that ribosomes are inactive in unfertilized eggs of echinoids, an idea that was initially unconfirmed by others. Various studies of ribosomes isolated from eggs and embryos either showed no differences in their activities (Stavy & Gross, 1967, 1969; Clegg & Denny, 1974; Ilan & Ilan, 1978) or tested their activity with synthetic templates that lacked initiation regions (Metafora et al., 1971). To determine whether ribosomes have a regulatory role in the post-fertilization activation of protein synthesis in sea urchin eggs, we measured the translational activity of ribosomes isolated from unfertilized eggs and embryos of S. purpuratus. Our studies differed in two significant ways from the previous ones: we used physiological salt solutions and nondenaturing conditions during isolation of the ribosomes, and we isolated embryonic ribosomes that we knew to be active -- ones derived from polyribosomes. We obtained ribosomes from polysomes by incubating polysomes in a ribosome-free reticulocyte lysate. In addition this lysate had little or no chain initiation activity, but sufficient elongation activity to run the ribosomes off of the messages. The resulting "runoff" ribosomes were thus freed of mRNA: they sedimented as 80S monoribosomes and will be referred to as polysome-derived ribosomes. Egg monoribosomes were simultaneously exposed to the ribosome-free reticulocyte lysate to control for the effect of lysate on the ribosomes. No significant difference was observed between treated and nontreated monoribosomes. Polysome-derived ribosomes from blastulae were active in translating reticulocyte mRNA in a ribosome-dependent cell-free system (Danilchik & Hille, 1981). Egg monoribosomes were always significantly less active than polysome-derived ribosomes at an incubation pH between 6.9 and 7.65. In addition, monoribosomes obtained from unfertilized eggs became fully active only after a characteristic, reproducible delay of up to 15 min at 26°C (Figs. 6 & 7). The delay varied with incubation pH, but not with concentrations of K^+, Mg^{2+}, initiation factors, or mRNA (Danilchik & Hille, 1981). One possible

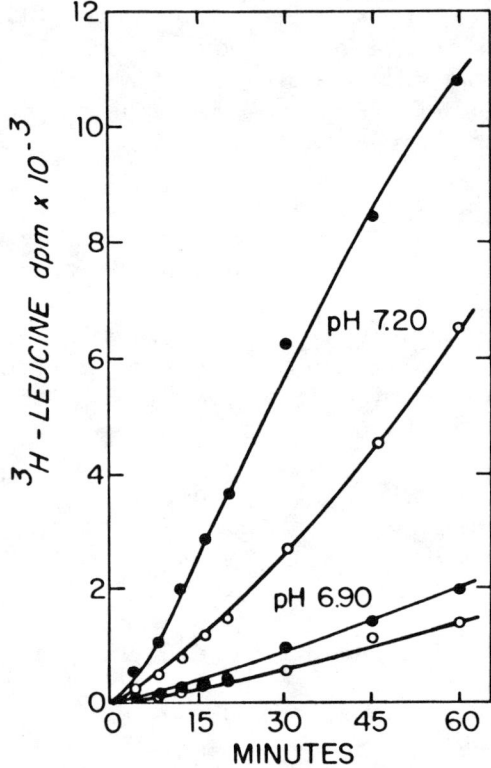

Fig. 6. The time course of protein synthesis by monoribosomes from unfertilized eggs (o) and polysome-derived ribosomes from blastulae (•) isolated at pH 6.8. Values refer to 3H-leucine counted in 5-µl samples of the reaction mixture which contained 100 mM KCl, 0.5 mM $MgCl_2$, 40 µg/ml reticulocyte poly A RNA, 3 mg/ml KCl-washed fraction from reticulocyte ribosomes containing initition factors, 4 A_{260} nm unit/ml ribosomes isolated by run off at pH 6.8. Details of the incubation conditions are given in Danilchik and Hille (1981). Incubation pH values are indicated in the figure.

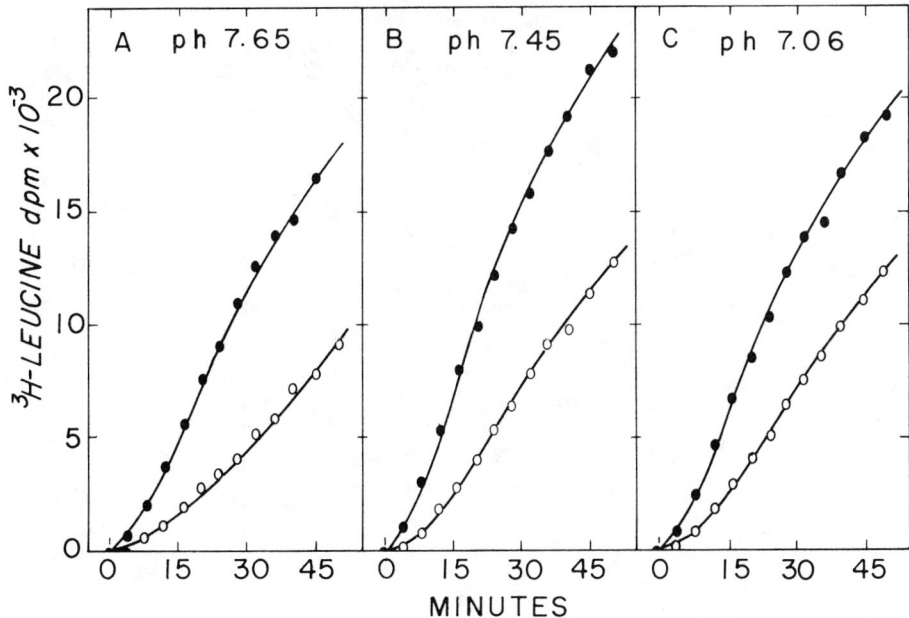

Fig. 7. Effect of incubation pH on the time course of protein synthesis by monoribosomes from unfertilized eggs (o) and by ribosomes derived from polysomes of hatched blastulae (●). Reaction mixtures contained 40 µg/ml reticulocyte mRNA, 3 mg/ml partially purified reticulocyte initiation factors, 100 mM KCl, 0.5 mM Mg(OAc)$_2$, and 6.1 A$_{260}$ nm unit/ml of ribosomes isolated and run off at pH 6.8. Incubation pH values in the figure refer to polysome-derived ribosomes and egg monoribosomes, respectively (Danilchik & Hille, 1981).

explanation for the variable delay is that egg monoribosomes are changed by incubation in reticulocyte lysates, and that the higher the pH of the lysate, the faster this change. Another explanation is that elongation rates as well as initiation rates are lower. A slower elongation rate would increase the time required for the incorporation of ^{3}H-amino acids into peptides to reach a steady state. Figure 8 suggests both a lower initiation rate and a slower elongation rate for monoribosomes: (1) fewer ribosomal subunits dissociate from egg monoribosomes incubated in reticulocyte lysates (Fig. 8d) as compared to those from polysome-derived ribosomes (Fig. 8c) suggesting a lower initiation rate for monoribosomes; and (2) the time required to complete an average peptide chain is 1.5 times longer for egg monoribosomes (about 17 min, Fig. 8d) as compared to polysome-derived ribosomes (11 min, Fig. 8a).

If ribosomes have increased activities after fertilization, do the ionic changes occurring participate in ribosome

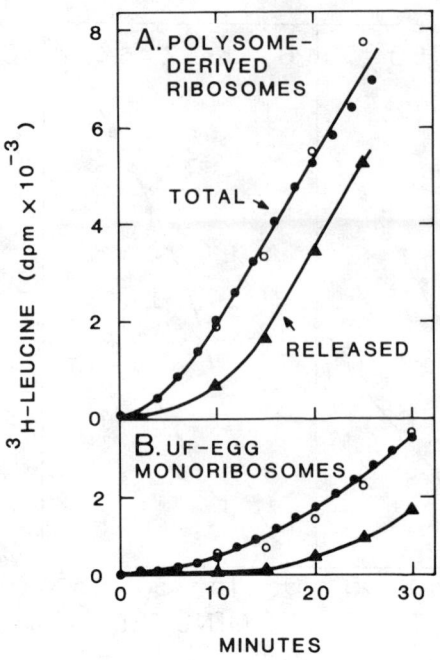
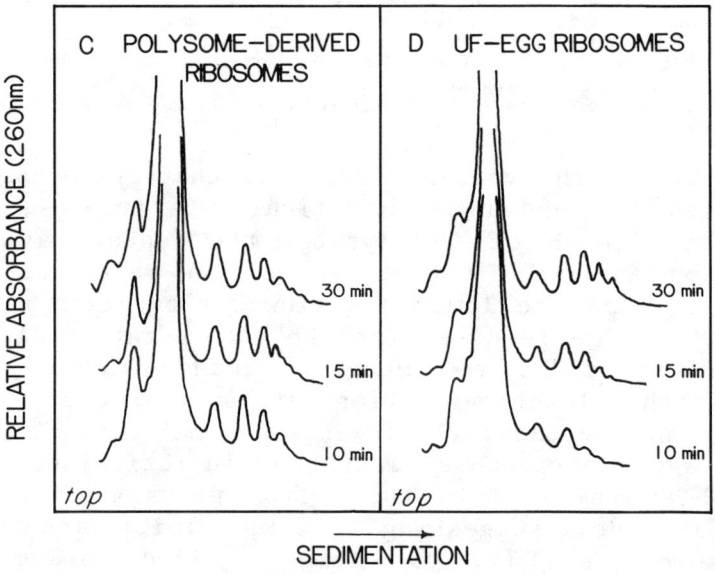

Fig. 8. Approach to steady state of ribosome activity in vitro: kinetics and polyribosomal profiles. A comparison was made of monoribosomes from unfertilized eggs and polysome-derived ribosomes from 16-cell embryos during the first 30 min of incubation in a ribosome-deficient reticulocyte lysate. Both ribosome preparations were pretreated in an initiation-dependent, ribosome-depleted reticulocyte lysate. Test incubations were with 100 mM KCl, 1.5 mM Mg(OAc)$_2$, 7.5 A$_{260}$ nm units/ml ribosomes, and 40 µg/ml globin mRNA at pH 7.3 and 26°C. Panels A

and B: Total incorporation of ^{3}H-valine into peptides determined by the trichloroacetic acid precipitation of aliquots of the reaction mixtures (●); and incorporation of ^{3}H-valine into total peptides (o) and released peptides (▲) determined by separation of aliquots of the reaction mixtures on sucrose gradients. Panel A: Kinetics for peptide synthesis with polysome-derived ribosomes form blastulae. The time of the synthesis of an average-size peptide is equal to twice the horizontal distance between the incorporation into total peptides and released peptides; it is 11 min for polysome-derived ribosomes. Panel B: Kinetics of peptide synthesis with unfertilized egg monoribosomes. The time for the synthesis of an average peptide is estimated to be about 17 min. Panel C: Profiles of polysome-derived ribosomes displayed as scans of absorbance at 260 nm of the centrifuged sucrose gradients. Panel D: Polyribosome profiles of unfertilized egg monoribosomes.

activation? Does the change occur rapidly (within minutes) or is it a gradual process? We have studied the possibility that either an increased concentration of Ca^{2+} ions or the alkalinization of the cytoplasm triggers the conversion of the monoribosomes to the more active polysome-derived ribosomes in embryos. For these studies, we measured the activity of monoribosomes and polysome-derived ribosomes isolated from eggs that had been treated either with NH_4Cl in Ca-free sea water or with the Ca-ionophore, A23187, in Na-free sea water. In the absence of Na^+, A23187 did not cause the formation of polyribosomes or active monoribosomes. However, when eggs were treated with NH_4Cl in Ca-free sea water, some ribosomes formed polyribosomes (Epel et al., 1974). Ribosomes derived from polysomes of these NH_4Cl-treated eggs were as active as ribosomes derived from polysomes of blastula embryos (Danilchik & Hille, unpubl.). Thus, the activation of monoribosomes by NH_4Cl-treatment, and not by A23187-treatment of sea urchin eggs, suggests that the conversion of ribosomes from an inactive state to a functional one is dependent on the pH change in the cytoplasm.

When are the less active monoribosomes converted to the polysomal-type ribosomes? To determine whether the change occurs immediately after fertilization or gradually, we compared the activities of monoribosomes and polysome-derived ribosomes from several developmental stages. We found that populations of monoribosomes from unfertilized eggs, 16-cell embryos, and hatched blastulae are similar in activity, and are all less active than polyribosome-derived ribosomes from embryos (Danilchik & Hille, unpubl.). These observations agree with those of others (Stavy & Gross, 1969; Clegg & Denny, 1974; Ilan & Ilan, 1978), which show that monoribosomes from eggs and zygotes have equivalent translational activities in vitro. Our new finding is that ribosomes from embryonic polyribosomes are more active than monoribosomes from either source. Any conversion from the less active monoribosomes to active polysomal ribosomes must be gradual since the less active form is found in embryos many hours after fertilization. Furthermore, we found that ribosomes released from polyribosomes in vivo during experimentally induced slow cooling of embryos retained their higher activity even though they had entered the pool of 80S monoribosomes (Danilchik & Hille, unpubl.). Our results suggest that, after fertilization, ribosomes are slowly changed to a more active form and that once active they remain so, even when they are released from polyribosomes. It is these ribosomes that are incorporated into polyribosomes under normal conditions.

Is the difference between monoribosomes and polysome-derived ribosomes a property specific to these eggs and embryos or is it a commonly observed property of other cells? In rabbit reticulocyte lysates a difference has been observed between the rate that ribosomes from polysomes reinitiate on messages and the rate that new monoribosomes enter polyribosomes. However, for these lysates, Howard et al. (1970) showed that due to the low concentrations of ribosomes and mRNAs, and consequently the distance between the molecules, the subunits released from polyribosomes were at least 100 times more likely to reinitiate on the same mRNA than any random subunit generated by the monoribosome pool. The difference in activity between monoribosomes and polyribosome-derived ribosomes of sea urchins cannot be explained by such statistical considerations, since the polysome-derived ribosomes and the monoribosomes were incubated at equal concentrations in the cell-free system and have an equal probability of encountering mRNAs and initiation factors. The difference in the behavior of ribosomes from sea urchins must, therefore, be attributed to their respective activities.

Ribosome-Associated Factors and Phosphorylation

The idea that egg ribosomes are "repressed" is supported by evidence that inhibitors of protein synthesis can be obtained by high salt extraction of 80S particles from unfertilized eggs (Metafora et al., 1971; Gambino et al., 1973; Hille, 1974). The inhibitory fraction described by Metafora et al. (1971) and Gambino et al. (1973) and isolated from Paracentrotus lividus has several properties: (1) it inhibits the translation of natural mRNAs in sea urchin lysates and reticulocyte lysates and decreases the runoff of polyribosomes in reticulocyte lysates. Thus, it probably

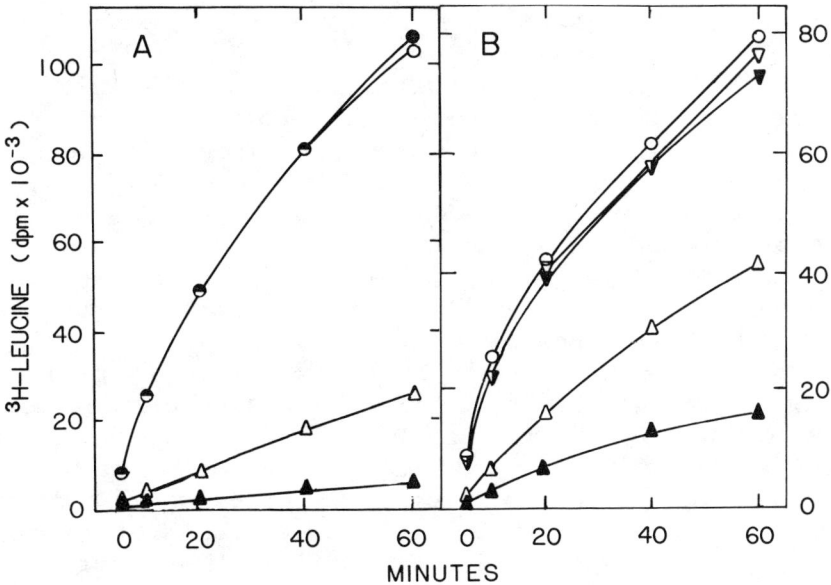

Fig. 9. Inhibition of translation by salt washes from unfertilized egg monoribosomes. Monoribosomes from unfertilized eggs and polysome-derived ribosomes from blastulae of S. purpuratus were successively treated with and centrifuged from solutions of 0.8 M KCl, 1.1 M KCl-NH$_4$Cl, and 1.4 M KCl-NH$_4$Cl containing 18 mM MgCl$_2$, 50 mM Tris-HCl, and 1 mM dithiothreitol. The ribosome-free salt solutions were dialyzed against 50 mM Tris-HCl, 0.1 mM EDTA, and 1 mM dithiothreitol (Hille, 1974). The proteins remaining in solution at the end of the dialysis (salt-wash proteins) were tested for their effect on translation in the heterologous reticulocyte-sea urchin system described in Figure 8. The salt-wash proteins added to the reaction mixture were isolated from twice as many ribosomes as those in the reaction mixture. No additions (o), salt-wash proteins from egg monosomes using concentrations of salt between 0.0 M to 0.8 M (●), 0.8 M to 1.1 M (▲), and 1.1 M to 1.4 M (Δ); and salt-wash proteins from polysome-derived ribosomes of blastulae using concentrations of salt between 0.8 M to 1.1 M (▼), and 1.1 M to 1.4 M (∇).

inhibits peptide bond formation. (2) It prevents the in vitro binding of poly U to ribosomes, the poly U-dependent binding of phe-tRNA to ribosomes, and reduces the in vitro occurrence of ribosomal subunits required for initiation. (3) It inhibits polyphenylalanine synthesis to a greater extent when tested with purified elongation factors than with postribosomal supernatants from lysates of sea urchin embryos, suggesting that the embryo lysate decreases the inhibitory activity. (4) The inhibitor is probably proteinaceous but not a ribonuclease.

Similar to Metafora and colleagues, we have observed that salt washes from egg monoribosomes of S. purpuratus are inhibitory while those from blastula polyribosomes are not (Hille, 1974). This inhibitory activity is not removed by salt at concentrations of less than 0.8 M (Fig. 9a & b).

Since the free mRNPs of eggs co-sediment with monoribosomes (Kaumeyer et al., 1978; Moon et al., 1980, 1983), it is not known if the inhibitor originates from the mRNP complexes or from the ribosomes. However, since poly A-containing RNPs purified by oligo-(dT)-cellulose chromatography under mild ionic conditions (0.25 M K^+) are translatable (Moon et al., 1982), it is unlikely that mRNPs contain the inhibitor. When the monoribosomes are dissociated in 0.8 M salt, the inhibitory activity remains with a pool of large ribosomal subunits and mRNPs, and not with the small ribosomal subunits (K. Moore & M.B. Hille, unpubl.).

Studies have been made to determine whether there are differences in the core proteins of sea urchin ribosomes. In many cells and organisms, an increase in the phosphorylation of the small ribosomal subunit protein, S6, parallels or precedes stimulation of protein synthesis (Kabat, 1970; Thomas et al., 1980; Gordon et al., 1982; Nielsen et al., 1982a & b). Ballinger and Hunt (1981) and Ballinger et al. (1984) observed in the sea urchin, Arbacia punctulata, that a decrease in phosphatase activity results in the increased phosphorylation of the S6 protein in monoribosome and polyribosome pools shortly after fertilization. Ward et al. (1983) found, however, that the post-fertilization increase in S6 phosphorylation does not occur in S. purpuratus or L. pictus. Takeshima and Nakano (1983) have found five egg proteins that undergo some form of rapid change at fertilization in several different Japanese sea urchins. One of these, tentatively identified as the sea urchin equivalent of the mammalian S6 protein, showed increased phosphorylation in three species. The other four peptides, two of which originated from the large subunit, underwent major shifts in electrophoretic positions. The significance of these post-fertilization changes in ribosomal proteins is not yet known. The observed phosphorylation of S6 is potentially important, since several groups studying other organisms have suggested that phosphorylation of S6 increases initiation rates (Thomas, 1980; Hanocq-Quertier & Baltus, 1981; Glover, 1982; Thomas et al., 1982) or elongation rates (Ballinger & Pardue, 1983). Some of the observations on sea urchins, however, suggest that phosphorylation may not be related to the recruitment of ribosomes and mRNA into polysomes, e.g., the absence of fertilization-induced S6 phosphorylation in S. purpuratus and L. pictus (Ward et al., 1983), and an absence of phosphorylation when eggs are activated with ammonia (Ballinger et al., 1984).

Is there evidence that translational control occurs at the ribosome level in other organisms? Translation may be regulated at this level during the response of Drosophila cells to heat shock. It is well known that heat shock causes a rapid decrease in the synthesis of normal peptides without degrading the normal 25°C mRNAs. This decrease is followed by a rise in the synthesis of the heat-shock proteins (McKenzie et al., 1975; Mirault et al., 1978; Lindquist, 1980). Heat shock decreases both the initiation and the elongation rates on mRNAs of normal peptides (Ballinger & Pardue, 1984). Scott and Pardue (1981) found that a component of the ribosomal pellet of 25°C-treated controls rescues the synthesis of functional mRNAs in heat-shocked lysates. Thus, the major regulatory factor may be associated with the ribosomes.

Regulation of Protein Synthesis by Initiation and Discriminatory Factors

In the last decade, great strides have been made in our understanding of how eukaryotic initiation factors regulate the translation of specific mRNAs. Our increase in knowledge has been facilitated by the purification of these factors (see methods in Enzymology, Vol. 60 H, 1979), and by the kinetic analysis of competition among different mRNAs for initiation factors (Lodish, 1974; Bergmann & Lodish, 1979), including competition for the message discriminatory factors eIF-4B and CPB-II (Godefroy-Colburn & Thach, 1981; Ray et al., 1983). Competition for discriminatory factors is now recognized as playing a central role in the regulation of viral synthesis (Walden et al., 1981; Rosen et al., 1982) and in the regulation of protein synthesis in cells such as rabbit reticulocytes (Kabat & Chappell, 1977), growing Vero cells (see analysis of Lee & Engelhardt, 1979, by Godefroy-Colburn & Thach, 1981), and chick embryo fibroblasts (Ignotz et al., 1981). In their model of protein synthesis, Godefroy-Colburn and Thach (1981) treat initiation as a multistep process in which mRNA must bind to a "discriminatory factor" prior to its recognition by the eIF-2-containing 40S preinitiation complex. Protein synthesis rates and mRNA selection, therefore, depend on concentrations of both factors.

Our experiments (Hille et al., 1981) suggest that initiation factors such as eIF-2, which recycle with each initiation step, are not rate limiting in sea urchins. The absence of a deficiency in recycling initiation factors was shown by examining polyribosome profiles of eggs and embryos incubated in the presence and absence of low amounts of

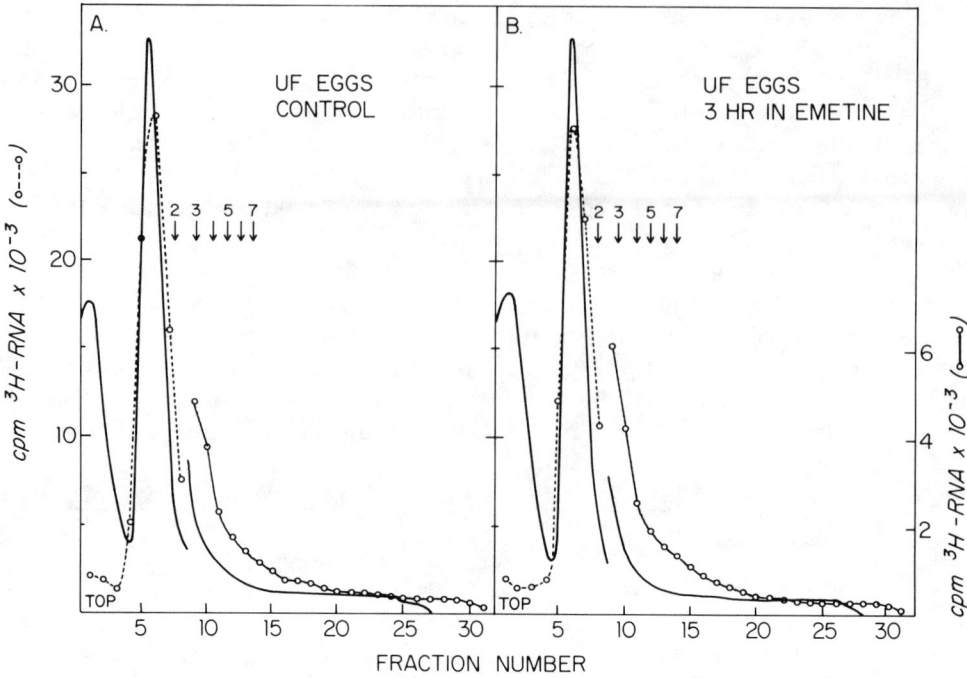

Fig. 10. Polyribosome profiles of unfertilized egg cultures incubated for 3.1 h in sea water (A) and for 3.2 h in 1.5 μm emetine in sea water (B). The inhibition of peptide synthesis by emetine was initially 90% then decreased to 70% after 2 h. The postmitochondrial supernatants were centrifuged at 200,000g_{ave} for 1.25 h at 2°C on 15% to 40% sucrose gradients. Centrifugation is from left to right. The first and second peaks of absorbance at the top of the gradient are supernatant and monoribosome peaks. The positions of polyribosomes with two to seven ribosomes are indicated by arrows. Absorbance at 260 nm (——) and tritium counts of oogenetically labeled RNA (o---o, o——o) are indicated. There is a scale change after fraction 9. The total counts in gradient A are 104,970 cpm and in gradient B are 102,190 cpm (Hille et al., 1981).

emetine, an antibiotic that inhibits polypeptide bond formation. The elongation of peptides is slowed relative to the initiation of new peptides in cells cultured in low concentrations of this inhibitor. Under these conditions, polyribosome sizes would increase if recycling initiation factors were previously rate limiting since the relative amount of initiation as compared to the elongation would increase (McCormick & Penman, 1969; Lodish, 1971, 1974; Christman, 1973). We determined the distribution of ribosomes found in polyribosomes by analyzing sucrose density gradient fractions for 3H-uridine that had been incorporated into rRNAs and mRNAs during oogenesis. The measurement of polyribosomes by radioactivity allowed an accurate determination of changes in the size and amounts of polyribosomes.

We found that concentrations of emetine that inhibited protein synthesis 40% to 90% did not change either the size of the polyribosomes or their numbers in unfertilized eggs (Fig. 10) or in 4- to 16-cell embryos (Fig. 11). Since the size of the polyribosomes did not increase, recycling initiation factors are probably not rate limiting in sea urchin eggs or cleavage-stage embryos. In addition, our observation that the total numbers of polyribosomes did not increase implies that few, if any, untranslated mRNAs were added to the polyribosomal pools. These results, by themselves, suggested that a block at initiation must occur because mRNA and/or ribosome are limiting, or because a factor necessary for the recruitment of messages is rate limiting. We know from the studies described above that the rate-limiting factor is probably not mRNA.

Concentrations of discriminatory factors could impose an absolute limit on the amount of protein synthesis in these or other eggs. Similar to echinoid eggs, postvitellogenic amphibian oocytes use only about 5% of their mRNAs and 1% of their ribosomes. Progesterone-stimulated maturation results in a 2-fold increase in protein synthesis; fertilization in a 5-fold increase by 8 h (Woodland, 1974; Adamson & Woodland, 1977). The difference is not due to a change in peptide elongation rates (Richter et al., 1982). As in echinoids, Xenopus oocytes have a limited capacity to translate mRNA. Injection of increasing amounts of globin mRNA results in an increase in globin synthesis, and a reciprocal decrease in endogenous synthesis; total protein synthesis does not increase (Laskey et al., 1977; Richter & Smith, 1981). Laskey et al. (1977) postulated that either there are limited amounts of recruitment factors in the cytoplasm for which all mRNAs compete (analogous to discriminatory factors), or limited amounts of ribosomes. They suggested that in the latter case, the binding of one ribosome to a mRNA would increase the probability of binding subsequent ribosomes by changing the secondary structure of the mRNA, and thus creating fully loaded polyribosomes. Importantly, Richter and Smith have observed that free and membrane-bound mRNAs injected into these oocytes probably do not compete with one another (Richter & Smith, 1981; Richter et al., 1983). These studies argue for the shortage of two message recruitment factors, one recognizing membrane-bound mRNAs and the other, free mRNAs. Thus, it is possible that a nonrecycling initiation factor, one that binds relatively tightly to the initiation region of the mRNA, limits translation in either amphibians or echinoids or both.

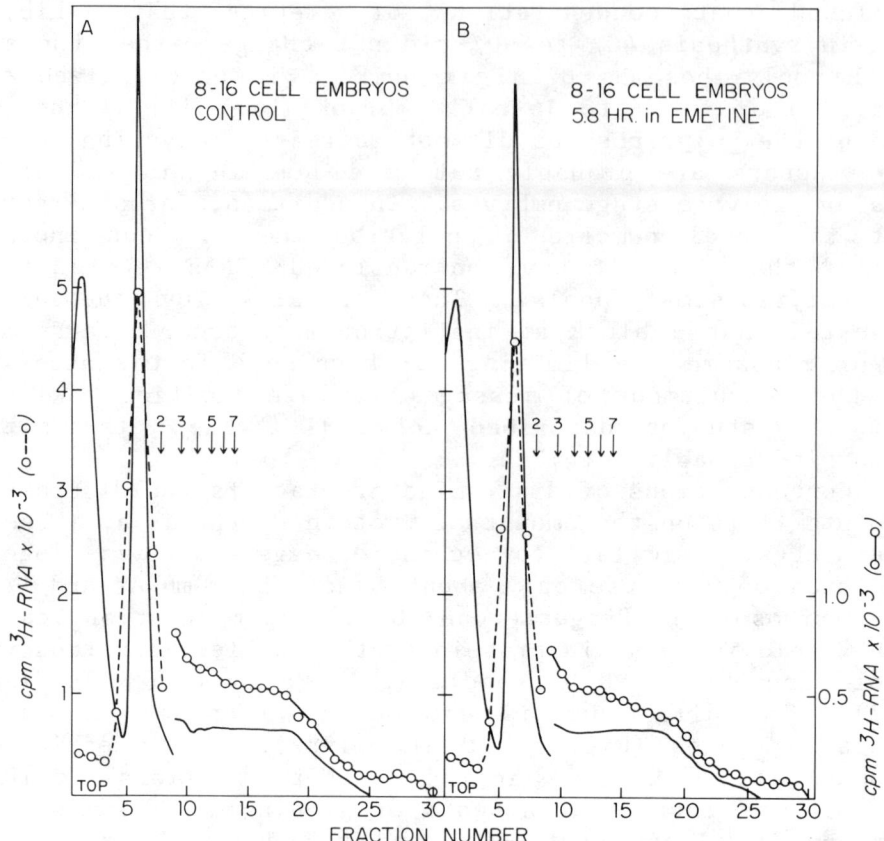

Fig. 11. Polyribosome profiles of 8- to 16-cell embryos incubated in sea water (A) and in 8 μM emetine in sea water (B). The inhibition of peptide synthesis by emetine was initially 75% and decreased to 40% after 2 h. The control culture taken 5.2 h after fertilization was 32% 8-cell and 52% 16-cell. The emetine-treated culture taken at 7.2 h after fertilization (6 h in emetine) was 83% 8-cell and 17% 16-cell. Notations and conditions are as in Figure 10. The total counts in gradient A are 21,390 cpm and in B are 19,320 cpm (Hille et al., 1981).

Regulation by Compartmentalization

The sequestration of α-histone mRNAs in echinoids is elegantly described by Venezky et al. (1981), Showman et al. (1982), DeLeon et al. (1983), Angerer et al. (1984), Cox et al. (1984) and by Showman in this symposium volume. They observed that stored α-histone mRNAs are transcribed after meiotic maturation of sea urchins and stored in the nucleus until the first post-fertilization cleavage when the nuclear membrane breaks down. Cleavage-stage histone mRNAs and poly A RNAs are, however, cytoplasmic in location and translated immediately after fertilization.

What evidence is there that ribosomes are compartmentalized in echinoids? Basophilic structures of 0.3 to 3 µm diameter have been observed in sea urchins that appear at the ultrastructural level to contain aggregations of electron-dense granules, approximately 150 Å in diameter, that are bound by annulate lamellae (Afzelius, 1957; Harris, 1967; Conway & Metz, 1974). These structures, termed heavy bodies, are absent during oogenesis (Afzelius, 1957; Conway, 1971); they originate after the maturation divisions of the oocyte and several hours after the formation of the egg pronuclear membrane. The cytological origin of these structures is obscure since they appear simultaneously as bleb-like outpouchings of the nuclear membrane and as membrane-bound entities in the cytoplasm. Similar membrane-bound granular aggregations have been induced in the cytoplasm of sea urchin eggs by treatment with hypertonic salt (Kallenbach, 1983, 1984). Hence, it is likely that in normal embryos, the origin of the heavy bodies is dual and that some structures originate from the cytoplasm and associate with condensations of membranes. The majority of the heavy bodies occur in the cytoplasm (about 1500). Those structures that are attached to the nucleus move into the cytoplasm during fusion of the sperm and egg pronucleus. Heavy bodies remain intact until nuclear membrane breakdown, at the time of the first cytoplasmic division, when they permanently disappear (Harris, 1967; Conway & Metz, 1974). There are other distinct electron-dense materials that remain attached to the inner nuclear membrane after pronuclear fusion (Harris, 1967). These structures, however, have a finer granularity than heavy bodies.

The granular structure and the basophilic nature of the heavy bodies are sensitive to RNase, but not to DNase, and hence, contain RNA (Conway, 1971). Though suspected of containing stored mRNA, the size and histochemical staining properties of heavy bodies are more similar to ribosomes (Afzelius, 1957; Conway, 1971). Additional observations, which suggest that these bodies contain ribosomes rather than messenger RNPs, are that 3H-poly U, which should detect poly A RNA, does not hybridize preferentially to heavy bodies (Venezky et al., 1981; Danilchik & Moon, unpubl.), and that ribosomal RNA binds to these structures (Steinert et al., 1984).

Despite the above arguments that heavy bodies may contain stored ribosomes, it is doubtful that a large enough portion of the ribosomes are stored in heavy bodies to significantly affect the rates of protein synthesis of the eggs. Also, free ribosomes are visible throughout the cytoplasm; and after centrifugation of the eggs into halves, both halves, whether or not they contain the heavy bodies, develop normally when fertilized (Harvey, 1933).

What evidence is there that ribosomes are stored in other organisms? The motile zoospores of Blastocladiella emersonii, an aquatic fungus, are dormant with regard to protein synthesis, and contain ribosomes, mRNA and polyribosomes (about 10% of the ribosomes) stored in a membrane-bound nuclear cap (Lovett, 1963, 1975; Gong & Lovett, 1977). During germination and rhizoid formation, the material in the nuclear cap disperses rapidly and polyribosomes accumulate. Actinomycin D at levels which inhibit mRNA synthesis has no effect on development until 40 min, or just before the newly formed primary rhizoid begins to branch (Lovett, 1968). Therefore, protein synthesis is activated during germination without the synthesis of new mRNAs as is seen following fertilization of sea urchin eggs. Stored ribosomes have also been observed in pre-vitellogenic oocytes of the hibernating lizard, Lacerta sicula (Taddei et al., 1973). In this case the ribosomes appear to be crystalline and not bound by membranes. Though inactive in vivo, they readily translate poly U in vitro. No such crystalline structures have been observed in echinoderm eggs.

SUMMARY

There is no overwhelming evidence that masking of mRNAs is the primary rate-limiting step for either post-vitellogenic oocytes of amphibians or for echinoid eggs. Therefore, we propose two alternatives. First, our mRNA microinjection experiments indicate that echinoid eggs contain no spare translational capacity, as previously found for ripe amphibian oocytes. Evidence is accumulating from other laboratories that the concentrations of message discriminatory factors limit the rate of translation in normal and injected cells. These findings, therefore, suggest that either message recruitment factors and/or ribosomes could be rate limiting in echinoid eggs. Second, in two cases, sequestration has been shown to be the mechanism of the synthetic block: the storage of α-histone mRNAs in the nuclei of sea urchin eggs, and the storage of ribosomes and mRNAs in nuclear caps of B. emersonii zoospores. Furthermore, annulate lamellae, reminiscent of nuclear membranes, surround ribosome-like particles in postmeiotic echinoid eggs. These examples suggest that, besides message recruitment factors and/or ribosomal inhibitors, sequestration may be a means of regulating translation in eggs. Finally, following the release of components from their sequestered state, polyribosomes could form and associate with cytoskeletal elements (Moon et al., 1983), further indicating that protein synthesis may be affected by macromolecular cellular structures. The complexity of the activation of protein synthesis in echinoids via the dual ionic triggers, the stimulation of both elongation and the recruitment of mRNAs and ribosomes, the association of an inhibitory molecule with ribosomes, and the delayed translation of sequestered α-histone mRNAs suggest that translation is indeed regulated at several levels in echinoid egg and embryos.

ACKNOWLEDGMENTS

This research was supported by National Institutes of Health Grants HD 11070 to M.B. Hille, National Institutes of Health Training Grants GM 07270 to M.V. Danilchik and A. Colin and HD 07183 to R.T. Moon. We thank Thomas Iberle for technical assistance.

LITERATURE CITED

Adamson, E.D. and H.R. Woodland. 1977. Changes in the rate of histone synthesis during oocyte maturation and very early development of Xenopus laevis. Dev. Biol. 57: 136-149.
Afzelius, B.T. 1957. Electron microscopy of the basophilic structures of the sea urchin egg. Z. Zellforsch 45: 660-675.
Anderson, D.M., J.D. Richter, M.E. Chamberlain, D.H. Price, R.J. Britten, L.D. Smith, and E.H. Davidson. 1982. Sequence organization of the poly A RNA synthesized and accumulated in lampbrush chromosome stage Xenopus laevis oocytes. J. Mol. Biol. 155: 281-309.
Angerer, L.M., D.V. DeLeon, R.C. Angerer, R.M. Showman, D.E. Wells, and R.A. Raff. 1984. Delayed accumulation of maternal histone mRNA during sea urchin oogenesis. Dev. Biol. 101: 477-487.
Aviv, H. and P. Leder. 1972. Purification of biologically active globin messenger RNA by chromatography on oligothymidylic acid-cellulose. Proc. Nat. Acad. Sci., USA 69: 1408-1412.

Ballinger, D.G. and T. Hunt. 1981. Fertilization of sea urchin eggs is accompanied by 40S ribosomal subunit phosphorylation. Dev. Biol. 87: 277-285.

Ballinger, D.G. and M.L. Pardue. 1983. The control of protein synthesis during heat shock in Drosophila cells involves altered polypeptide elongation rates. Cell 33: 103-114.

Ballinger, D.G., S.J. Bray, and T. Hunt. 1984. Studies of the kinetics and ionic requirements for the phosphorylation of ribosomal protein S6 after fertilization of Arbacia punctulata eggs. Dev. Biol. 101: 192-200.

Bergmann, J.E. and H.F. Lodish. 1979. A kinetic model of protein synthesis. J. Biol. Chem. 254: 11927-11937.

Brachet, J., M. Deeroly, A. Ficq, and J. Quertier. 1963. Ribonucleic acid metabolism in unfertilized and fertilized sea urchin eggs. Biochim. Biophys. Acta 72: 660-662.

Brandis, J.W. and R.A. Raff. 1978. Translation of oogenetic mRNA in sea urchin eggs and early embryos. Demonstration of a change in translational efficiency following fertilization. Dev. Biol. 67: 99-113.

Brandis, J.W. and R.A. Raff. 1979. Elevation of protein synthesis is a complex response to fertilization. Nature 278: 467-468.

Chambers, E.L. 1975. Na^+ is required for nuclear and cytoplasmic activation of sea urchin eggs by sperm and divalent ionophores. J. Cell Biol. 67: 60a.

Christman, J. 1973. Effect of elevated potassium level and amino acid deprivation on polysome distribution and rate of protein synthesis in L cells. Biochim. Biophys. Acta 294: 138-152.

Clegg, K.B. and P.C. Denny. 1974. Synthesis of rabbit globin in a cell-free protein synthesis system utilizing sea urchin egg and zygote ribosomes. Dev. Biol. 37: 263-272.

Costantini, F.D., R.J. Britten, and E.H. Davidson. 1980. Message sequences and short repetitive sequences are interspersed in sea urchin egg $poly(A)^+$ RNAs. Nature 287: 111-117.

Conway, C.M. 1971. Evidence for RNA in the heavy bodies of sea urchin eggs. J. Cell Biol. 51: 889-893.

Conway, C.M. and C.B. Metz. 1974. In vitro maturation of Arbacia punctulata oocytes and initiation of heavy body formation. Cell Tissue Res. 599: 1-9.

Cox, K.H., D.V. DeLeon, L.M. Angerer, and R.C. Angerer. 1984. Detection of mRNAs in sea urchin embryos by in situ hybridization using asymmetric RNA probes. Dev. Biol. 101: 485-502.

Crystal, R., N. Elson, and W. Anderson. 1974. Initiation of globin synthesis: Assays. pp. 101-127. In: Methods in Enzymology, Vol. 30. K. Moldave and L. Grossman (eds.). Academic Press, N.Y.

Danilchik, M.V. and M.B. Hille. 1981. Sea urchin egg and embryo ribosomes: Differences in translational activity in a cell-free system. Dev. Biol. 84: 291-298.

Denny, P.C. and A. Tyler. 1964. Activation of protein biosynthesis in non-nucleate fragments of sea urchin eggs. Biochim. Biophys. Res. Commun. 14: 245-249.

DeLeon, D.V., K.H. Cox, L.M. Angerer, and R.C. Angerer. 1983. Most early-variant histone mRNA is contained in the pronucleus of sea urchin eggs. Dev. Biol. 100: 197-206.

Epel, D. 1967. Protein synthesis in sea urchin eggs: a "late" response to fertilization. Proc. Nat. Acad. Sci., USA 57: 899-905.

Epel, D. 1974. An analysis of the partial metabolic derepression of sea urchin eggs by ammonia; The existence of independent pathways. Dev. Biol. 40: 245-255.

Epel, D., R.A. Steinhardt, T. Humphreys, and D. Mazia. 1974. An analysis of the partial metabolic derepression of sea urchin eggs by ammonia: The existence of independent pathways. Dev. Biol. 40: 245-255.

Gambino, R., S. Metafora, L. Felicetti, and J. Raisman. 1973. Properties of the ribosomal salt wash from unfertilized and fertilized sea urchin eggs and its effect on natural mRNA translation. Biochim. Biophys. Acta 312: 377-391.

Gilbert, J. and W. Anderson. 1970. Cell-free hemoglobin synthesis. II. Characteristics of the transfer ribonucleic acid-dependent assay system. J. Biol. Chem. 245: 2342-2349.

Glover, C.V.C. 1982. Heat shock induces rapid dephosphorylation of a ribosomal protein in Drosophila. Proc. Nat. Acad. Sci., USA 79: 1781-1785.

Godefroy-Colburn, T. and R.E. Thach. 1981. The role of mRNA competition in regulating translation. J. Biol. Chem. 256: 11762-11773.

Gong, C-S. and J.S. Lovett. 1977. Regulation of protein synthesis in Blastocladiella zoospores: Factors for synthesis in nonsynthetic spores. Exp. Mycol. 1: 138-151.

Gordon, J., P.J. Nielsen, K.L. Manchester, H. Towbin, L. Jimenez de Asua, and G. Thomas. 1982. Criteria for establishment of the biological significance of ribosomal protein phosphorylation. Current Topics in Cell Reg. 21: 89-99.

Goustin, A.S. and F.H. Wilt. 1981. Protein synthesis, polyribosomes, and peptide elongation in early development of Strongylocentrotus purpuratus. Dev. Biol. 82: 32-40.

Grainger, J.L., M.M. Winkler, S.S. Shen, and R.A. Steinhardt. 1979. Intracellular pH controls protein synthesis rate in the sea urchin egg and early embryo. Dev. Biol. 68: 396-406.

Gross, P.R. and G.H. Cousineau. 1963. Effects of actinomycin-D on macromolecular synthesis and early development of sea urchin eggs. Biochem. Biophys. Res. Commun. 10: 321-326.

Gross, P.R. and G.H. Cousineau. 1964. Macromolecular synthesis and the influence of actinomycin on early development. Exp. Cell Res. 33: 368-395.

Gross, P.R., L.I. Malkin, and W.A. Moyer. 1964. Templates for the first proteins of embryonic development. Proc. Nat. Acad. Sci., USA 51: 407-414.

Hanocq-Quertier, J. and E. Baltus. 1981. Phosphorylation of ribosomal proteins during maturation of Xenopus laevis oocytes. Europ. J. Biochem. 120: 351-355.

Harris, P. 1967. Structural changes following fertilization in the sea urchin egg: Formation and dissolution of heavy bodies. Exp. Cell Res. 48: 569-581.

Harvey, E.B. 1933. Development of the parts of sea urchin eggs separated by centrifugal force. Biol. Bull. 64: 125-148.

Hille, M.B. 1974. Inhibitors of protein synthesis isolated from ribosomes of unfertilized eggs and embryos of sea urchins. Nature 249: 556-558.

Hille, M.B. and A.A. Albers. 1979. Efficiency of protein synthesis after fertilization of sea urchin eggs. Nature 278: 469-471.

Hille, M.B., D.C. Hall, Z. Yablonka-Reuveni, M.V. Danilchik, and R.T. Moon. 1981. Translational control in sea urchin eggs and embryos: Initiation is rate limiting in blastula stage embryos. Dev. Biol. 86: 241-249.

Howard, G.A., S.D. Adamson, and E. Herbert. 1970. Subunit recycling during translation in a reticular cell-free system. J. Biol. Chem. 245: 6237-6239.

Hultin, T. 1964. On the mechanism of ribosomal activation in newly fertilized sea urchin eggs. Dev. Biol. 10: 305-328.

Humphreys, T. 1971. Measurements of messenger RNA entering polysomes upon fertilization of sea urchin eggs. Dev. Biol. 26: 201-208.

Ignotz, G.G., S. Hokari, R.M. Dephilip, K. Tsukaclay, and L. Lieberman. 1981. Lodish model and regulation of ribosomal protein synthesis by insulin-deficient chick embryo fibroblasts. Biochemistry 20: 2550-2557.

Ilan, J. and J. Ilan. 1978. Translation of maternal message ribonucleoprotein particles from sea urchin in a cell-free system and product analysis. Dev. Biol. 66: 375-385.

Infante, A.A. and M. Nemer. 1967. Heterogeneous ribonucleoprotein particles in the cytoplasm of sea urchin embryos. Proc. Nat. Acad. Sci., USA 58: 681-688.

Jenkins, N.A., J.F. Kaumeyer, E.M. Young, and R.A. Raff. 1978. A test for masked message: The template activity of messenger ribonucleoprotein particles isolated from sea urchin eggs. Dev. Biol. 63: 279-298.

Johnson, J.D., D. Epel, and M. Paul. 1976. Intracellular pH and activation of sea urchin eggs after fertilization. Nature 262: 661-664.

Johnson, C.H. and D. Epel. 1981. Intracellular pH of sea urchin eggs measured by the dimethyloxazolidinedione (DMO) method. J. Cell Biol. 89: 284-291.

Johnson, C.H. and D. Epel. 1982. Starfish oocyte maturation and fertilization: Intracellular pH is not involved in activation. Dev. Biol. 92: 461-469.

Kabat, D. 1970. Phosphorylation of ribosomal proteins in rabbit reticulocytes. Characterization and regulatory aspects. Biochemistry 9: 4160-4175.

Kabat, D. and M.R. Chappell. 1977. Competition between globin messenger ribonucleic acids for a discriminating initiation factor. J. Biol. Chem. 252: 2684-2690.

Kallenbach, R.J. 1983. The induction of de novo centrioles in sea urchin eggs: A possible common mechanism for centriolar activation among parthenogenetic procedures. Eur. J. Cell Biol. 30: 159-166.

Kallenbach, R.J. 1984. Endoplasmic reticulum whorls as a source of membranes for early cytaster formation in parthenogenetically stimulated sea urchin eggs. Cell Tissue Res. 236: 237-244.

Kaumeyer, J.F., N.A. Jenkins, and R.A. Raff. 1978. Messenger ribonucleoprotein particles in unfertilized sea urchin eggs. Dev. Biol. 63: 266-278.

Laemmli, U.K. 1970. Cleavage of structural proteins during the assembly of the head of bacteriophage. Nature 227: 680-685.

Laskey, R.A. and A.D. Mills. 1975. Quantitative film detection of ^{3}H and ^{14}C in polyacrylamide gels by fluorography. Eur. J. Biochem. 56: 335-341.

Laskey, R.A., A.D. Mills, J.B. Gurdon, and G.A. Partington. 1977. Protein synthesis in oocytes of Xenopus laevis is not regulated by the supply of messenger RNA. Cell 11: 345-351.

Lee, G.T.-Y. and D.L. Engelhardt. 1979. Growth-related fluctuations in messenger RNA utilization in animal cells. J. Mol. Biol. 129: 221-233.

Lee, S.C. and R.A. Steinhardt. 1981. pH changes associated with meiotic maturation in oocytes of Xenopus laevis. Dev. Biol. 85: 358-369.

Lindquist, S. 1980. Varying patterns of protein synthesis in Drosophila during heat shock: Implications for regulation. Dev. Biol. 77: 463-479.

Lodish, H.F. 1971. Alpha and beta globin messenger ribonucleic acid. J. Biol. Chem. 246: 7131-7138.

Lodish, H.F. 1974. Model for the regulation of mRNA translation applied to haemoglobin synthesis. Nature 251: 385-388.

Lovett, J.S. 1963. Chemical and physical characterization of "nuclear caps" isolated from Blastocladiella zoospores. J. Bacteriol. 85: 1235-1246.

Lovett, J.S. 1968. Reactivation of ribonucleic acid and protein synthesis during germination of Blastocladiella zoospores and the role of the ribosomal nuclear cap. J. Bacteriol. 96: 962-969.

Lovett, J.S. 1975. Growth and differentiation of the water mold Blastocladiella emersonii: Cytodifferentiation and the role of ribonucleic acid and protein synthesis. Bacteriol. Rev. 39: 345-404.

McCormick, W. and S. Penman. 1969. Regulation of protein synthesis in HeLa cells: Translation at elevated temperatures. J. Mol. Biol. 39: 315-333.

McKenzie, S.L., S. Henikoff, and M. Meselson. 1975. Localization of RNA from heat-induced polysomes at puff sites in Drosophila melanogaster. Proc. Nat. Acad. Sci., USA 72: 1117-1121.

Metafora, S., L. Felicetti, and R. Gambino. 1971. The mechanism of protein synthesis activation after fertilization of sea urchin eggs. Proc. Nat. Acad. Sci., USA 68: 600-604.

Mirault, M.-E., M. Goldschmidt-Clermont, L. Moran, A.P. Arrigo, and A. Tissieres. 1978. The effect of heat shock on gene expression in Drosophila melanogaster. Cold Spring Harbor Symp. Quant. Biol. 42: 819-827.

Monroy, A., R. Maggio, and A. Rinaldi. 1965. Experimentally induced activation of the ribosomes of the unfertilized sea urchin egg. Proc. Nat. Acad. Sci., USA 54: 107-111.

Moon, R.T. 1983. Poly(A)-containing messenger ribonucleoprotein complexes from sea urchin eggs and embryos: Polypeptides associated with native and UV-crosslinked mRNPs. Differentiation 24: 13-23.

Moon, R.T., K.D. Moe, and M.B. Hille. 1980. Polypeptides of nonpolyribosomal messenger ribonucleoprotein complexes of sea urchin eggs. Biochemistry 19: 2723-2730.

Moon, R.T., M.V. Danilchik, and M.B. Hille. 1982. An assessment of the masked message hypothesis: Sea urchin egg messenger ribonucleoprotein complexes are efficient templates for in vitro protein synthesis. Dev. Biol. 93: 389-403.

Moon, R.T., R.F. Nicosia, C. Olsen, M.B. Hille, and W.R. Jeffery. 1983. The cytoskeletal framework of sea urchin eggs and embryos: Developmental changes in the association of messenger RNA. Dev. Biol. 95: 447-458.

Murray, A. and R. Sosnowski. 1980. The status of mRNA in eggs and early embryos. Biol. Bull. 159: 476.

Nielsen, P.J., K.L. Manchester, H. Towbin, J. Gordon, and G. Thomas. 1982a. The phosphorylation of ribosomal protein S6 in rat tissues following cycloheximide injection, in diabetes, and after denervation of diaphragm. J. Biol. Chem. 257: 12316-12321.

Nielsen, P.J., G. Thomas, and J.L. Maller. 1982b. Increased phosphorylation of ribosomal protein S6 during meiotic maturation of Xenopus oocytes. Proc. Nat. Acad. Sci., USA 79: 2937-2941.

Palmiter, R. 1974. Magnesium precipitation of ribonucleoprotein complexes. Expedient techniques for the isolation of undegraded polysomes and messenger ribonucleic acid. Biochemistry 13: 3606-3615.

Palmiter, R. 1973. Ovalbumin messenger ribonucleic acid translation: Comparable rates of polypeptide initiation and elongation on ovalbumin and globin messenger ribonucleic acid in a rabbit reticulocyte lysate. J. Biol. Chem. 248: 2095-2106.

Paul, M. and R.N. Johnston. 1978. Absence of a Ca response following ammonia activation of sea urchin eggs. Dev. Biol. 67: 330-335.

Payan, P., J. Girard, and B. Ciapa. 1983. Mechanisms regulating intracellular pH in sea urchin eggs. Dev. Biol. 100: 29-38.

Raff, R.A. 1980. Masked messenger RNA and the regulation of protein synthesis in eggs and embryos. pp. 107-136. In: Cell Biology: A Comprehensive Treatise, Vol. 4. D.M. Prescott and L. Goldstein (eds.). Academic Press, N.Y.

Raff, R.A., J.W. Brandis, C.J. Huffman, A.L. Koch, and D.E. Leister. 1981. Protein synthesis as an early response to fertilization of the sea urchin egg: A model. Dev. Biol. 86: 265-271.

Raff, R.A. and R.M. Showman. In Press. Maternal messenger RNA: Quantitative, qualitative, and spatial control of its expression in embryos. In: The Biology of Fertilization. C.B. Metz and A. Monroy (eds.). Academic Press, N.Y.

Ray, B.K., T.G. Brendler, S. Adya, S. Daniels-McQueen, J.K. Miller, J.W.B. Hersey, J.A. Grifo, W.C. Merrick, and R.E. Thach. 1983. Role of mRNA competition in regulating translation: Further characterization of mRNA discriminatory initiation factors. Proc. Nat. Acad. Sci., USA 80: 663-667.

Regier, J.C. and F.C. Kafatos. 1977. Absolute rate of protein synthesis in sea urchins with specific activity measurements of radioactive leucine and leucyl-tRNA. Dev. Biol. 57: 270-283.

Richter, J.D. and L.D. Smith. 1981. Differential capacity for translation and lack of competition between mRNAs that segregate to free and membrane-bound polysomes. Cell 27: 182-191.

Richter, J.D., W.J. Wasserman, and D.L. Smith. 1982. The mechanism for increased protein synthesis during Xenopus oocyte maturation. Dev. Biol. 89: 159-167.

Richter, J.D., D.C. Evers, and L.D. Smith. 1983. The recruitment of membrane-bound mRNAs for translation in microinjected Xenopus oocytes. J. Biol. Chem. 258: 2614-2620.

Richter, J.D., D.M. Anderson, E.H. Davidson, and L.D. Smith. 1984. Interspersed poly(A)RNAs of amphibian oocytes are not translatable. J. Mol. Biol. 173: 227-241.

Rosen, H., G. Disegni, and R. Kaempfer. 1982. Translational control by messenger RNA competition for eukaryotic initiation factor 2. J. Biol. Chem. 257: 946-951.

Scott, M.P. and M.L. Pardue. 1981. Translational control in lysates of Drosophila melanogaster cells. Proc. Nat. Acad. Sci., USA 78: 3353-3357.

Shen, S.S. and R.A. Steinhardt. 1978. Direct measurement of intracellular pH during metabolic depression at fertilization and ammonia activation of the sea urchin egg. Nature 272: 253-255.

Showman, R.M. and C. Foerder. 1979. Removal of the fertilization membrane of sea urchin embryos employing aminotriazole. Exp. Cell Res. 120: 253-255.

Showman, R.M., D.E. Wells, J. Anstrom, D.A. Hursch, and R.A. Raff. 1982. Message-specific sequestration of maternal histone mRNA in the sea urchin egg. Proc. Nat. Acad. Sci., USA 79: 5944-5947.

Spirin, A. 1966. On "Masked" forms of messenger RNA in early embryogenesis and in other differentiating systems. pp. 1-38. In: Current Topics in Developmental Biology, Vol. 1. A.A. Moscona and A. Monroy (eds.). Academic Press, N.Y.

Stavy, L. and P.R. Gross. 1967. The protein synthesis lesion in unfertilized eggs. Proc. Nat. Acad. Sci., USA 57: 735-742.

Stavy, L. and P.R. Gross. 1969. Protein synthesis in vitro with fractions of sea urchin eggs and embryos. Biochim. Biophys. Acta 182: 193-202.

Steinert, G., A. Felsani, R. Kettmann, and J. Brachet. 1984. Presence of rRNA in the heavy bodies of sea urchin eggs. Exp. Cell Res. 154: 203-212.

Steinhardt, R.A., L. Lundin, and D. Mazia. 1971. Bioelectric responses of the echinoderm egg to fertilization. Proc. Nat. Acad. Sci., USA 68: 2426-2430.

Steinhardt, R.A., R. Zucker, and G. Schatten. 1977. Intracellular calcium release at fertilization in the sea urchin egg. Dev. Biol. 58: 185-196.

Taddei, C., R. Gambino, S. Metafora, and A. Monroy. 1973. Possible role of protein synthesis in pre-vitellogenic oocytes of the lizard Lacerta sicula. Exp. Cell Res. 78: 159-167.

Takeshima, K. and E. Nakano. 1983. Modification of ribosomal proteins in sea urchin eggs following fertilization. Eur. J. Biochem. 137: 437-443.

Thomas, G. 1980. Regulation of 40S ribosomal protein S6 phosphorylation during the G_0/G_1 transition of the cell cycle. pp. 102-110. In: Protein Phosphorylation and Bio-Regulation. G. Thomas, E.J. Podesta, and J. Gordon (eds.). FMI-EMBO Workshop, Basel, 1979, Karger, Basel.

Thomas, G., M. Siegmann, A. Kulber, J. Gordon, and L. Jimenez de Asua. 1980. Regulation of 40S ribosomal protein S6 phosphorylation in swiss mouse 3T3 cells. Cell 19: 1015-1023.

Thomas, G., J. Martin-Perez, M. Siegmann, and A.M. Otto. 1982. The effect of serum, EGF, $PGF_{2\alpha}$ and insulin on S6 phosphorylation and the initation of protein and DNA synthesis. Cell 30: 235-242.

Venezky, D.L., L.M. Angerer, and R.C. Angerer. 1981. Accumulation of histone repeat transcripts in the sea urchin egg pronucleus. Cell 24: 385-391.

Walden, W.E., T. Godefroy-Colburn, and R.E. Thach. 1981. The role of mRNA competition in regulating translation. J. Biol. Chem. 256: 11739-11746.

Ward, G.E., V.D. Vacquier, and S. Michel. 1983. The increased phosphorylation of ribosomal protein S6 in Arbacia punctulata is not a universal event in the activation of sea urchin eggs. Dev. Biol. 95: 360-371.

Winkler, M.M. and R.A. Steinhardt. 1981. Activation of protein synthesis in a sea urchin cell-free system. Dev. Biol. 84: 432-439.

Winkler, M.M., R.A. Steinhardt, J.L. Grainger, and L. Minning. 1980. Dual ionic controls for the activation of protein synthesis at fertilization. Nature 287: 558-560.

Woodland, H.R. 1974. Changes in the polysome content of developing Xenopus laevis embryos. Dev. Biol. 40: 90-101.

Zucker, R.S., R.A. Steinhardt, and M. M. Winkler. 1978. Intracellular calcium release and mechanisms of parthenogenetic activation of the sea urchin egg. Dev. Biol. 65: 285-295.

PATTERNS OF MATERNAL mRNA DISTRIBUTION AND THEIR ROLE IN EARLY DEVELOPMENT
William R. Jeffery

ABSTRACT

Current progress in elucidating the patterns, mechanisms, and developmental significance of maternal mRNA distributions in invertebrate eggs and early embryos is reviewed. Three patterns of poly A^+ RNA distribution have been observed using in situ hybridization with a ^{3}H-labeled poly U probe. Poly A^+ RNA is concentrated in the germinal vesicle (GV) plasm of oocytes and the ectoplasm of ascidian eggs. The ectoplasmic poly A^+ RNA is segregated into the animal hemisphere (AH) blastomeres of early embryos. Poly A^+ RNA is concentrated in a distinctly staining perinuclear region in oocytes of the annelid Nereis. The perinuclear poly A^+ RNA appears to diffuse into the plasm released from the GV at the time of maturation and is eventually delivered to the AH cells of the early embryo. Poly A^+ RNA is concentrated in the cortex in eggs of the annelid Chaetopterus and a large proportion of these molecules also enter the AH cells of the early embryo. The different patterns of poly A^+ RNA distribution in ascidians and annelids each result in the segregation of maternal mRNA into the ectoderm cell lineages. The spatial distribution of poly A^+ RNA in ascidian eggs appears to be mediated by the association of these molecules with specific domains of the cytoskeletal framework. Poly A^+ RNA appears to be present in a cortical complex in Chaetopterus eggs which may also be associated with the cytoskeletal framework. The high specific gravity of the poly A^+ RNA associated with cortical complexes results in its displacement to the centrifugal pole region when eggs are stratified by centrifugation. The association of poly A^+ RNA with the cortical complex also results in its displacement to the centrifugal fragment when eggs are fragmented by stronger centrifugation. The distinctive behavior of poly A^+ RNA during the fragmentation of Chaetopterus eggs provides a test to determine whether the mRNA and associated cortical complexes are required for development. It was found that centripetal fragments depleted in poly A^+ RNA do not develop after fertilization, even though they contain the egg nucleus. Likewise, if the cortical complexes with their associated poly A^+ RNA molecules are displaced into one of the two blastomeres by centrifugation at first cleavage there is interference with normal development. These results are consistent with the idea that maternal mRNA localized in the cortex of Chaetopterus eggs plays an important role in early development.

INTRODUCTION

A central issue of developmental biology is how different embryonic cell lineages are formed from the totipotent, apparently-undifferentiated egg. The egg is not really an undifferentiated cell, however. From its beginning it possesses a well-defined polarity that is visibly expressed as a radially-symmetric distribution of organelles along the animal-vegetal axis. An axis of bilateral symmetry also becomes apparent during early development. The phenomena that underlie the expression of the earliest egg axes are probably followed by other, more subtle polarizations that may in part be responsible for the origin of the different embryonic cell lineages. The purpose of this article is to review our current understanding of the polarization of invertebrate eggs at the molecular level. In recent years it has been discovered that both RNA and proteins can be unevenly distributed in eggs and early embryos. This article is focused on the spatial distribution of maternal mRNA because there is evidence that these molecules may be important in some cases of embryonic determination (Kalthoff, 1979; Brown & Kalthoff, 1982; see Jeffery, 1983 for review). In this paper, I describe the spatial distribution of poly A^+ RNA. Poly A^+ RNA is considered to be mRNA for the purpose of this discussion, although all mRNAs may not be polyadenylated and some poly A^+ RNAs may not be typical messages in that they resemble unprocessed or partially processed nuclear transcripts (Richter et al., 1984). Spatial patterns of mRNA distribution are compared in the eggs of ascidians and annelids. A possible mechanical basis for the distinct patterns of mRNA distribution observed in these eggs is also presented. Finally, the question of whether the pattern of maternal mRNA distribution has an important role in early development is considered.

Mapping The Distribution of mRNA

In recent years the spatial distribution of mRNA molecules has been mapped in sections of eggs and embryos by in situ hybridization with nucleic acid probes (Capco & Jeffery, 1978; Angerer & Angerer, 1981; Cox et al., 1984). The studies reported in this article were conducted using the in situ hybridization techniques developed by Capco and Jeffery (1978), and Jeffery et al. (1983). The method for mapping the distribution of poly A^+ RNA consists of the following steps. Eggs or embryos are fixed with Petrunkewitsch's fluid. This fixes poly A^+ RNA in the specimen and thus the distributions of poly A^+ RNA that are observed cannot be due to differential extraction of hybridizable RNA from the ooplasmic regions (Jeffery & Wilson, 1983). The specimens are dehydrated, cleared, embedded in paraffin, and sectioned after fixation. The sections are briefly treated with DNase I and proteinase K, which serve to enhance the efficiency of poly A^+ RNA detection, and are then annealed with an excess of ^{3}H-labeled poly U. Hybridization is carried out in 10 mM Tris-HCl - 200 mM NaCl - 5 mM $MgCl_2$ for 4 h at 50°C. These hybridization criteria were selected to avoid a possible interaction of the poly U probe with oligo(dA) sequences in DNA (Riley et al., 1966). Following the annealing step, unhybridized poly U is removed by RNase hydrolysis and cold trichloroacetic acid extraction and the location of the poly A:poly U hybrids is determined by autoradiography.

Several lines of evidence suggest that the poly U probe interacts specifically with poly A sequences preserved in the sections (Capco & Jeffery, 1978). The hybridization signal is sensitive to pretreatment of the sections with alkali, excluding a possible interaction with DNA. The signal is also sensitive to pretreatment of the sections with RNase T_2, a nuclease which hydrolyzes poly A. Treatment of the sections with pancreatic RNase A markedly reduces the signal, but only when administered in low ionic strength media: the same enzyme in media of higher ionic strength does not affect the signal. The sensitivity of poly A to RNase A at low, but not at high, ionic strength is also observed in vitro (Beers, 1960), suggesting that the hybridization signal originates from an interaction of poly U and poly A in the sections. The strongest evidence for a poly A target, however, is provided by microinjection experiments. The hybridization signal can be enhanced by microinjecting poly A into eggs, but is not affected by microinjecting similar amounts of poly dA:dT (Capco & Jeffery, 1978). These controls validate the use of in situ hybridization with poly U to map the ooplasmic distribution of poly A^+ RNA during early development.

Patterns of Poly A^+ RNA Distribution

The spatial distribution of poly A^+ RNA has been mapped in four different invertebrate eggs. Poly A^+ RNA is distributed unevenly in three of these. The exception is the sea urchin egg where we, like Angerer and Angerer (1981), observe a uniform distribution of poly A^+ RNA up to the blastula stage (Fig. 1). In contrast, ascidian and annelid eggs exhibit three different patterns of poly A^+ RNA distribution.

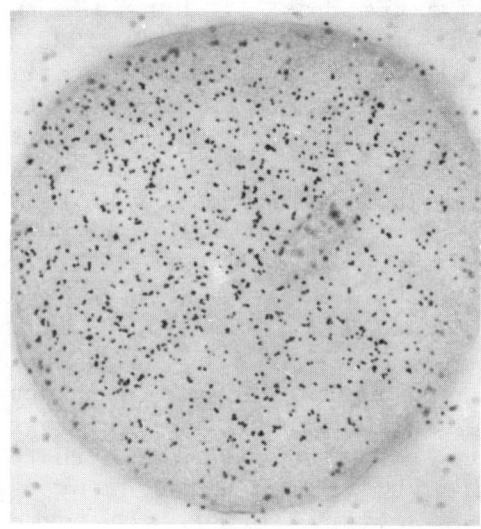

Fig. 1. The distribution of poly A^+ RNA is even in a fertilized Strongylocentrotus purpuratus egg as shown by in situ hybridization with poly U.

The GV Plasm-Ectoplasm Pattern. A large proportion of the oocyte poly A^+ RNA accumulates in the germinal vesicle (GV) during oogenesis in the ascidian Styela plicata (Jeffery & Capco, 1978). These molecules become a part of the cytoplasmic RNA population at maturation when the large GV breaks down and releases its nucleoplasm into the animal hemisphere of the egg. The GV material remains segregated from the rest of the cytoplasm as a special ooplasm known as the ectoplasm.

The ectoplasm primarily enters the ectodermal cell lineages during embryogenesis (Conklin, 1905). The precise histological distinctions between the GV plasm-ectoplasm and the other ooplasms have been exploited in mapping of the distribution of poly A^+ RNA (Jeffery & Capco, 1978; Jeffery et al., 1983). According to these studies, the GV plasm exhibits a poly A concentration about 25-30 fold higher than that of the myoplasm, an ooplasm at the periphery of the egg that enters the mesodermal cells, and about 10-fold higher than that of the endoplasm, a yolk-filled region (located between the myoplasm and the GV) that is partitioned largely to the endodermal cells during embryogenesis (Fig. 2A). The magnitude of these regional differences (and also those reported

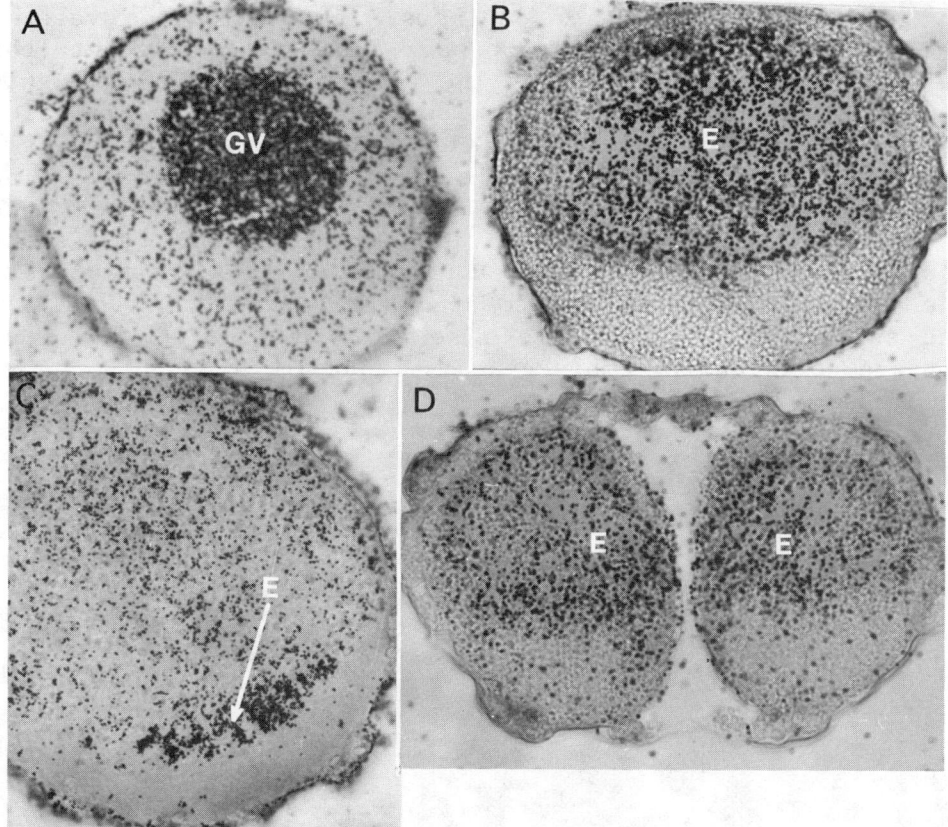

Fig. 2. The GV plasm-ectoplasm pattern of poly A^+ RNA distribution in oocytes, eggs, and early embryos of the ascidian Styela plicata as shown by in situ hybridization with poly U. A. A mature oocyte showing most of the grains over the GV. B. An unfertilized egg showing most of the grains over the ectoplasm (E). C-D. One and two-cell zygotes showing grains over the ectoplasm (Jeffery et al., 1983).

below for other eggs) makes it unlikely that they are the result of regional variations in the length of poly A tracts on different ooplasmic mRNAs. The GV appears to contain about 50% of the mass of maternal poly A^+ RNA. The poly A^+ RNA of the GV does not diffuse throughout the egg at maturation and remains highly concentrated in the ectoplasm throughout early development (Fig. 2B).

Ooplasmic segregation occurs shortly after fertilization in ascidian eggs (see Jeffery, 1984a, for a review). This event begins when the myoplasm and ectoplasm flow into the vegetal hemisphere. Simultaneously, the endoplasm is displaced into the animal hemisphere. Ooplasmic segregation ends with a partial reversal of the initial movements. The ectoplasm returns to the animal hemisphere, and the endoplasm is displaced back into the vegetal hemisphere. The myoplasm, in contrast to the other ooplasms, remains in the vegetal region of the egg where it spreads out into a crescent (the yellow crescent in Styela). The poly A^+ RNA remains highly concentrated in the ectoplasm during ooplasmic segregation (Fig. 2C) suggesting that these molecules are tenaciously associated with a localized structural framework. The ectoplasmic poly A^+ RNA is equally divided between the blastomeres at the first cleavage (Fig. 2D). In subsequent cleavages, however, ectoplasmic poly A^+ RNA is partitioned to the animal hemisphere blastomeres (Jeffery & Capco, 1978). The animal blastomeres give rise to most of the ectodermal lineage cells during embryogenesis (Conklin, 1905).

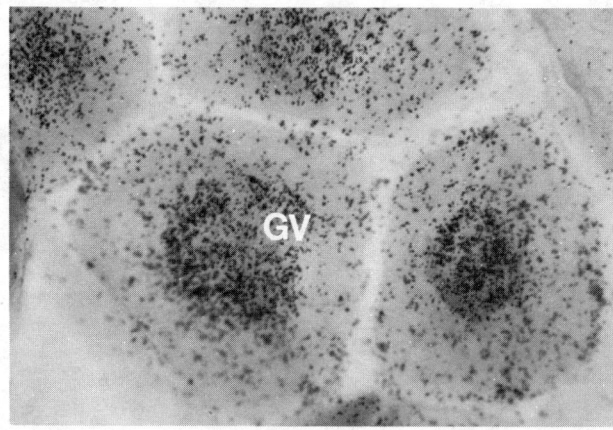

Fig. 3. The localization of poly A^+ RNA in the germinal vesicle (GV) of the chaetognath Sagitta as shown by in situ hybridization with poly U.

The GV-ectoplasm pattern of poly A^+ RNA distribution may not be unique to ascidians. Angerer et al. (1984) have shown that poly A^+ RNA accumulates in sea urchin oocyte GVs, but is absent from the haploid pronuclei of ootids. A search for other examples of poly A^+ RNA localization in the GV is in progress. Thus far, an enrichment of poly A^+ RNA has also been observed in the GV of chaetognath eggs (Fig. 3), but its developmental fate is unknown. It is tempting to speculate that the GV-localized poly A^+ RNAs of invertebrate eggs may be unprocessed or partially processed transcripts (Richter et al., 1984) which are subsequently matured during early development.

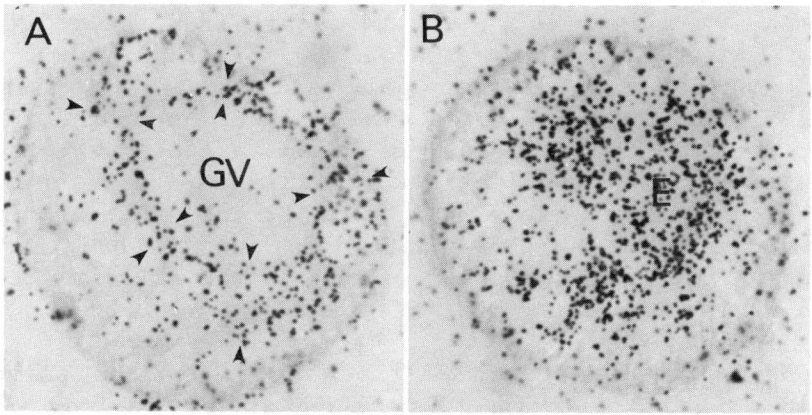

Fig. 4. The perinuclear plasm-ectoplasm pattern of poly A^+ RNA distribution in oocytes and fertilized eggs of the polychaete Nereis limbata as shown by in situ hybridization with poly U. A. A mature oocyte showing grains over the perinuclear plasm (outlined by arrowheads). B. A fertilized egg after germinal vesicle (GV) breakdown showing grains over the cytoplasm formed by the mixture of GV plasm and perinuclear plasm (E).

The Perinuclear Plasm-Ectoplasm Pattern. A variation in the GV plasm-ectoplasm pattern of poly A^+ RNA distribution occurs in the polychaete annelid Nereis limbata (Jeffery, unpubl.). Nereis eggs are shed at the primary oocyte stage. Unlike the situation in ascidian eggs, however, little poly A^+ RNA is localized in the GV plasm. When oocytes are mapped by in situ hybridization with poly U, most of the poly A^+ RNA is present in a basophilic perinuclear region (Fig. 4). The extent of the perinuclear region is limited peripherally by an area containing large oil droplets and yolk granules that fill the remainder of the cytoplasm. Similar to the case of the ectoplasm and the other ooplasmic regions of ascidian eggs, there is limited diffusion of poly A^+ RNA between the perinuclear and the peripheral ooplasms.

Fertilization triggers GV breakdown in Nereis, which is accompanied by the segregation of the ooplasm containing oil droplets and yolk to the vegetal hemisphere of the egg. When the GV ruptures, its contents mix with the perinuclear plasm and these constituents move into the animal hemisphere of the egg as a distinct ooplasm. The poly A^+ RNA molecules of the perinuclear plasm appear to mix with the GV components after maturation, but there is still no diffusion into the vegetal hemisphere cytoplasm. Like the situation discussed above for ascidian eggs, this behavior suggests that the RNA molecules are bound to a structural framework. As the perinuclear poly A^+ RNA mixes with the GV plasm there is also a two-fold increase in the hybridization signal. The enhancement of signal is likely to result from the addition of new poly A tails to maternal mRNA molecules or from the lengthening of pre-existing poly A sequences, as has been described in sea urchin eggs (Wilt, 1973; Angerer & Angerer, 1981). The poly A polymerase required for this process might be released from the GV at the beginning of maturation. The ooplasm containing poly A^+ RNA is distributed during cleavage to the animal hemisphere blastomeres and these in turn are responsible for the formation of the larval ectoderm cells in Nereis (Wilson, 1892).

The Cortical Pattern. Although eggs of the polychaete annelid Chaetopterus are also shed at the primary oocyte stage, there is little poly A^+ RNA within or near the GV (Jeffery & Wilson, 1983). Most of these molecules are found in a specialized ooplasmic region called the ectoplasm (Fig. 5) that lies in the animal two-thirds of the egg cortex (Lillie, 1906). The ectoplasm is filled with large, distinctly-staining granules. These granules along with surrounding nuage-like material form a pavement-like complex (Eckberg, 1981). Nuage are fibrillar-granular organelles present in many different kinds of germ cells that are thought to contain RNA (see Jeffery, 1983, for review). As described in the next section of this article, the cortical complex is a likely repository for maternal poly A^+ RNA. Poly A^+ RNA is also observed in regions of cortical complexes, which are scattered among lipid droplets and yolk platelets in the endoplasm (Fig. 5), an extensive area between the ectoplasm and GV of primary oocytes (Jeffery & Wilson, 1983).

The GV of Chaetopterus oocytes breaks down shortly after shedding and its clear plasm migrates into the animal pole region where the maturation divisions occur. Maturation also

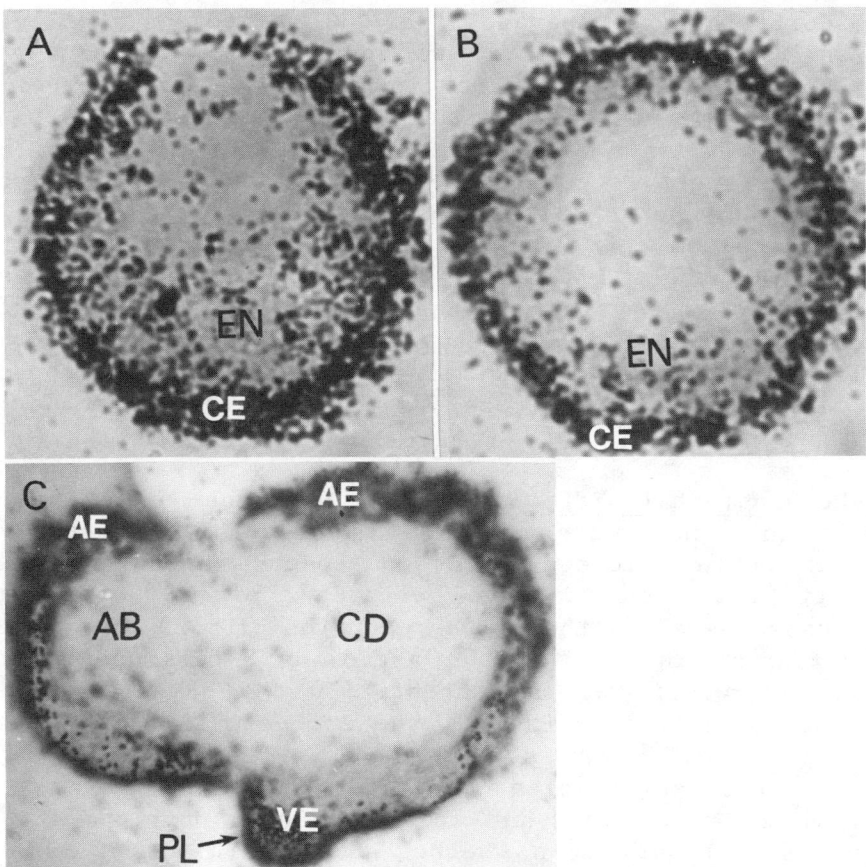

Fig. 5. The cortical pattern of poly A⁺ RNA distribution in oocytes, eggs, and early embryos of the polychaete Chaetopterus pergamentaceus as shown by in situ hybridization with poly U. Mature oocytes (A) and egg (B) showing grains over the cortical-ectoplasm (CE) and islets of cortical complexes in the endoplasm (EN). Trefoil stage embryo showing AB and CD cells, the small polar lobe (PL), and grains over the animal (AE) and vegetal (VE) ectoplasms (Jeffery & Wilson, 1983).

involves a vegetal movement of cortical complexes and poly A^+ RNA, and migration of internal poly A^+ RNA to the cortex. The latter joins with the main localization of poly A^+ RNA in the cortex (see Lillie, 1906, and Jeffery & Wilson, 1983, for more detailed discussions of ooplasmic movements in Chaetopterus eggs). When the cytoplasmic movements accompanying maturation are complete, about 95% of the total egg poly A^+ RNA is present in the cortex (Jeffery & Wilson, 1983). As noted above, the retention of poly A^+ RNA in a distinct region during ooplasmic segregation suggests that these molecules are bound to a structural framework in Chaetopterus eggs.

Shortly before cleavage, the cortical poly A^+ RNA localization is divided into animal and vegetal portions by an endoplasmic constriction which forms near the equator of the egg. The animal ectoplasm, the larger of the two portions, is partitioned to both the AB and CD blastomeres (Fig. 5C). The smaller vegetal ectoplasm becomes restricted to the polar lobe and is shunted almost entirely to the CD cell. The animal ectoplasm enters animal hemisphere blastomeres, which give rise to the epidermal cells during early embryogenesis, while much of the vegetal ectoplasm enters the D quadrant of the embryo and eventually the mesodermal cell lineages. The localization of poly A^+ RNA in the cortex of Chaetopterus eggs also results in its preferential delivery to ectodermal cell lineages.

Developmental Effect of mRNA Patterns. The mapping studies show that maternal mRNA is distributed unevenly in ascidian and annelid eggs. The consequence of these patterns is that certain embryonic cell lineages obtain more maternal mRNA than others. Although the patterns of mRNA distribution are different in eggs of the ascidians Nereis and Chaetopterus, their effects are essentially the same: to deliver most of the maternal mRNA to the ectodermal cell lineages. The use of different strategies of mRNA localization to achieve this segregation suggests that the presence of maternal mRNA is important for ectodermal development. At present it is not known whether the importance of maternal mRNA segregation to the ectoderm lies in its quantitative or possible qualitative aspects. It is conceivable, for instance, that the type of mRNA species segregated to the animal hemisphere blastomeres controls the spectrum of proteins synthesized by the ectodermal cells. Further comparative data will be necessary to determine the strategies of mRNA distribution used by other invertebrates and whether they also result in a preferential segregation of maternal mRNA to the ectodermal cell lineages.

Mechanism of Poly A^+ RNA Distribution

The existence of unique patterns of poly A^+ RNA distribution and the maintenance of these patterns during ooplasmic segregation strongly suggest that maternal poly A^+ RNA is linked to a regionalized structural framework in the egg. Intracellular membrane systems and cytoskeletal domains are excellent candidates for structural frameworks involved in mRNA localization. It has been appreciated for some time that certain mRNAs can be associated with intracellular membranes, and poly A^+ RNA has also been shown to be associated with non-membrane cytoskeletal elements in the nucleus and cytoplasm (Lenk et al., 1977; Long et al., 1979; Jeffery, 1982). Studies are reviewed below that demonstrate an association of poly A^+ RNA with the cytoskeletal framework.

Association of Poly A^+ RNA with a Cytoskeletal Framework in Ascidian Eggs. The association of mRNA with a cytoskeletal framework has been demonstrated in somatic cells by extraction with non-ionic detergents (Lenk et al., 1977). These detergents efficiently remove membrane lipids, soluble proteins and nucleic acids, but not cytoskeletal systems or their associated components. A large proportion of the cellular poly A^+ RNA is usually resistant to detergent extraction and this attribute is considered as evidence for its association with the cytoskeleton. The cytoskeletal framework of ascidian eggs has been examined by biochemical and microscopic methods after Triton X-100 extraction (Jeffery & Meier, 1983). About 70% of the total poly A is left in the detergent-insoluble fraction under conditions in which most of the lipids, proteins, and other types of RNA are extracted. These results suggest that egg mRNA molecules are specifically retained in the detergent-insoluble fraction. Most of these mRNA molecules are likely to be associated with the cytoskeleton and/or its associated organelles rather than systems of intracellular membrane, because Triton X-100 extracts membrane lipids. Despite the extraction of most of the cellular proteins and lipids, detergent-treated eggs retain an amazing amount of their structural detail, including cytoskeletal domains corresponding to the ectoplasm, the endoplasm, and the myoplasm. The cytoskeletal framework may be the structure referred to by early embryologists as the cytoplasmic ground substance.

The role of the cytoskeletal framework in organizing the pattern of mRNA distribution was tested by subjecting sections of detergent-extracted eggs and embryos to in situ hybridization with poly U (Jeffery, 1984b). The different cytoskeletal domains, like the ooplasms that they underlie, can be histologically distinguished by differential staining properties. In a typical experiment, a population of eggs or embryos is divided into two parts. One part is extracted with Triton X-100 while the other remains untreated. The detergent-treated and intact specimens are then mixed, fixed, and processed together for in situ hybridization. In this way the hybridization signal in the ooplasms and their corresponding cytoskeletal domains can be compared quantitatively. When an experiment of this nature is conducted during early development, 70-80% of the poly A^+ RNA is found in the detergent-insoluble residue at all stages of early development (Jeffery, 1984b). Moreover, poly A^+ RNA molecules are concentrated in the same regions of this residue (the GV matrix and the ectoplasmic cytoskeletal domain) as they are in the intact egg (Fig. 6). The cytoskeletal domains appear to retain their characteristic mRNA concentrations during ooplasmic segregation and during the period when the ectoplasm is partitioned to the animal hemisphere blastomeres. This result suggests that mRNA localization is based on the association of these molecules with cytoskeletal elements or their associated organelles and explains the inability of mRNA molecules to diffuse through the entire cytoplasm during GV breakdown or ooplasmic segregation.

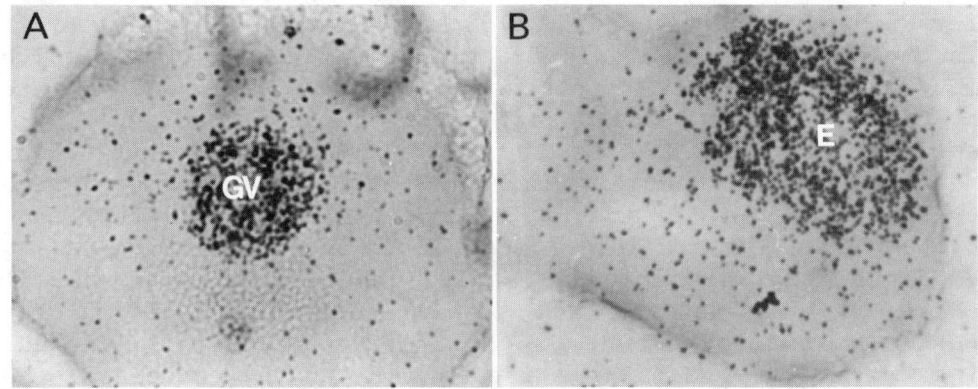

Fig. 6. Association of poly A^+ RNA with the cytoskeletal framework of Triton X-100 extracted Styela plicata oocytes and eggs as shown by in situ hybridization with poly U. A. An oocyte showing grains over the remnant of the germinal vesicle (GV). B. An egg showing grains over the ectoplasmic cytoskeletal domain (E) (Jeffery, 1984b).

The identity of the cytoskeletal recognition sites in mRNA or its associated protein moieties is currently unknown. It has been shown, however, that DNase I treatment does not affect the amount of poly A^+ RNA detected in detergent-extracted eggs by in situ hybridization (Jeffery, 1984b). This suggests that the actin cytoskeleton is not responsible for mRNA localization, since F-actin is depolymerized and extracted by the combination of DNase I and Triton X-100 (Jeffery & Meier, 1983). The nuclear matrix, a proteinaceous structural framework underlying the nucleoplasm (Berezney &

Coffey, 1974) and associated with poly A$^+$ RNA in mammalian cells (Long et al., 1979; van Eekelen & van Venrooij, 1981), would be an excellent candidate for the ectoplasmic localization framework. If poly A$^+$ RNA is bound to the matrix of the GV, then it is possible that this structure, or some remnant of it, may persist in the cytoplasm where it would continue to serve as a binding site for ectoplasmic mRNA during early development. This feature could promote the delivery of ectoplasmic messages, or their precursors originally sequestered in the GV, into the epidermal cell lineages. Efforts are now underway to test the predicted association of poly A$^+$ RNA with the matrix of the GV and to determine whether the elements of this matrix constitute what is perceived as the ectoplasmic cytoskeletal domain during early embryonic development.

Association of Poly A$^+$ RNA With the Cortical Complex in Chaetopterus Eggs. As described above, the distribution of poly A$^+$ RNA appears to be correlated with that of a complex containing distinctly-staining granules and associated nuage-like particles in the cortex of Chaetopterus eggs. It is possible that this complex is responsible for mRNA localization. Evidence for the presence of poly A$^+$ RNA in the cortical complex was obtained by coupling in situ hybridization and centrifugation. Relatively low centrifugal forces stratify the contents of Chaetopterus eggs into visible zones according to their specific gravities (Lillie, 1909). The contents of unfertilized Chaetopterus eggs are stratified into four major zones after centrifugation (Fig. 7). From the centripetal to the centrifugal pole, the zones are composed of lipid droplets, ribosome-sized granules and membranes, yolk granules, and cortical complexes. The displacement of the cortical complexes to the centrifugal pole can be exploited to determine whether poly A$^+$ RNA is associated with these structures. If poly A$^+$ RNA is associated with the complexes, it should also be displaced into the most centrifugal zone by centrifugal force. If it is not associated with the complex, it may be displaced into another zone or it may not be stratified at all. Poly A$^+$ RNA molecules were quantitatively displaced to the most centrifugal zone of stratified eggs as shown by in situ hybridization with poly U (Fig. 8). It is unlikely that mRNA or polysomes would exhibit such a high specific gravity unless they were associated with very dense structures like the cortical complexes. Further evidence for the association of mRNA with cortical complexes was obtained by centrifuging fertilized

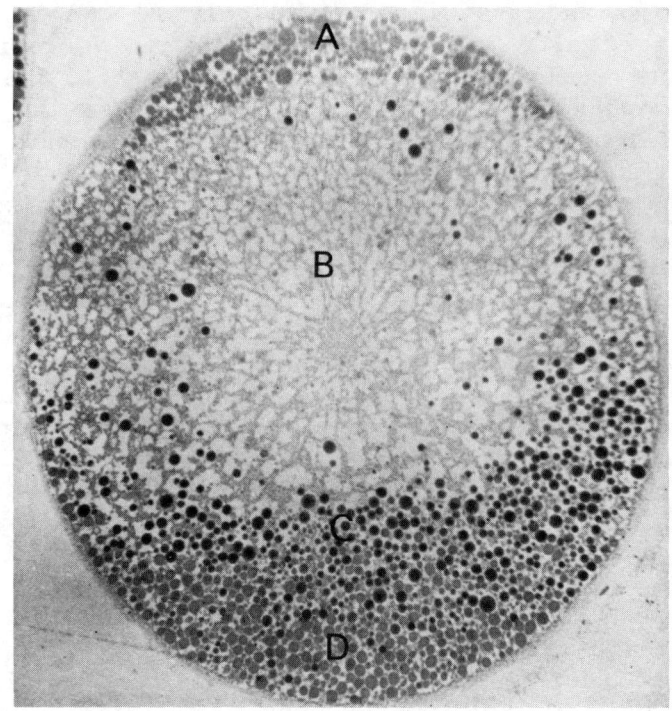

Fig. 7. Transmission electron micrograph of a section through the centripetal-centrifugal axis of a stratified Chaetopterus egg. A. Lipid zone. B. Zone of small granules and membranes. C. Zone of yolk granules. D. Zone of cortical complexes.

Chaetopterus eggs. Fertilized eggs are apparently more highly structured than their unfertilized counterparts and consequently their cortical complexes are not as easily displaced by low centrifugal forces. In situ hybridization shows that poly A^+ RNA molecules are also not easily displaced from the cortex of fertilized eggs by low centrifugal forces.

The centrifugal behavior of poly A^+ RNA suggests that it is a part of the cortical complex, an attribute which would explain its localization in the ectoplasm. The structure of the nuage-like component of the cortical complex is similar to that described for Drosophila polar granules, which have also been suggested to contain mRNA (Mahowald, 1971).

Role of the Cytoskeleton in mRNA Distribution. Although mRNA molecules at first appear to be localized by different mechanisms in ascidian and Chaetopterus eggs, the cytoskeleton may determine the distribution of mRNA in both. The cortical complex has recently been shown to be a part of the cytoskeleton in detergent-extracted Chaetopterus eggs (Eckberg, pers. comm.). The cytoskeleton may be an important matrix for maternal mRNA localization in both ascidian and Chaetopterus eggs. The interaction of mRNA with the cytoskeleton may be via the cortical complex in Chaetopterus eggs, and by a direct association with cytoskeletal elements in ascidian eggs.

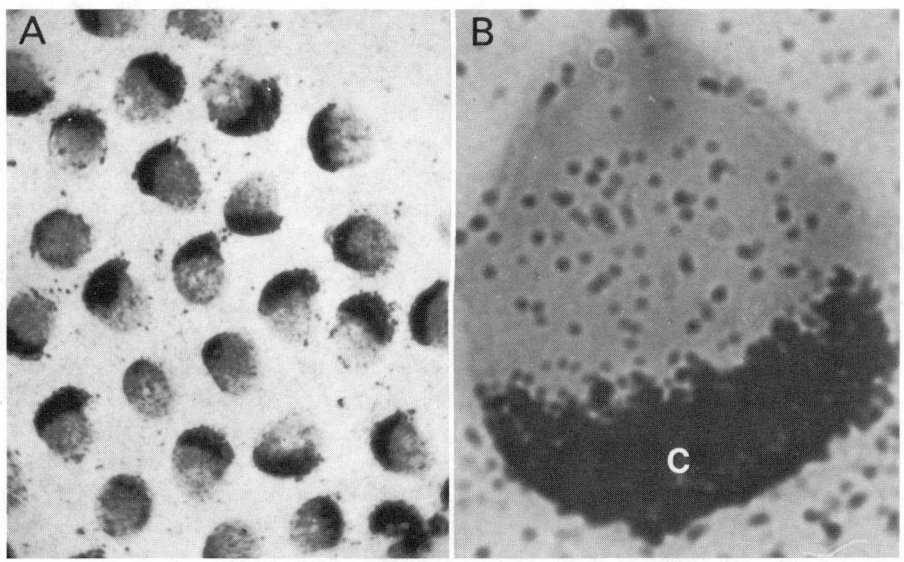

Fig. 8. Displacement of poly A⁺ RNA to the centrifugal zone (C) of stratified Chaetopterus eggs as shown by in situ hybridization with poly U. A. Field of eggs showing grains over one pole. The dark areas represent unresolved grains. B. An egg sectioned through the centripetal-centrifugal axis showing grains over the centrifugal zone.

Developmental Significance of Maternal mRNA

The results described above suggest that some invertebrates have specialized mechanisms to direct maternal mRNA into particular cell lineages during early embryogenesis. Do maternal mRNAs play a crucial developmental role in these cell lineages? If the titer of maternal messages in a given part of the embryo were reduced would the zygotic nuclei be able to compensate by new transcription, or is there something special about the mRNA complement that is synthesized during oogenesis and stored in a particular location in the egg? These questions have been difficult to approach in most embryos. The unique behavior of mRNA in centrifuged Chaetopterus eggs, in which egg fragments lacking mRNA and embryos with mRNA moved to atypical locations can be produced and challenged to develop, has been exploited to obtain answers to some of these questions.

Development of Egg Fragments Depleted in mRNA. Strong centrifugation of unfertilized Chaetopterus eggs through sucrose step gradients eventually results in their fragmentation into two parts. The centripetal fragment contains the nucleus, lipid granules, and most of the clear cytoplasm, while the centrifugal fragment contains yolk granules and cortical complexes. To determine whether mRNA remains associated with the cortical complex and becomes restricted to the centrifugal fragment during fragmentation, unfertilized eggs were centrifuged for varying lengths of time and the distribution of poly A^+ RNA was determined by in situ hybridization with poly U. As the eggs were centrifuged they first became stratified, then elongated, and were finally separated into two fragments (Fig. 9A). Because the eggs become oriented along their animal-vegetal axes by the centrifugal force, the fragments are usually derived from the original animal and vegetal regions of the egg. In situ hybridization of eggs fixed during the centrifugation shows that most of the detectable poly A^+ RNA molecules remain with the cortical complex during the elongation and fragmentation of the eggs and are distributed to the anucleate centrifugal fragment (Fig. 9B). The centripetal fragment contains much less poly A^+ RNA than the centrifugal fragment.

Fragmentation of fertilized eggs by the same centrifugation regime gives a different result. As discussed earlier, the cortical complexes and their associated mRNA molecules are not as easily displaced by centrifugation after fertilization, probably because of the polymerization of cytoskeletal elements that harness these structures in the egg cortex. Extensive centrifugation also pulls fertilized eggs into two fragments; however, many of the centripetal fragments still contain cortical complexes and significant amounts of poly A^+ RNA (data not shown). Thus, centrifugation of unfertilized or fertilized Chaetopterus eggs produces nucleated centripetal fragments with different amounts of maternal mRNA.

Are the centripetal fragments depleted in mRNA and cortical complexes able to regulate and develop normally? To answer this question, the centripetal fragments produced from unfertilized eggs were inseminated and cultured in parallel with uncentrifuged controls. Experiments similar to these have been reported previously by Harvey (1939) who generated

Chaetopterus egg fragments by centrifugation through step gradients and studied their developmental potential. She reported that many of the centripetal fragments exhibited normal cleavages and some formed small clear blastulae after insemination. Wilson (1929) also investigated the development of fragmented Chaetopterus eggs, using a fragmentation method much cruder than centrifugation through step gradients. It consisted of centrifuging eggs to the bottom of a tube of sea water where they were broken into heterogeneous-sized fragments by further application of centrifugal force. He reported the development of swimming larva from some of his fragments.

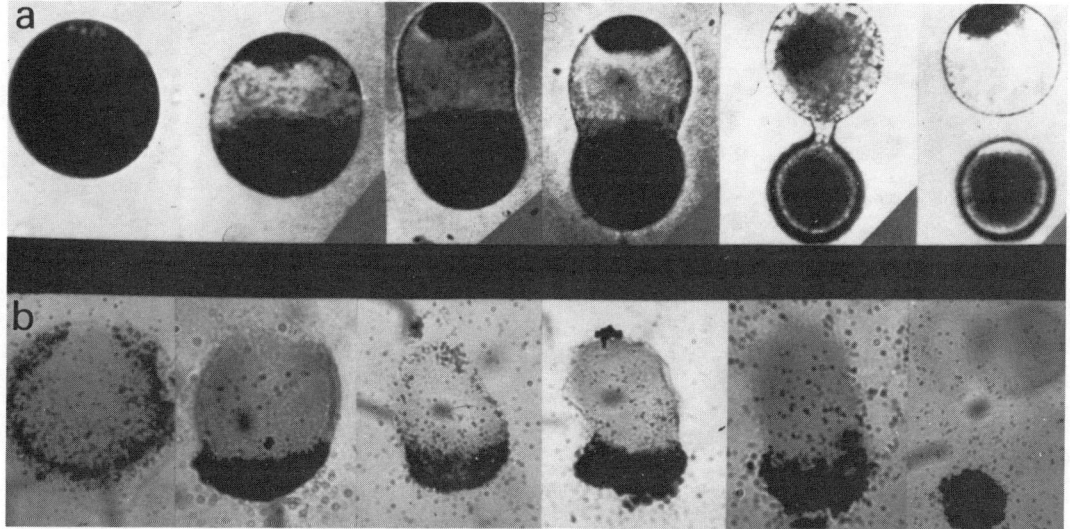

Fig. 9. Displacement of poly A⁺ RNA to the most centrifugal fragment after fragmentation of unfertilized Chaetopterus eggs by extended centrifugation. Frame a. Gradual fragmentation of eggs during centrifugation (from Harvey, 1939). Frame b. In situ hybridization with poly U during fragmentation of eggs by centrifugation showing grains over the most centrifugal zone and the centrifugal fragment.

The results we have obtained are quite different from those of Harvey (1939) or Wilson (1929). Less than 3% of the centripetal fragments derived from unfertilized eggs cleave and develop into larva (Fig. 12). Instead of cleaving, most of the fragments exhibit a curious behavior that could perhaps be mistaken for cleavage. They form large lobar pseudopods which are separated from the remainder of the cell by a thin isthmus of cytoplasm (Fig. 10). The pseudopods never become entirely separate from the body of the cell and are caused to retract into the main body of the cell by focusing the light from a microscope lamp on the pseudopod. The number of pseudopods increase with time, eventually making the centripetal fragments appear like a small blastula. At no time, however, is there nuclear multiplication or cleavage in the centripetal fragments. It is possible that pseudopod formation is what Harvey (1939) interpreted as cleavage and development in her centripetal fragments.

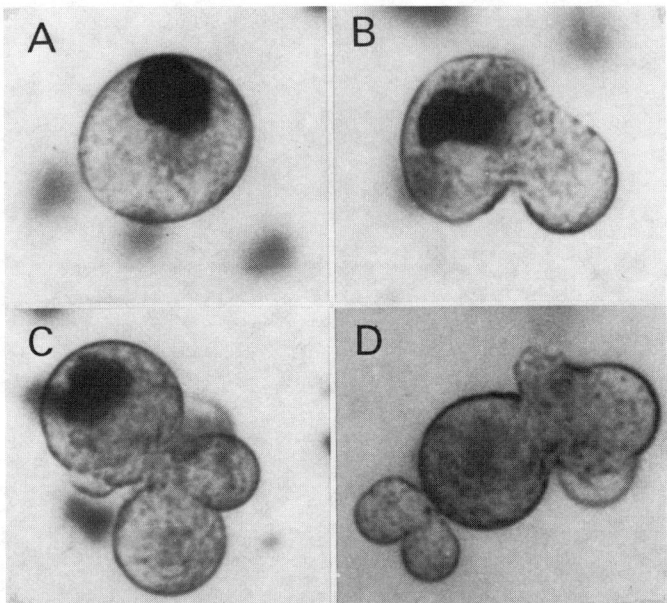

Fig. 10. Pseudopod formation by inseminated centripetal fragments of fragmented, unfertilized Chaetopterus eggs. A. A centripetal fragment immediately after its removal from the centrifuge. The dark spot represents the lipid zone. B. A large pseudopod forms in many centrifugal fragments at about the same time as controls are completing first cleavage. C-D. Numerous pseudopods occur in cultured centripetal fragments at later stages.

Wilson's (1929) fragmentation experiments have also been repeated. Consistent with his findings, some of the irregular fragments produced by his method cleave and develop into swimming larvae. When checked by in situ hybridization with poly U, however, all of the fragments produced by Wilson's method are found to contain poly A^+ RNA (data not shown), suggesting that this method does not displace cortical complexes to the centrifugal end of the egg prior to fragmentation. Wilson admits that most of the fragments produced by his method contain all of the visible organelles of the egg, including what we now recognize as the cortical complexes.

It is concluded that egg fragments depleted in mRNA do not cleave or develop after fertilization, even though they contain nuclei. It is not known whether the centrifugal fragments attempt to replace their maternal mRNA complement by new transcription.

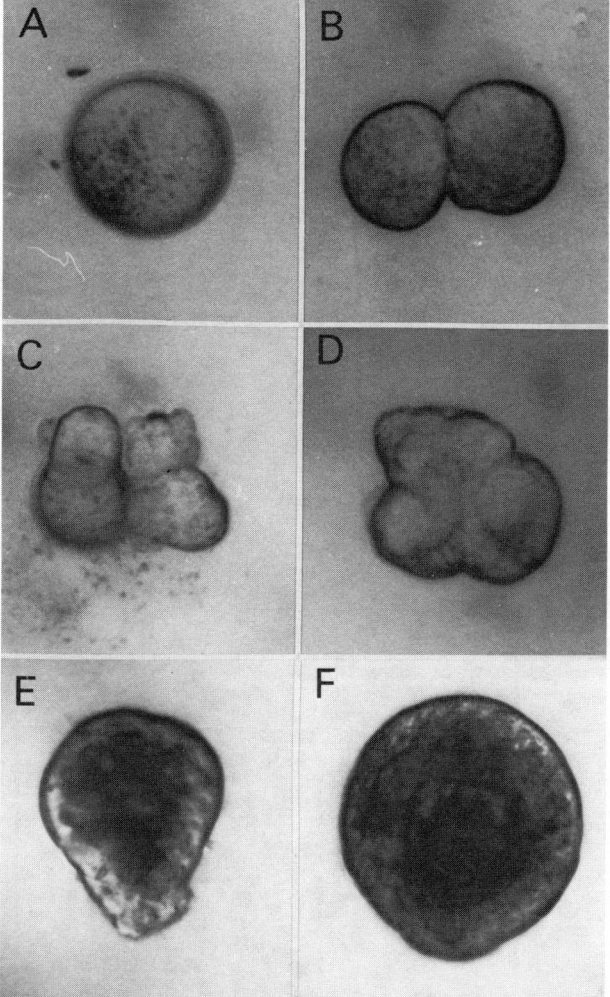

Fig. 11. Development of centripetal fragments of fertilized Chaetopterus eggs. A. A fragment immediately after removal from the centrifuge. B. A first cleavage. C. A four-cell stage. D. An eight-cell stage. E. One-day-old larvae formed from a centripetal fragment. F. One-day-old larvae formed from a control embryo.

It cannot be concluded from the experiments described above that the centripetal fragments of Chaetopterus eggs are developmentally arrested because they are depleted in mRNA or cortical complexes. It is possible, for example, that they are not properly activated by insemination or that they may lack essential materials besides mRNA and cortical complexes which are also displaced to the centrifugal fragment. Fortunately, there are ways to explore some of these alternatives. As mentioned earlier, fragmentation of fertilized Chaetopterus eggs produces a significant number of nucleated fragments depleted in cortical complexes and mRNA. The centripetal fragments from fertilized eggs are identical to those obtained from unfertilized eggs with respect to stratified components except that many of them also contain cortical complexes and are less depleted in mRNA. When centripetal fragments from fertilized eggs are cultured, about 20% of them cleave on schedule and eventually form miniature swimming larvae (Figs. 11 & 12). These larva die before feeding, probably due to a deficiency in yolk. The ability of some of the centripetal fragments of fertilized eggs to develop suggests an experiment which could distinguish be-

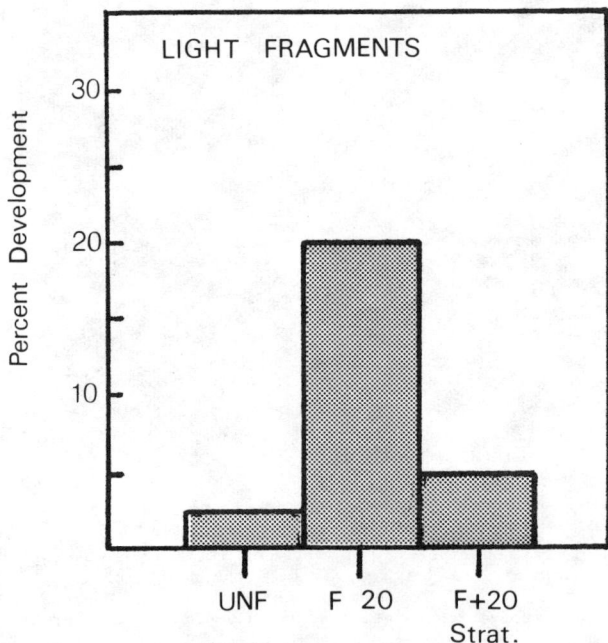

Fig. 12. The proportion of centripetal fragments prepared from unfertilized Chaetopterus eggs (UNF), fertilized eggs (F+20), and pre-stratified fertilized eggs (F+20 Strat.) that develop into one-day-old larvae.

tween developmental arrest because of deficiencies in activation by sperm or the presence of maternal mRNA and cortical complexes. A clutch of fertilized eggs was separated into two groups. The first group was fragmented as described above, producing some centripetal fragments containing poly A^+ RNA. The second group was initially stratified by very strong centrifugation over a pad of ficoll before being fragmented by further centrifugation through step gradients and, similar to unfertilized eggs, almost all of the centripetal fragments were deficient in poly A^+ RNA. When the two groups were compared to controls, it was found that many of the centripetal fragments that contained significant levels of poly A^+ RNA were able to develop, whereas, those that were relatively deficient in poly A^+ RNA remained arrested (Fig. 12). The results suggest that the inability of centrifugal fragments to develop is not due to improper activation. Since the only detectable difference between fragments that develop and those that do not is the presence of the cortical complex, it seems likely that development is dependent on a component of the complex, possibly maternal mRNA.

Development of Embryos with Displaced mRNA. Although the experiments discussed above provide evidence that maternal mRNA may be important for early development, they do not indicate whether the proper segregation of these molecules between the AB and CD cells is also required. As discussed earlier, the cortical poly A^+ RNA localization is divided into two parts prior to the first cleavage. The animal ectoplasm is distributed to the AB and CD cells, while the vegetal ectoplasm enters the polar lobe and is distributed mainly to the CD cells. The role of maternal mRNA segregation in early development has been studied by moving these molecules to atypical regions of the cleaving embryo. The movement of poly A^+ RNA can be accomplished only at times when the cortical complexes are responsive to stratification by low centrifugal forces. As we have seen, one of the favorable times is just prior to fertilization. Another is during the first cleavage.

If early cleavage-stage embryos are centrifuged, their contents are often stratified independently of their animal-vegetal axis. As a consequence, some of these embryos will contain cortical complexes and poly A^+ RNA in both cells, while others will have cortical complexes and poly A^+ RNA in only one cell after the completion of first cleavage (Fig. 13). Embryos that exhibit these features can be identified by light microscopy and are selected for culturing. During the period between the first and second cleavages, the stratified cortical complexes spread throughout the AB and CD cells of the selected embryos and are partitioned to either two or four blastomeres at the second cleavage. The embryos with cells depleted in maternal mRNA on one side of the animal-vegetal axis can be exploited to determine whether the normal distribution of maternal mRNA is important for embryonic development. The embryos with cells containing maternal mRNA on both sides of the axis serve as controls for the possible effects of centrifugation on development. As shown in Figure 14, a large proportion of the control embryos develop into normal swimming larva. Most of the embryos which were originally depleted in maternal mRNA in one of the two cells form grossly abnormal larvae. Some of the features missing from the abnormal larvae include apical tufts, eyespots, and guts. The overall shapes of the latter were also abnormal; often one side of the larvae appeared to be more highly differentiated than the other. Presumably, the most abnormal larval side forms from cells depleted in maternal mRNA.

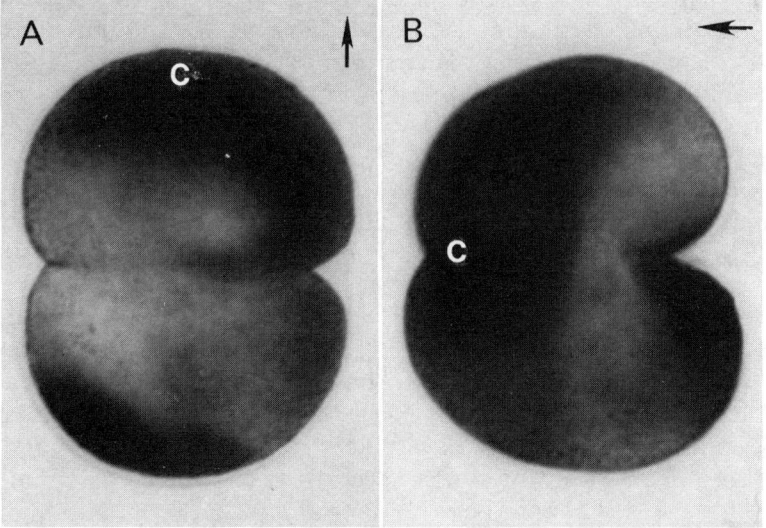

Fig. 13. Cleaving Chaetopterus eggs centrifuged so that the cortical complexes (C) enter one (A) or two (B) blastomeres. Arrows at upper right-hand side of each frame indicate the direction of centrifugal force.

The present results suggest that normal distribution of cortical complexes and maternal mRNA between the AB and CD blastomeres of the two-cell embryo is necessary for the formation of complete larva. Some embryos, however, can form relatively normal larvae despite having most of their cortical complexes centrifuged into one blastomere at first cleavage (Fig. 14). These embryos could have been selected as containing mRNA in only one cell, but could actually have had a critical titer of mRNA in both cells. It is difficult to rule out this possibility because centrifuged embryos cannot be simultaneously evaluated for the presence of mRNA and developmental potential.

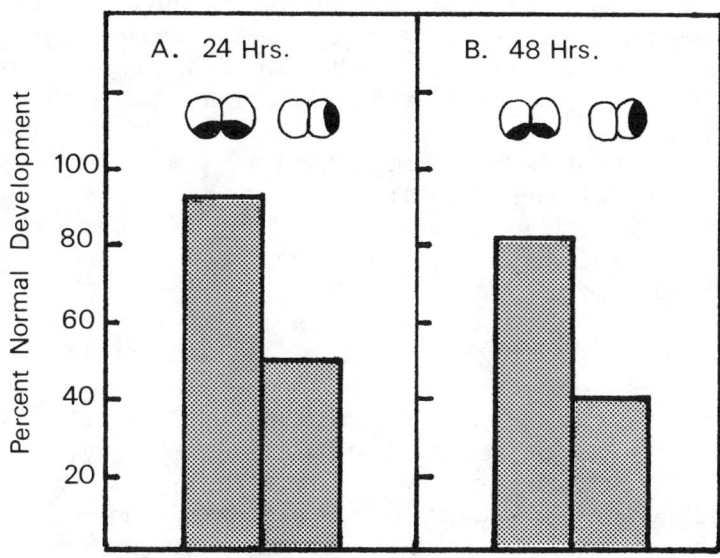

Fig. 14. The proportion of embryos centrifuged at first cleavage to move cortical complexes into one or both blastomeres that develop into normal larvae after one (A) or two (B) days of culture.

CONCLUSIONS

It is concluded that maternal mRNA sequences are subject to uneven spatial distributions in ascidian and annelid eggs. The patterns of mRNA distribution have been classified into groups in which most of the maternal mRNAs are originally concentrated in the oocyte GV, the perinuclear cytoplasm, or the cortical cytoplasm. Despite the different initial localizations, most of the maternal mRNA sequences are subsequently partitioned into the animal hemisphere blastomeres and eventually to the ectodermal cell lineages. It is emphasized that these patterns result in more maternal mRNA in the ectodermal lineages, but that the other cell lineages also obtain mRNA from the maternal stockpile. It should also be recalled that sea urchins do not use the strategy of uneven mRNA distribution to shunt maternal mRNA to particular cell lineages, even though these mRNAs are also associated with the cytoskeleton (Moon et al., 1983). The significance of the difference in mRNA distribution between sea urchin eggs and the other invertebrate eggs is obscure.

The retention of egg mRNA by Triton X-100-extracted ascidian eggs implies that the cytoskeletal framework is instrumental in stabilizing the distribution of these molecules. Although comparable detergent extraction studies have not been done in Chaetopterus eggs, the cytoskeletal framework also seems to be important in mRNA localization in these cells. It has been shown that Chaetopterus mRNA is associated with distinct cortical organelles and that these structures are likely to be associated with the cytoskeletal framework. Cytoskeletal localization may be an important factor in controlling the distribution of maternal mRNA in eggs. Ascidian eggs are known to contain different cytoskeletal domains that underlie the cytoplasmic regions of unique morphogenetic fate (Jeffery & Meier, 1983). It is possible that mRNAs bind differently to these domains and that this constitutes the basis for their uneven distribution. Further experiments will be necessary to test this possibility and to accurately define the cytoskeletal elements that are associated with mRNA.

Exploitation of the unique stratification behavior of maternal mRNA associated with cortical organelles in Chaetopterus eggs shows that egg fragments depleted in mRNA or embryos in which mRNA was displaced to atypical locations are incapable of normal development. This suggests that the zygotic nucleus is unable to compensate for the loss of maternal mRNA by producing new transcripts. These results are also consistent with the possibility that the segregation of localized maternal mRNA sequences or their associated components (i.e., cortical organelles) is required for early development. It will be an important endeavor to determine the nature of this requirement.

ACKNOWLEDGMENTS

This research was supported by NIH Grant HD-13970 and a grant from the Muscular Dystrophy Association. Some of these studies were conducted in the Embryology course at the Marine Biological Laboratory, Woods Hole, MA supported by NIH training grant 232-HDO-7098. Technical assistance was provided by Bonnie Brodeur, Priscilla Kemp, Dianne McCoig, and Linda Wilson. I wish to thank my colleagues and students

Richard Brodeur, David Capco, Stephen Meier, Randall Moon, Billie Swalla, and Craig Tomlinson who participated in many aspects of these studies.

LITERATURE CITED

Angerer, L.M. and R.C. Angerer. 1981. Detection of poly A^+-RNA in sea urchin eggs and embryos by quantitative in situ hybridization. Nuc. Acids Res. 9: 2819-2840.

Angerer, L.M., D.V. DeLeon, R.C. Angerer, R.M. Showman, D.E. Wells, and R.A. Raff. 1984. Delayed accumulation of maternal histone mRNA during sea urchin oogenesis. Develop. Biol. 101: 477-484.

Beers, R.F. 1960. Hydrolysis of polyadenylic acid by pancreatic ribonuclease. J. Biol. Chem. 235: 2393-2398.

Berezney, R. and D.S. Coffey. 1974. Identification of a nuclear protein matrix. Biochm. Biophys. Res. Commun. 60: 1410-1417.

Brown, M.P. and K. Kalthoff. 1982. Inhibition by ultraviolet light of pole cell formation in Smittia sp. (Chironomidae, Diptera): Action spectrum and photoreversibility. Develop. Biol. 97: 113-122.

Capco, D.G. and W.R. Jeffery. 1978. Differential distribution of poly A-containing RNA in the embryonic cells of Oncopeltus fasciatus: Analysis by in situ hybridization with an [^{3}H]-poly U probe. Develop. Biol. 67: 137-151.

Conklin, E.G. 1905. The organization and cell-lineage of the ascidian egg. J. Acad. Nat. Sci. Phila. 13: 1-119.

Cox, K.H., D.V. DeLeon, L.M. Angerer, and R.C. Angerer. 1984. Detection of mRNAs in sea urchin embryos by in situ hybridization using asymmetric RNA probes. Develop. Biol. 101: 485-502.

Eckberg, W.R. 1981. An ultrastructural analysis of cytoplasmic localization in Chaetopterus pergamentaceus. Biol. Bull. 160: 228-239.

Harvey, E.B. 1939. Development of half-eggs of Chaetopterus pergamentaceus with special reference to parthenogenetic merogony. Biol. Bull. 76: 384-404.

Jeffery, W.R. 1982. Messenger RNA in the cytoskeletal framework: Analysis by in situ hybridization. J. Cell Biol. 95: 1-7.

Jeffery, W.R. 1983. Maternal RNA and the embryonic localization problem. pp. 73-144. In: Control of Embryonic Gene Expression. M.A.Q. Saddiqui (ed.). CRC Press, Boca Raton, FL.

Jeffery, W.R. 1984a. Pattern formation by ooplasmic segregation in ascidian eggs. Biol. Bull. 166: 277-298.
Jeffery, W.R. 1984b. Spatial distribution of messenger RNA in the cytoskeletal framework of ascidian eggs. Develop. Biol. 103: 482-492.
Jeffery, W.R. and D.G. Capco. 1978. Differential accumulation and localization of poly A-containing RNA during early development of the ascidian Styela. Develop. Biol. 67: 151-166.
Jeffery, W.R. and S. Meier. 1983. A yellow crescent cytoskeletal domain in ascidian eggs and its role in early development. Develop. Biol. 96: 125-143.
Jeffery, W.R., C.R. Tomlinson, and R.D. Brodeur. 1983. Localization of actin messenger RNA during early ascidian development. Develop. Biol. 99: 408-417.
Jeffery, W.R. and L. Wilson. 1983. Localization of messenger RNA in the cortex of Chaetopterus eggs and early embryos. J. Embryol. Exp. Morphol. 75: 225-239.
Kalthoff, K. 1979. Analysis of a morphogenetic determinant in an insect embryo (Smittia sp., Chironomidae, Diptera). pp. 97-126. In: Determinants of Spatial Organization. S. Subtelny and I.R. Konigsberg (eds.). Academic Press, N.Y.
Lenk, R., L. Ransom, Y. Kaufmann, and S. Penman. 1977. A cytoskeletal structure with associated polyribosomes obtained from HeLa cells. Cell 10: 67-78.
Lillie, F.R. 1906. Observations and experiments concerning the elementary phenomena of embryonic development in Chaetopterus. J. Exp. Zool. 3: 153-268.
Lillie, F.R. 1909. Polarity and bilaterality of the annelid egg. Experiments with centrifugal force. Biol. Bull. 16: 54-79.
Long, B.H., C.Y. Huang, and A.O. Pogo. 1979. Isolation and characterization of the nuclear matrix in Friend erythroleukemia cells: Chromatin and hnRNA interactions with the nuclear matrix. Cell 18: 1079-1090.
Mahowald, A.P. 1971. Polar granules of Drosophila. IV. Cytochemical studies showing loss of RNA from polar granules during early stages of embryogenesis. J. Exp. Zool. 176: 345-352.
Moon, R.T., R. Nicosia, C. Olsen, M. Hille, and W.R. Jeffery. 1983. The cytoskeletal framework of sea urchin eggs and embryos: Developmental changes in the association of messenger RNA. Develop. Biol. 95: 447-458.

Richter, J., L.D. Smith, D. Anderson, and E.H. Davidson. 1984. Interspersed poly A RNAs of amphibian oocytes are not translatable. J. Mol. Biol. 173: 227-241.

Riley, M., B. Maling, and M.J. Chamberlain. 1966. Physical and chemical characterization of the two and three-stranded adenine-thymine and adenine-uracil homopolymer complexes. J. Mol. Biol. 20: 359-389.

van Eekelen, C.A.G. and W.J. van Venrooij. 1981. HnRNA and its attachment to the nuclear matrix. J. Cell Biol. 88: 554-563.

Wilson, E.B. 1892. The cell lineage of Nereis. A contribution to the cytogeny of the annelid body. J. Morphol. 6: 361-480.

Wilson, E.B. 1929. The development of egg fragments in annelids. W. Roux Archiv. 117: 179-210.

Wilt, F. 1973. Polyadenylation of maternal RNA of sea urchin eggs after fertilization. Proc. Nat. Acad. Sci. USA 70: 2345-2349.

MATERNAL MESSENGER RNA: SYNTHESIS AND LOCALIZATION OF HISTONE mRNAs IN THE SEA URCHIN OOCYTE

R.M. Showman, D.E. Wells, J.A. Anstrom, D.A. Hursh, D.S. Leaf, and R.A. Raff

ABSTRACT

The synthesis of protein during the early stages of sea urchin embryogenesis occurs using stored maternally synthesized messenger RNAs. Here we discuss the role of these maternal mRNAs, particularly those coding for the alpha subtype histones. We have previously examined the delayed recruitment of these mRNAs and demonstrated their localization to the egg pronucleus. Here we examine the fact that these same mRNAs are not synthesized until after the meiotic maturation of the oocyte is complete and consider the problem of post-transcriptional processing and transport of the mRNAs within the egg. We also discuss the biological significance of these stored maternal mRNAs and the consequences of their removal from the egg.

INTRODUCTION

Because they are informational molecules, messenger RNAs have long been considered prime candidates for embryonic determinants. RNA synthesis studies, enucleation experiments, and interspecies hybridization studies have all served to demonstrate that mature oocytes of many organisms contain significance stores of RNA. Generally termed maternal mRNA, these sequences have been the subject of much study and speculation. Even now, however, little is known about either the structure or function of these RNAs.

Maternal mRNAs and the Problem of Cytoplasmic Localization

By 1900 the work of Wolff and his successors had marked the end of the preformationist doctrine and a re-emphasis of the interactive nature of development. This shift did not, however, preclude or eliminate the concept of prelocalization of components within an egg that could later be segregated via cleavage into specific cell lines. Thus His in 1874 speaks of "organ forming germ-regions" that are present in the egg and which contain material "not yet morphologically marked off and hence not directly recognizable". This concept of localized elements or molecules that are segregated and responsible for later patterns of differentiation was supported by Lankester, Whitman and others and is well summarized by Wilson in his classic text on the cell in development and heredity (Wilson, 1925).

Demonstration of the phenomena of localized morphological determinants has taken many forms during the past century, ranging from early cell separation studies in molluscs and annelids (cf. Wilson, 1904) to simple visual observation of segregation of subcellular components. These observations continue today in the recent studies of localized determinants in the ctenophores and the annelids (Freeman, 1976, 1979; Render, 1983). This cytoplasmic organization can conceivably take two forms. First, a message can be restrained in its movement to a specific region of the cytoplasm, perhaps through association with the cytoskeleton. There are clear hints that such a pattern of localization exists in organisms such as ascidians and Chaetopterus (cf. Jeffery & Wilson, 1983). Second, RNA can be localized by its retention inside the nucleus or some other organelle following synthesis and processing. This is the case for the maternal alpha histone mRNAs in sea urchins. The consequences of these two types of localization can, in general, be quite different. Regional restriction allows for partitioning of the message into discrete cell lines where they may be expressed in a tissue or cell-line specific manner. Restriction to the nucleus or other subcellular organelle allows for a delay in timing of translation without the need for message-specific modifications of the RNA. Both these alternatives have been exploited by organisms where appropriate. With the advent of recombinant DNA and refinements in in situ localization technology, we are now capable of following the segregation of even low to moderate prevalence messenger RNA molecules in individual eggs and embryos (see Jeffery, this volume).

Synthesis and Utilization of Maternal mRNAs

For a long time it has been argued that maternal RNAs are storage pools of messenger RNA that can be segregated and utilized during early development. Perhaps the strongest argument for this RNA having a defined biological role can be made in dipteran embryos. In Smittia, the condition "double abdomen" seems to be caused by an RNAse sensitive component in the oocyte's cytoplasm (Kandler-Singer & Kalthoff, 1976). Similarly, in Drosophila, injection of poly A^+ RNA can restore pole cell formation in UV-sterilized embryos (Togashi & Okada, 1983). Probably the best evidence to date is the role of poly A^+ RNA in establishing the dorsal-ventral axis in Drosophila embryos. Anderson and Nüsslein-Volhard (1984) have been able to demonstrate that stored maternal poly A^+ RNAs are responsible for establishing dorsal-ventral polarity. Injection of normal egg poly A^+ RNA into mutant embryos which show no polarity in the dorsal-ventral axis results in complete phenotypic rescue. Thus it seems that, at least for the diptera, maternal RNAs play a major role in establishing the early patterns of morphogenesis.

Less is known of either the pattern of synthesis or the purpose of the maternal mRNA in sea urchins. Hough-Evans et al. (1977, 1979) have studied the relationship between single copy transcripts of the egg and the RNA transcripts present in both the nucleus and cytoplasm of immature and mature oocytes. For Strongylocentrotus purpuratus, the egg RNA includes transcript from roughly six percent of the total single copy DNA (Galau et al., 1976). When Hough-Evans et al. examined the single copy gene transcripts in the cytoplasm of previtellogenic oocytes, they found roughly 44% of the complexity of the mature egg. They concluded that as much as 56% of the total RNA complexity of the mature oocyte is synthesized by vitellogenic stage oocytes in a stage-specific manner.

A major difficulty in interpreting these results is that although measurements of complexity tell us a good deal about the sequence relationship between various populations of RNA, they do not address the problem of what percentage of that total complexity is functionally significant RNA. In both sea urchins and amphibians, as much as 70% of the poly A+ RNA of the mature oocyte is linked to transcripts from repetitive DNA (Constantini et al., 1980; Anderson et al., 1982). Because the complexity of the interspersed repeat-containing maternal RNA is much lower than the heterogeneous nuclear RNA in both sea urchin (Hough-Evans et al., 1979; Constantini et al., 1980) and Xenopus oocytes (Davidson & Hough-Evans, 1971; Rosbash & Ford, 1974), it has been argued that this RNA is not simply nuclear RNA that has leaked randomly from the germinal vesicle, but rather exists to perform some specific biological role.

Richter et al. (1984) have assayed the translatability of this fraction of the maternal poly A+ RNA. In amphibians, these transcripts apparently fail to assemble into polysomes and are translationally inactive. While similar experiments have not yet been done using sea urchin oocyte RNA, if one looks at polysome-associated RNA, one finds few repeat-containing transcripts (Scheller et al., 1978; Thomas et al., 1981). Rudensey and Infante (1979) also report a lower transcriptional efficiency in vitro of non-polysomal poly A+ RNA from sea urchin blastulae, a fraction expected to have repeat-containing transcripts. Thus it would appear that much of the maternal RNA of the oocytes may well be inactive. The sequence organization of the maternal RNA may itself be the critical factor. Posakony et al. (1983) find stop signals in all reading frames of some of sea urchin repeat elements. Although it is not yet known where these repeat elements are located, relative to the protein-coding regions of the RNA, it is distinctly possible that such stop signals are at least in part responsible for any failure of the RNA to be translated.

That these maternal RNAs are not simply random transcripts can be deduced from the conservation of their DNA during evolution. Moore (1984) has assessed the extent of evolutionary divergence of the DNA coding for mature oocyte RNA from two species of sea urchin. He estimates the divergence somewhere between 4.8 and 7.4%. This is markedly less than that seen in total single copy DNA in these species and suggests that there are selective constraints on the divergence of the DNA coding for most of the oocyte RNA.

It also seems clear that these maternal RNAs are not simply pools of messenger RNAs that are released at random into the polysomes immediately following fertilization (Raff & Showman, in press; Hille et al., this volume, for reviews of translational regulation of these RNAs). Cabrara et al. (1984) have examined the stability of sea urchin maternal RNAs and find that while many of these RNAs turn over at a rapid rate following fertilization, there is no single stage at which all the maternal transcripts are lost. At least one of these transcript persists in the cytoplasm well beyond the gastrula stage, with the maternal contribution remaining the major portion of the total messenge present. Thus it would seem that while much of the maternal RNA pool in sea urchins may be translationally inactive, a biologically significant portion does exist.

Localization in the Sea Urchin

The sea urchin has, since the original blastomere separation studies of Driesch, been considered to have a classically "regulative" egg. Thus, division of either the two- or four-cell embryo into individual blastomeres yields complete, albeit diminutive, embryos. It is only with the third cleavage, bisecting the egg or animal-vegetal axis, that one sees restriction in the developmental potential of the blastomeres. Extensive studies by Horstadius and others argue strongly for the presence of a pair of gradients along this axis, each containing either an "animalizing" or a "vegetalizing" factor that determines cell fates. Yet beyond postulating the need for its existence, little has been done to characterize the gradient or define its physical/chemical nature (see Hörstadius, 1973, for a review). The only other studies of possible segregation of components in the sea urchin egg have been studies of the distribution of mRNAs in the micromeres at the 16-cell stage (Rodgers & Gross, 1978; Tufaro & Brandhorst, 1979; Ernst et al., 1980). With these exceptions, little work has been done demonstrating cytoplasmic segregation of informational molecules in the eggs of these organisms. This is somewhat ironic since the most striking case of organelle-associated localization of a messenger RNA in an egg is that of the alpha histone mRNA in the echinoids. It is now clear from both in situ studies (Venezky et al., 1981; DeLeon et al., 1983), RNA Northern blot analysis (Showman et al., 1982) and pronuclear isolation experiments (Showman et al., 1984) that essentially all of the maternal alpha histone mRNA of the sea urchin is restricted to the egg pronucleus (Fig. 1). This message remains inside the nucleus through syngamy and is not released into polysomes until the time of nuclear envelope breakdown just prior to first division (Wells et al., 1981a & b) (Fig. 2). (See Raff, 1983, and Showman et al., 1984 for a more complete review of recent literature on localization and translational regulation of echinoid maternal alpha histone mRNAs.)

It is not clear why the alpha histone mRNA should be sequestered in such a unique manner or why its release should be so effectively delayed until after first cleavage has occurred. One possibility that has been suggested is that the presence of an alpha histone protein in the cytoplasm

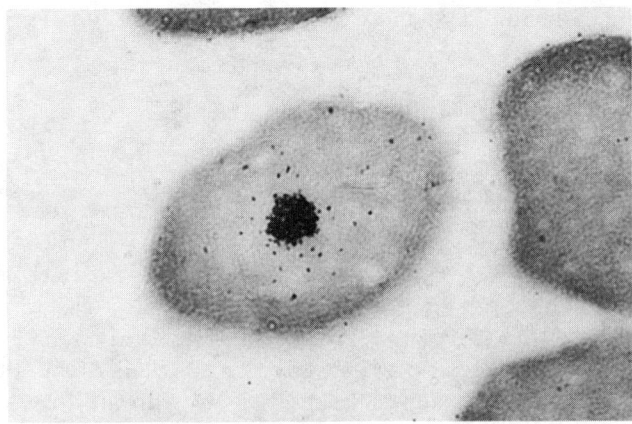

Fig. 1. In situ autoradiograph of sea urchin eggs prepared as described by Venezky et al. (1981). Note the concentration of silver grains over the female pronucleus indicating binding of the ^{32}P-labeled histone mRNA specific DNA probe.

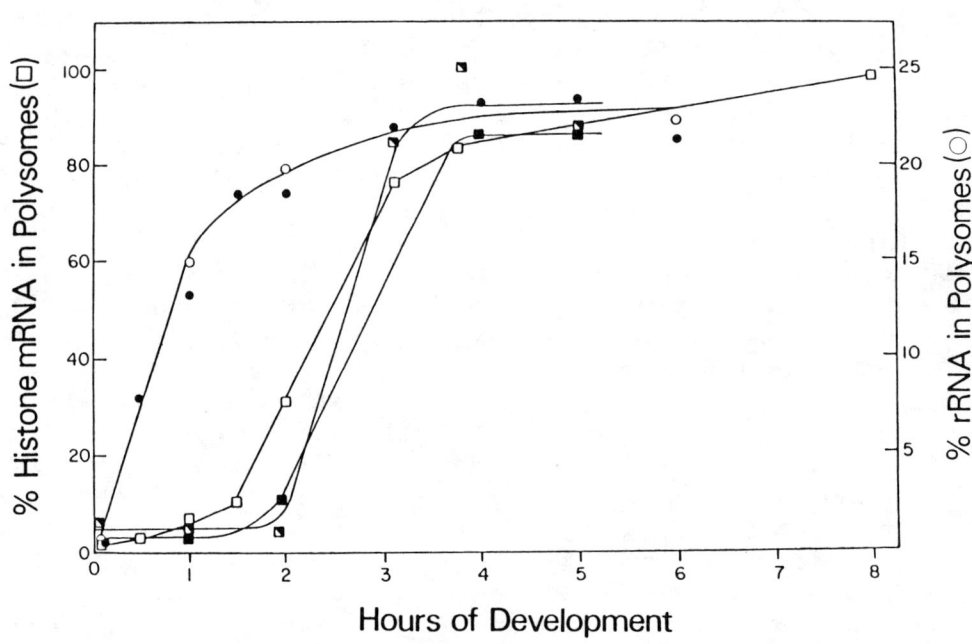

Fig. 2. Time of appearance of rRNA and histone H1, H2B, and H3 mRNAs in the polysomes of S. purpuratus embryos at 13.5°C. (●) % rRNA in the polysomes as determined by A_{253}; (o) % rRNA in polysomes as determined by hybridization with rDNA probe; % of histone H1 (□), H2B (▨), and H3 (■) mRNA in polysomes as determined by cloned DNA probes. (Reproduced from Wells et al., 1981a & b; Showman et al., 1984.)

would result in an incorrect restructuring of the chromatin of the male pronucleus (Raff, 1983; Showman et al., 1984). It is clear that as the male pronucleus transverses the newly fertilized egg, it undergoes a rapid change in the protein complement of its chromatin, exchanging its sperm-specific proteins for cleavage stage (CS) histones that are found stored in large quantities in the egg (Poccia et al., 1981; Savic et al., 1981). By the time the female and male pronuclei fuse, their chromatin is presumably in the same "ground state", consisting primarily of cleavage-stage histones. What might happen if the cleavage-stage histones on the male chromatin were to be replaced by a prematurely synthesized alpha histone? Despite several attempts to dissociate alpha histone synthesis from first cleavage, and hence possibly get premature alpha histone synthesis, no such experiment has been successful to date (Showman et al., 1984). Nonetheless, the egg does seem to preclude premature alpha histone protein synthesis. Thus the possibility remains that premature incorporation of alpha histone protein into the zygotic chromatin might have deleterious consequences for embryonic development.

Synthesis of the Alpha Histone mRNAs

We have already commented on the unique manner in which translation of the maternal alpha histone component is regulated. It should not be surprising, therefore, that when one asks not how the maternal alpha message translation is regulated, but rather how the pronuclear localization is achieved initially, that the answer is different from most messenger RNAs found in the egg. Fundamentally there are two alternatives that should be considered. In the first, the alpha histone messenger RNA could be synthesized prior to oocyte maturation along with the majority of the egg messenger RNAs. Then following maturation of the oocyte and reformation of the female pronucleus, the alpha histone message would be selectively resegregated into the pronucleus via some yet to be discovered mechanism. Although at first glance awkward, this resegregation of specific mRNAs is not inconceivable considering the efficiency with which some RNA protein complexes can be efficiently concentrated inside the nucleus (Mattaj & DeRoberts, 1985). This model would result in the pronuclear localization of the alpha histone message without the necessity of having some specific transcription-regulating mechanism for the alpha histone gene complex. The second possibility is for the alpha histone mRNA to accumulate only after the oocyte has undergone its meiotic maturation divisions. This would preclude the need for a special mechanism to achieve post-transcriptional segregation, but requires specific control of the time of transcription and/or significant modifications in the stability of the alpha histone messenger mRNA. It also requires that any alpha histone messenger RNA be retained within the pronucleus following its synthesis.

To distinguish between these two possibilities, we have, in collaboration with Drs. Robert and Lynn Angerer, determined the relative amounts of alpha histone messenger RNA present in the full-size (80μm) 4N germinal vesicle stage oocyte and mature (1N) egg of the sea urchin Strongylocentrotus purpuratus (Angerer et al., 1984). Both in situ measurements and direct titration via Northern blot analysis were used. As can be seen in Figure 3, the in situ hybridization studies show no silver grains (above background) located over the germinal vesicle stage oocytes (Fig. 3a). At the same time, the mature egg shows its typical high concentration of silver grains over the nucleus (Fig. 3b). The two-cell section (Fig. 3c) demonstrates that one can easily detect the alpha histone messenger RNAs even after they have been distributed throughout the cytoplasm following first cleavage. These in situ results have been confirmed by RNA Northern gel blot analysis as well. Equal numbers of full-size (80 μm) germinal vesicle stage oocytes and mature eggs (typically 200-500 of each) were isolated by hand and the total RNA in each sample extracted, run on an agarose gel, blotted to nitrocellulose paper and hybridized with ^{32}P-labeled probes specific for the alpha histone gene cluster. As a control, the same filters were washed to remove the alpha histone probe and hybridized with a ^{32}P-labeled probe specific for mitochondrial ribosomal RNA. Use of this second probe allowed us to control for variables of RNA recovery between preparations. The results from seven separate pairs of extraction can be seen in Table 1. In no case were we able to detect any alpha histone messenger RNA in the germinal vesicle stage oocytes. In each case however, the mitochondrial RNAs were detectable. Since our assay procedure is sufficiently sensitive to detect considerably less than one percent of the signals seen in the control egg samples, it can be safely concluded that there is little if any accumulation of alpha histone messenger RNA prior to the completion of the maturation division. The possibility does still exist that the alpha subtype genes are active during oogenesis, but that the message half-life is sufficiently short that the

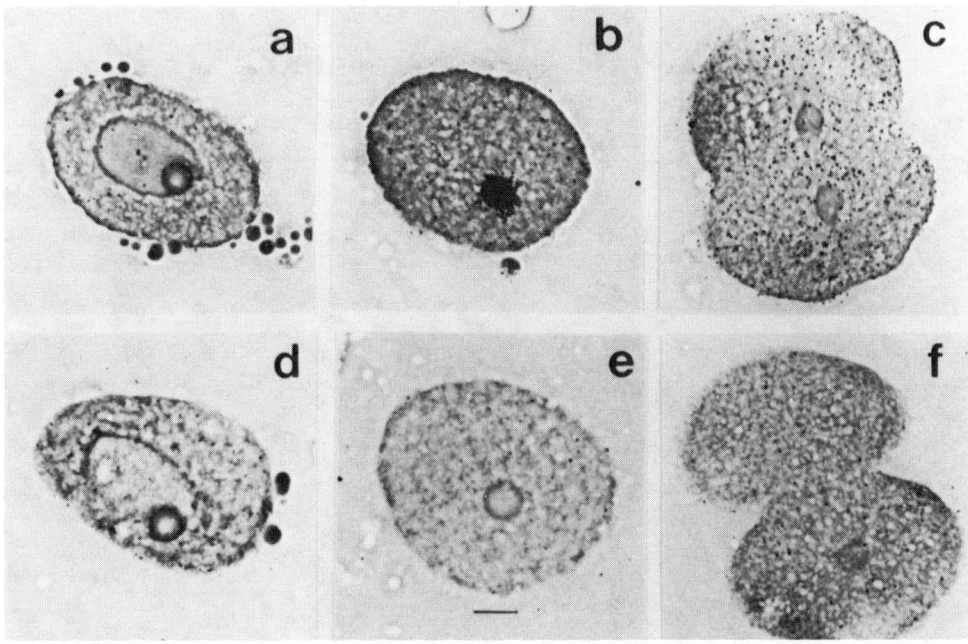

Fig. 3. In situ hybridization of alpha subtype mRNAs. Sections of large oocytes (a & d), unfertilized eggs (b & e), and two-cell embryos (c & f) on the same slides were hybridized with either ^{3}H-RNA transcripts of pCO2R$^+$ (a-c) to detect alpha subtype histone mRNA or with transcripts of the vector R7Δ7 supercoiled DNA (d-f) as a control of nonspecific binding. The specific activity of both probes was 5.9×10^7 dpm/ug and the exposure time was 27 h. Note the even distribution of grains in the two-cell embryo (c), the concentration of grains over the female pronucleus in (b), and the absence of grains in the large 4N oocyte (a) and the controls (d-f). (Reproduced from Angerer et al., 1985.)

effective concentration is below the sensitivity of either the in situ hybridization or the RNA Northern blot techniques. By adjusting the half-life following maturation, it would then be possible to start accumulating alpha histone mRNA inside the pronucleus. While this possibility can not be rigorously discounted, the other alternative of onset of alpha histone mRNA synthesis following reformation of the pronucleus is the simpler hypothesis.

One of the characteristics of the maternal alpha histone messenger RNA pool is the large number of copies of each message that exist. Mauron et al. (1982) estimate there are approximately 10^6 messages per egg for the core histones and somewhat less (800,000) for H1. The question can, therefore, be asked if there is sufficient time following maturation to accumulate these messenger RNAs before shedding and fertilization. Following the meiotic maturation divisions, the sea

Table 1. RNA-blot analysis of relative α-histone and mitochondrial RNA contents in large oocytes and unfertilized eggs.

Experiment	Large oocytes/ unfertilized eggs	α-Histone mRNA[b]		Mitochondria rRNA[c]	
		Oocytes	Eggs	Oocytes	Eggs
1	200	<0.01	1.0	0.67	0.33
2	800	<0.01	1.0	0.46	0.54
3	620	<0.01	1.0	0.39	0.61
4	200	<0.01	1.0	0.32	0.68
5	150	<0.01	1.0	0.70	0.30
6	500	<0.01	1.0	0.34	0.66
7	550	<0.01	1.0	0.47	0.53
	Average	<0.01	1.0	0.48	0.52

[a]The relative content of each RNA in samples from eggs and large oocytes was determined from densitometry of the autoradiographs. The values shown are the fraction of the sum of the optical density in both the egg and oocyte lanes.
[b]Hybridized with nick-translated DNA of the cloned α-histone repeat (pCO_2).
[c]Hybridized with nick-translated DNA of the cloned mitochondrial rRNA sequence (pSpm9).
Reprinted from Angerer et al., 1985.

urchin oocyte continues to synthesize RNA at an active pace (see Brandhorst, in press, for a review). Furthermore, all available evidence seems to indicate that the alpha histone messenger RNAs are highly stable prior to release from the pronucleus (Wells et al., 1981a). If this is so, then considering a total gene copy number of approximately 300-400 per haploid genome (Maxson et al., 1983) and a transcription rate equivalent to the maximum value per haploid nucleus observed in post-fertilization development (about 0.15×10^{-3} pg/min, Maxson & Wilt, 1982), it should only require six days to accumulate the typical egg complement of 1.3 pg of alpha histone message (Mauron et al., 1982). Considering the length of time the mature eggs are retained in the ovary before shedding (as long as six months for S. purpuratus) there is clearly more than sufficient time to accumulate the pool of stored maternal alpha histone message, even if reasonable turnover levels and slower transcription rates are postulated (Angerer et al., 1984).

Post-Transcriptional Processing and Transport

Since there is evidence that the maternal alpha histone mRNAs are synthesized and stored in the female pronucleus, an interesting point is raised. Brandhorst and others have argued that the egg pronucleus is not quiescent at all. Rather, the pronucleus is transcribing messenger at normal rates. Since, by using large amounts of label, it is possible to detect polysome-associated labeled RNAs in eggs, this would seem to argue that there is transcription, processing and transport of at least some messenger RNAs during this period. Yet no alpha histone messenger RNAs are found in the egg cytoplasm, in spite of very high concentrations of these messages in the female pronucleus. This difference is even more striking if one compares the alpha subtype histones with the cleavage-stage subtypes. Both Brandhorst (1980) and Ruderman and Schmidt (1981) report the rapid translation of newly synthesized 9S mRNAs. Since it has not been possible to detect any stored alpha histone protein in the egg, this is presumably cleavage-stage histone that is being synthesized. Thus, not only are the histone messenger RNAs processed in a manner different from the majority of the egg messages, but even within the histone multigene family, one subset (the cleavage-stage histones) are transported and translated, while another (the alphas) are kept inside the pronucleus.

The conclusion one draws from this is that transportation of the maternal alpha histone messages across the nuclear membrane is somehow prevented. Three reasons make it unlikely that the retention of the alpha histone messenger RNA in the egg is due to incomplete processing of the message. First, the size of the message, as assayed on high resolution gels, does not shift following release from the pronucleus (Lifton & Kedes, 1976). Second, release of the message following first cleavage is accompanied by the very rapid loading of the messages on polysomes (Wells et al., 1981a). Third, extracted RNA from sea urchin oocytes translated in vitro gives faithful products (Showman et al., 1984). If processing is occurring, it must be happening very rapidly, right after release of the message, and it must effect only a very small portion of the message or its associated proteins. It will be of interest to compare the structure of the pronuclear alpha histone ribonucleoprotein particles with those synthesized by the embryo to see if small differences do exist.

The assumption has been made that the maternal RNA pool is present to provide a "reserve" of informational molecules from which the embryo can draw while it is undergoing its rapid cleavage divisions. Although this is an appealing model, there is little evidence to either support or refute it. Unfortunately the crucial experiment, that of following embryonic development after deletion of the entire maternal messenger RNA pool, is not yet technically feasible. Such an experiment is, however, possible for the alpha histone mRNA subpopulation of the maternal pool and, in fact, was done, albeit unintentionally, in the 1950s by E.B. Harvey (Harvey, 1956). Because the female pronucleus contains the total store of maternal alpha histone messenger RNA of the egg, the generation by centrifugation of an anucleate egg half results in a cell without any alpha histone message (Showman et al., 1982). It is possible to fertilize such an anucleate ("heavy") half and get development of a small haploid embryo. We have measured such markers of development as cleavage time, time of onset of alpha histone mRNA synthesis, and cell number, and find no significant differences in any parameter studied (Wells et al., in prep.). Furthermore, since transcription of new alpha histone messenger RNAs is underway by the 16-cell stage (Wells et al., 1981a) and since alpha histone proteins synthesized from new alpha histone mRNA can be detected in the chromatin as early as the 16-cell stage (Showman & Murray, unpubl. obs.) this means that the point at which an absence of alpha protein might possibly have an effect on development is restricted to the second and third cleavage divisions. Thus one is forced to consider the possibility that, whatever the biological basis for its appearance, the accumulation of at least the maternal alpha histone messenger RNA may not be crucial to the overall development of the sea urchin, at least up to the pluteus stage.

It is of course possible that sequestration and delayed translation of the alpha histone mRNAs is an indirect and developmentally inconsequential result of a distant, neutral evolutionary event that turned on alpha histone gene expression following oocyte maturation, or that laboratory experiments have not yet been able to assess the adaptive significance of this mechanism in natural populations. As has been previously mentioned, deletion of the maternal alpha histone mRNA pool appears not to harm the embryo. This suggests that the key aspect of the process might be "delayed translation" rather than "onset of transcription". Although we tend to think in terms of the positive value of a gene product, it is equally likely that, rather than needing the alpha histone messenger RNA at a key time, it is critical that that same message not be present in the cytoplasm (or its protein product be absent) until after some time-critical events occurs. By delaying the appearance of the mRNA in the cytoplasm until after that key time, the embryo avoids a potentially detrimental interaction.

We have examined the eggs of a variety of echinoderms for the presence or absence of maternal alpha histone mRNA pools to determine if such a pool is critical for all echinoderm species (Raff et al., 1984). Among the major extant echinoid super orders, the three euechinoids, the Diadematacea, Irregularia and Echinacea (the latter including most of the most commonly used sea urchins such as Arbacia and Strongylocentrotus spp.), all possess an alpha histone maternal mRNA pool in their eggs. In contrast, the more ancient Cidaracea, represented by the Caribbean urchin, Eucidaris tribuloides, do not contain such a pool, although Eucidaris does exhibit the same early turn on of its alpha subtype genes following fertilization. This pattern of accumulation during cleavage, but of having no maternal pool of alpha histone message, is the same as that seen in asterioids and holothurians which diverged from the echinoid line approximately 500 million years ago. Because of the brief time span (geologically speaking) in which the cidaroids gave rise to the euechinoids and they in turn subdivided into their three major subdivisions, it is possible to assign a date of approximately 200 million years before present when the mechanism for storing the pool of maternal alpha histone messenger RNAs first appeared. Thus we are faced with the observation that at least one species belonging to a primitive sea urchin order develops without any stored maternal pool of alpha histone message, while species belonging to the advanced orders have retained this baroque regulatory mechanism as part of their developmental repertoire for over 190 million years.

CONCLUSIONS

Although it is widely accepted, albeit not proven, that maternally synthesized RNAs stored in the egg play a central role in the process of determination, we still know few specific details concerning the synthesis, storage or segregation of these molecules. The best studied cases to date are the sea urchin alpha histone messenger RNAs which are synthesized in large amounts after the final maturation divisions are complete. They accumulate in the female pronucleus where they remain until released into the cytoplasm coincident with breakdown of the pronuclear envelope at first cleavage. One consequence of this is the delayed translation of these messages. Although the significance of this segregation and delayed translation remains elusive, we now understand the basic mechanism behind at least one form of cytoplasmic localization. As time goes on and techniques continue to improve we should be able to continue to unravel the organizational complexity of the eggs and hopefully begin to understand the physical/biochemical bases behind at least a few of the patterns of transcriptional and translational regulation seen during oogenesis and following fertilization.

LITERATURE CITED

Anderson, D.M., T.D. Richter, M.E. Chamberlin, D.H. Price, R.J. Britten, L.D. Smith, and E.H. Davidson. 1982. Sequence organization of the Poly(A)RNA synthesized and accumulated in lampbrush chromosome stage Xenopus laevis oocytes. J. Mol. Biol. 155: 281-309.

Anderson, K.V. and C. Nüsslein-Volhard. 1984. Information for the dorsal-ventral pattern of the Drosophila embryo is stored as maternal mRNA. Nature 311: 223-227.

Angerer, L.M., D.V. DeLeon, R.C. Angerer, R.M. Showman, D.E. Wells, and R.A. Raff. 1985. Delayed accumulation of maternal histone mRNA during sea urchin oogenesis. Devel. Biol. 101: 477-484.

Brandhorst, B.P. 1980. Simultaneous synthesis, translation and storage of mRNA, including histone mRNA in sea urchin eggs. Devel. Biol. 79: 139-148.

Brandhorst, B.P. In press. The informational content of the echinoderm egg. In: Developmental Biology: A Comprehensive Synthesis. I. Oogenesis. L. Browder (ed.). Plenum, N.Y.

Cabrera, C.V., T.J. Lee, T.W. Ellinson, R.J. Britten, and E.H. Davidson. 1984. Regulation of cytoplasmic mRNA prevalance in sea urchin embryos: Rates of appearance and turnover for specific sequences. J. Mol. Biol. 174: 85-111.

Constantini, F.D., R.J. Britten, and E.H. Davidson. 1980. Message sequences and short repetitive sequences are interspersed in sea urchin poly(A)+ RNAs. Nature (London) 287: 111-117.

Davidson, E.H. and B.R. Hough. 1971. Genetic information in oocyte RNA. J. Mol. Biol. 56: 491-506.

DeLeon, D.V., K.H. Cox, L.M. Angerer, and R.C. Angerer. 1983. Most early varient histone mRNA is contained in pronuclei of sea urchin eggs. Devel. Biol. 100: 197-207.

Ernst, S.G., B.R. Hough-Evans, R.J. Britten, and E.H. Davidson. 1980. Limited complexity of the RNA in micromeres of 16-cell sea urchin embryos. Devel. Biol. 79: 119-127.

Freeman, G. 1976. The role of cleavage in the localization of developmental potential in the ctenophore Mnemiopsis leidyi. Devel. Biol. 49: 143-177.

Freeman, G. 1979. The multiple roles which cell division can play in the localization of developmental potential. pp. 53-76. In: Determinants of spatial organization. S. Subtelny and I.R. Konigsberg (eds). Academic Press, NY.

Galau, G.A., W.H. Klein, M.M. Davis, B.T. Wold, R.T. Britten, and E.H. Davidson. 1976. Structural gene sets active in embryos and adult tissues of the sea urchin. Cell 7: 487-505.

Harvey, E.B. 1956. The American Arbacia and Other Sea Urchins. Princeton University Press, Princeton.

Hille, M.B., M.V. Danilchik, A.M. Colin, and R.T. Moon. 1985. Translational control in echinoid eggs and early embryos. In: The Cellular and Molecular Biology of Invertebrate Development. Roger H. Sawyer and Richard M. Showman (eds.). University of South Carolina Press, Columbia.

Hörstadius, S. 1973. Experimental Embryology of Echinoderms. Clarendon Press, Oxford.

Hough-Evans, B.R., S.G. Ernst, R.J. Britten, and E.H. Davidson. 1979. RNA complexity in developing sea urchin oocytes. Devel. Biol. 69: 258-269.

Hough-Evans, B.R., B.T. Wold, S.G. Ernst, R.J. Britten, and E.H. Davidson, E.H. 1977. Appearance and persistence of maternal mRNA sequences in sea urchin development. Devel. Biol. 60: 258-277.

Jeffery, W.R. 1985. Patterns of maternal mRNA distribution and their role in early development. In: The Cellular and Molecular Biology of Invertebrate Development. Roger H. Sawyer and Richard M. Showman (eds.). University of South Carolina Press, Columbia.

Jeffery, W.R. and L.J. Wilson. 1983. Localization of messenger RNA in the cortex of Chaetopterus eggs and early embryos. J. Emb. Exp. Morph. 75: 225-239.

Kandler-Singer, I. and K. Kalthoff. 1976. RNase sensitivity of an anterior morphogenetic determinant in an insect egg (Smittia spec. Chironomidae, Diptera). Proc. Nat. Acad. Sci. USA 73: 3739-3743.

Lifton, R.P. and L.H. Kedes. 1976. Size and sequence homology of masked maternal and embryonic histone messenger RNAs. Devel. Biol. 48: 47-55.

Mattaj, I.W. and E.M. DeRoberts. 1985. Nuclear segregation of U2 snRNA requires binding of specific snRNP proteins. Cell 40: 111-118.

Mauron, A., L.H. Kedes, B.R. Hough-Evans, and E.H. Davidson. 1982. Accumulation of individual histone mRNAs during embryogenesis of the sea urchin Strongylocentrotus purpuratus. Devel. Biol. 94: 425-434.

Maxson, R., R. Cohn, and L. Kedes. 1983. Expression and organization of histone genes. Ann. Rev. Genet. 17: 239-277.

Maxson, R.E. and F.H Wilt. 1982. Accumulation of the early histone messenger RNAs during development of Strongylocentrotus purpuratus. Devel. Biol. 94: 435-440.

Moore, G.P. 1984. Evolutionary conservation of DNA coding for maternal RNA in sea urchins. Biochem. Biophys. Res. Comm. 123: 278-285.

Poccia, D., J. Salik, and G. Krystal. 1981. Transitions in histone variants of the male pronucleus following fertilization and evidence for a maternal store of cleavage stage histones in the sea urchin egg. Devel. Biol. 82: 287-296.

Posakony, J.W., C.N. Flytzanis, R.J. Britten, and E.H. Davidson. 1983. Interspersed sequence organization and developmental representation of cloned poly(A)RNAs from sea urchin eggs. J. Mol. Biol. 157: 361-389.

Raff, R.A. 1983. Localization and temporal control of expression of maternal histone mRNA in sea urchin embryos. pp. 65-86. In: Time, Space and Pattern in Embryonic Development. W.R. Jeffery and R.A. Raff (eds.). A.R. Liss, NY.

Raff, R.A., T.A. Anstrom, C.J. Huffman, D.S. Leaf, J.-H. Loo, R.M. Showman, and D.E. Wells. 1984. Evolutionary change in developmental expression of highly conserved histone genes in echinoderms. Nature 310: 312-314.

Raff, R.A. and R.M. Showman. In press. Maternal messenger RNA: Quantitative, qualitative and spatial control of its expression in embryos. In: The Biology of Fertilization. C.B. Metz and A. Monroy (eds.). Academic Press, NY.

Render, J.A. 1983. The second polar lobe of the Sabellaria cementarium embryo plays an inhibitiory role in apical tuft formation. W. Roux's Arch. Devel. Biol. 192: 120-129.

Richter, J.D., L.D. Smith, D.M. Anderson, and E.H. Davidson. 1984. Interspersed poly(A)RNAs of amphibian oocytes are not translatable. J. Mol. Biol. 173: 227-241.

Rodgers, W.A. and P.R. Gross. 1978. Inhomogenous distribution of egg RNA sequences in the early embryo. Cell 14: 279-288.

Rosbash, M. and P.J. Ford. 1974. Polyadenylic acid-containing RNA in Xenopus laevis oocytes. J. Mol. Biol. 85: 87-101.

Rudensey, L.M. and A.A. Infante. 1979. Translational efficiency of cytoplasmic non-polysomal messenger ribonucleic acid from sea urchin embryos. Biochem. 18: 3056-3063.

Ruderman, J.V. and M.R. Schmidt. 1981. RNA transcription and translation in sea urchin oocytes and eggs. Devel. Biol. 81: 220-228.

Savic, A., P. Richman, P. Williamson, and D. Poccia. 1981. Alterations in chromatin structure during early sea urchin embryogenesis. Proc. Nat. Acad. Sci. USA 78: 3706-3710.

Scheller, R.H., F.D. Constantini, M.R. Kozlowski, R.J. Britten, and E.H. Davidson. 1978. Specific representation of cloned repetitive DNA sequences in sea urchin RNA. Cell 15: 189-203.

Showman, R.M., D.E. Wells, J.A. Anstrom, D.A. Hursch, D.S. Leaf, and R.A. Raff. 1984. Subcellular localization of maternal histone mRNAs and the control of histone synthesis in the sea urchin embryo. pp. 109-130. In: Molecular Aspects of Early Development. G.M. Malacinski and W.H. Klein (eds.). Plenum, NY.

Showman, R.M., D.E. Wells, J. Anstrom, D.A. Hursh, and R.A. Raff. 1982. Message-specific sequestration of maternal histone mRNA in the sea urchin egg. Proc. Nat. Acad. Sci. USA 79: 5944-5947.

Thomas, T.L., J.W. Posakony, D.M. Anderson, R.J. Britten, and E.H. Davidson. 1981. Molecular structure of maternal RNA. Chromosoma (Berl.) 84: 319-335.

Togashi, S. and M. Okada. 1983. Poly(A)$^+$ RNA extracted from <u>Drosophila</u> embryos restores pole-cell-forming ability in U.V.-sterilized embryos. Devel., Growth and Diff. 25: 423.

Tufaro, F. and B.P. Brandhorst. 1979. Similarity of proteins synthesized by isolated blastomeres of early sea urchin embryos. Develop. Biol. 72: 390-397.

Venezky, D.L., LM. Angerer, and R.C. Angerer. 1981. Accumulation of histone repeat transcripts in the sea urchin egg pronucleus. Cell 24: 385-391.

Wells, D.E., R.M. Showman, W.H. Klein, and R.A. Raff. 1981a. Delayed recruitment of maternal mRNA in sea urchin embryos. Nature 292: 477-478.

Wells, D.E., R.M. Showman, W.H. Klein, and R.A. Raff. 1981b. Translational regulation in sea urchin embryos. Biol. Bull. 161: 322.

Wilson, E.B. 1904. Experimental studies on germinal localization. I. The germ-regions in the egg of <u>Dentalium</u>. J. Exp. Zool. 1: 1-72.

Wilson, E.B. 1925. The Cell in Development and Heredity. 3rd ed. Macmillan, NY.

EXPRESSION AND APPEARANCE OF GERM LAYER-SPECIFIC ANTIGENS ON THE SURFACE OF EMBRYONIC SEA URCHIN CELLS

David R. McClay, Valeria Matranga, and Gary Wessel

ABSTRACT

A group of monoclonal antibodies are described which define a number of patterns in sea urchin development. The cell surface antigens identified by the antibodies appeared at different times and became localized to subsets of cells in the gastrula and pluteus larva. Some of the antigens became germ layer-specific secondarily after being present on all germ layers. Other antigens appeared de novo and were located in a single germ layer from first transcription onward. Finally, some antigens were highly localized but not restricted to a single germ layer. These were expressed by portions of two germ layers. Thus, at gastrulation there is a dramatic appearance and redistribution of cell-surface molecules at the time of origin of the three germ layers.

INTRODUCTION

Two major events occur to signal the onset of gastrulation in the sea urchin. First, the embryonic genome begins to take full control of the succession of morphogenetic events and second, cell movements and rearrangements establish the three germ layers which mark the basic lineages from which all differentiated tissues arise. These generalizations of gastrulation are based on two types of studies. First, early drug studies established that development could proceed without transcription in many organisms until the onset of gastrulation (Gross & Cousineau, 1964). Second, descriptive studies provided detailed analyses of the morphological origins of the three germ layers (Gustafson & Wolpert, 1967).

Much remains to be learned about these components of gastrulation. At the genomic level, major efforts are underway in many laboratories to learn how the embryonic genome is selectively activated and how gene expression becomes distinct in the three germ layers. At the cellular level, very little is known about the mechanisms of cell rearrangements that establish the three germ layers.

In their classic paper on sea urchin gastrulation, Gustafson and Wolpert (1967) suggested that cell rearrangements occurred through a combination of motility and molecular cues from the cell surface. This suggests that the cell surface acquires molecular markers that somehow provide information necessary for the correct direction of movement, and for the appropriate cell-cell, and cell-substrate associations. Implicit in their model is that the selective genomic activities that will determine the germ layer lineages provide, as part of their information, cell-surface proteins that have morphogenetic roles in the cell rearrangements. Several predictions follow from this model. First, gastrulation should mark the beginning of some dramatic cell-surface changes in that new molecules should be inserted in subsets of cells to provide the necessary molecular cues for morphogenetic movements. Second, the insertions should coincide with the visible separation of the three germ layers. And third, the proteins should be products of the newly activated embryonic genome.

Early attempts to study the cell-surface changes indicated that some changes did occur. For example, in an immunochemical study, McClay and Chambers (1978) showed that antigenic changes occurred at gastrulation, and that these changes seemed to be subdivided into several germ layer specificities. As that study used polyclonal antibodies, it was difficult to make specific conclusions because the activity of the antisera was against many antigenic determinants.

In order to test the prediction that the cell surface is a mosaic starting at gastrulation, we adapted protocols for isolating monoclonal antibodies. These reagents are products of cloned cells growing in culture such that an antibody with a single specificity is produced (Kohler & Milstein, 1975). Thus it was possible to follow the distribution of single protein species on the cell surface. When one makes monoclonal antibodies, there is a problem of selection. In practice, splenic lymphocytes from immunized mice are fused with myeloma cells in culture. The hybridomas that result must be screened by selecting for clones producing antibodies of interest. In our case, we were looking for germ layer-specific antigens. Thus we screened the hybridomas with membranes from ectoderm and endoderm. From such screens on a number of cell fusions, we collected a number of germ layer-selective antibodies. These have been studied and are presented here to show the variety of patterns exhibited by the cell surface at gastrulation. As will be seen below, there are antigen distributions as predicted from the Gustafson and Wolpert model. In the course of the study it was found that germ layer-specific antigens are a category of molecules and that there are a variety of ways for the antigens to arrive at a germ layer-specific position. These patterns will be described. Some of the data described below have been reported previously (McClay et al., 1983; McClay & Wessel, 1984).

Antigens in the Egg that Later Become Confined to a Single Germ Layer

Although membranes of late gastrula stages were used for the immunization, a number of antigens were found that were intracellular prior to fertilization. They were synthesized during oogenesis and deposited into intracellular granules or vesicles that could be subdivided experimentally by centrifugation (Wessel et al., 1984). They became associated with membranes or with the extracellular matrix at different times after fertilization. Initially all the antigens were secreted to cover every cell without germ layer preference. A number of antigens continued to be associated with all plasma membranes, without germ layer preference. The antigens described below, however, were present in the egg and became associated with, or compartmentalized to, one germ layer secondarily.

Egg Antigens that Become Localized to Ectoderm

At fertilization or shortly thereafter, presumptive ectodermal antigens are released to the surface of the embryo (Fig. 1) (the term surface is used because at the current level of resolution it is not possible to distinguish between antigens that are part of an extracellular matrix and antigens that are membrane-associated). An area in the egg

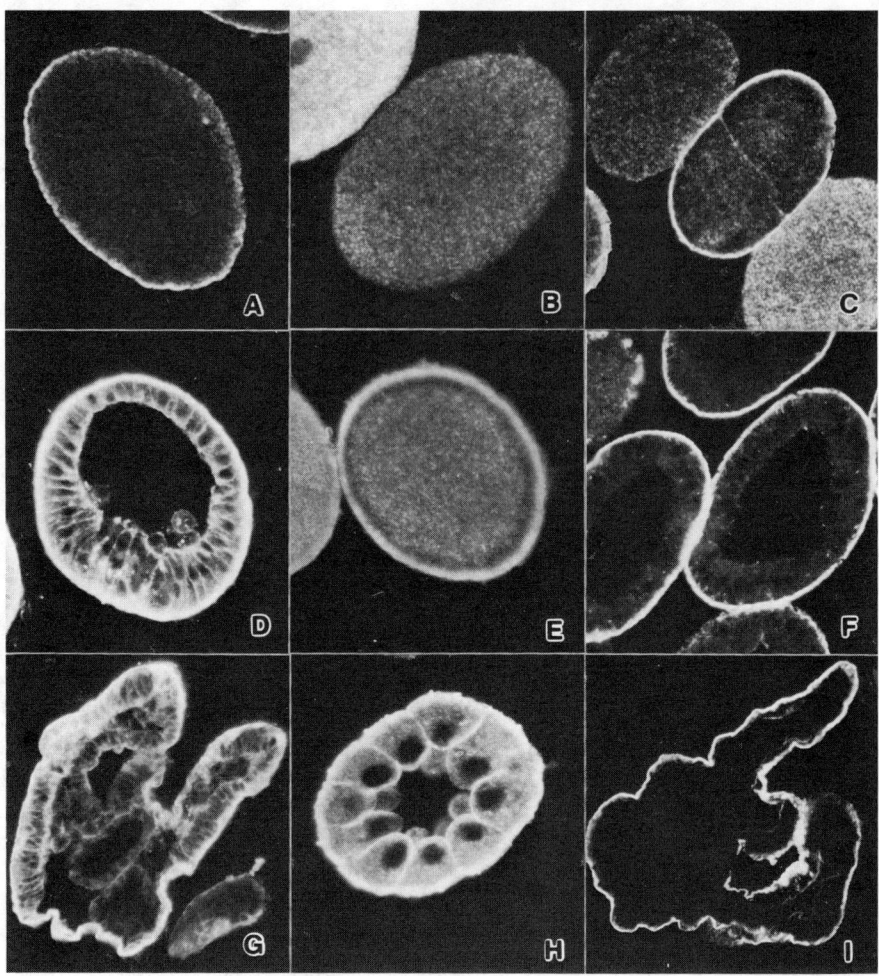

Fig. 1. Ectodermal antigens originally found in the egg. Sections of Lytechinus were stained with two monoclonal antibodies and compared with hyalin to show the diversity of this group. (a) Polyclonal antibody to hyalin. (b,d,e,g,h) ectoderm monoclonal LL1c10. (c,f,i) ectoderm monoclonal De27e9. Both ectodermal antigens are stored in the egg (e.g., LL1c10 in b) as granules that are distinct from cortical granules (stained with anti-hyalin in a). De27e9 is released at fertilization and is predominantly on the cell surface by the two-cell stage (c). At the blastula stage, De27e9 surrounds the embryo (f). At the pluteus stage, this antigen covers the embryo and the lining of the stomodaeum (i). Some of LL1c10 is released from the intracellular granules at fertilization (e), but this antigen continues to be within all cells through the blastula (h) and gastrula stages (d). In the pluteus this antigen is found in the ectoderm; the staining is much reduced in endoderm, and the mesoderm is negative (g). All the embryos shown in Figures 1-7 were fixed in the same way (1% paraformaldehyde). Controls in each case included the following which were found to be negative for fluorescence: parent myeloma supernatant substituted for primary antibody, no primary antibody, and no secondary antibody. In addition, a polyclonal anti-sea urchin antibody was used as a positive control, and other monoclonal antibodies served as internal positive controls for nonspecific staining. All of the figures are at about the same magnification. For reference, each egg has a diameter of ~120 μm.

cortex just beneath the surface loses fluorescent material, suggesting that the surface antigen originally comes from this area (Fig. 1e). A portion of each of the presumptive ectodermal antigens remains in intracellular compartments through early development. These are released, or turned over, at different rates, depending on the individual antigen. Some antigens are almost entirely extracellular by the two-cell stage (Fig. 1c), while others are still detectable within blastula cells (Fig. 1d). At the gastrula stage some ectodermal antigens initially follow along the invaginating archenteron and then are lost secondarily from the endodermal surface to become associated predominantly or exclusively with the ectodermal surface (Fig. 1g & i).

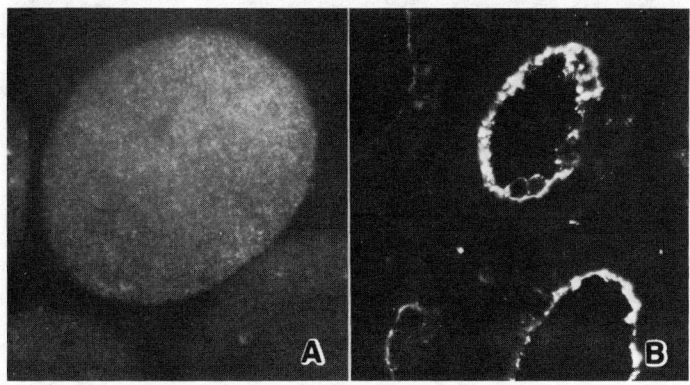

Fig. 2. An antigen in the egg that becomes localized to the basal lamina. In (a) antigen LL1b10 was detected, but at reduced amounts when compared to the lining of the blastocoel at the mesenchyme blastula stage (b).

Egg Antigens that Localize to the Basal Lamina

The antigens described as "ectodermal" are first secreted toward the apical surface of all cells, then become associated only with ectoderm. The secretion of basal laminar antigens is initially a function of all blastomeres. Later, some of the antigens are secreted only by the primary mesenchyme cells (Fig. 2). The antigens localized in the region of the basal lamina all come from the same class of egg granules (Wessel et al., 1984). Many of these are released into the blastocoel early in cleavage, others are released later at the mesenchyme blastula stage. As with the ectodermal antigens, the term mesodermal most likely results from continued synthesis of these antigens only in the mesoderm.

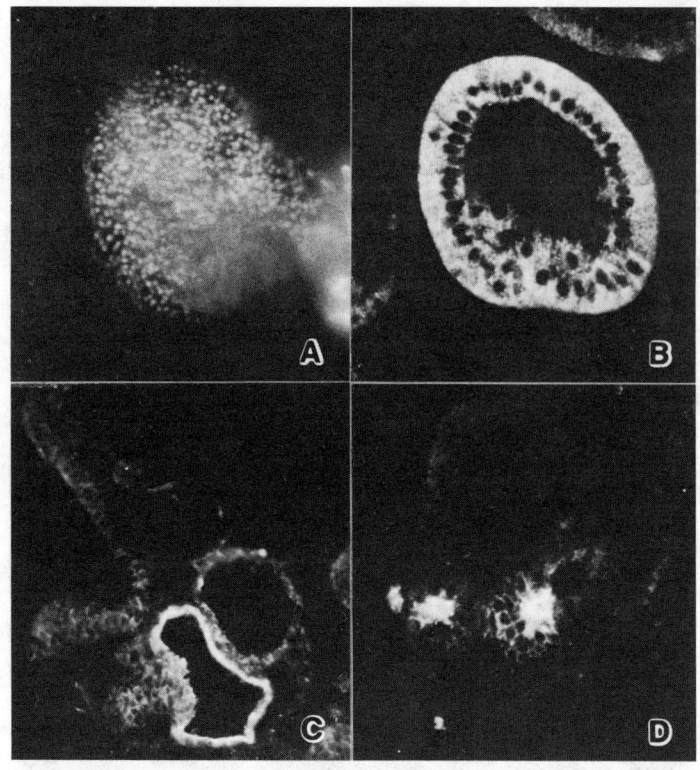

Fig. 3. Antigens in the egg that later become endodermal. (a-c) AA1a3, (d) LL5f7. (a) shows a section of an unfertilized egg to demonstrate the intracellular granules characteristic of antigens in the egg. Bar = 10 µm. In the mesenchyme blastula the antigen remains intracellular and in all blastomeres (b), but by the pluteus stage the antigen is concentrated in the midgut and absent or reduced elsewhere (c). Another antigen in this class becomes confined to cell surfaces in the midgut and hindgut (d). (b-d) magnification same as in Figure 1.

Egg Antigens that Later Become Endoderm-Specific

Like many other antigens in the egg, endoderm-specific antigens segregate in a centrifugal field, in a pattern that is distinct from the segregation of yolk, the basal laminar antigens, ectodermal antigens, and cortical granular antigens. At fertilization there is no obvious change in intracellular distribution of the presumptive endodermal antigens. In fact, through the blastula stage until the late gastrula-prism stage, many of these antigens often appear to be equally distributed in intracellular compartments in all blastomeres (Fig. 3b). As the pluteus stage approaches, these antigens become increasingly confined to the endoderm, often to a small segment of the gut (Fig. 3c & d). The endoderm antigens do not appear to be found ultimately in the stomodaem, an area that was previously considered to be endodermal. Instead this area contains antigens that are continuous with the ectoderm (Fig. 1i).

It is possible that some of the antigens that are being described as endoderm are yolk proteins, but four facts argue against that possibility. First, yolk proteins are abundant, yet the endodermal antigens, as seen on Western blots, are not associated with gel bands of abundant proteins. Second, the molecular weights are not comparable (Ozaki, 1982). Third, the antigens ultimately are localized at the cell surface which is not a location where yolk is usually thought to be. Fourth, an antibody against yolk does not stain in a pattern that resembles any of the antibodies in this class.

Antigens that Appear *De Novo* at Gastrulation

The term de novo is used to describe antigens that first appear at some time during development. Without additional information it cannot be assumed that these antigens are newly synthesized, since their antigenicity could be a secondary modification of a preexisting protein. Our screens have picked up de novo antigens in the mesoderm and endoderm, and these will be described below. Carpenter et al. (1984) have detected several ectodermal antigens that appear de novo.

Mesodermal Antigens

One very striking pattern is seen in a group of mesoderm monoclonals, all of which begin to stain the embryo at the same developmental time, but consist of a heterogeneous group of antigens on Western blots. Of those, AA1g8 is the best characterized and will be described here. The AA1g8 antigen is not present prior to the mesenchyme blastula stage (Fig. 4a). The antigen appears at a very precise time during the mesenchyme blastula stage. This embryonic stage is defined by the delamination of primary mesenchyme cells from the blastula wall. Cells located at the vegetal pole move into the blastocoel one by one until a certain specific number come to lie on the floor of the blastocoel. The AA1g8 antigen appears just at the time a cell completes delamination (Fig. 4b & c). A number of sections capture a situation

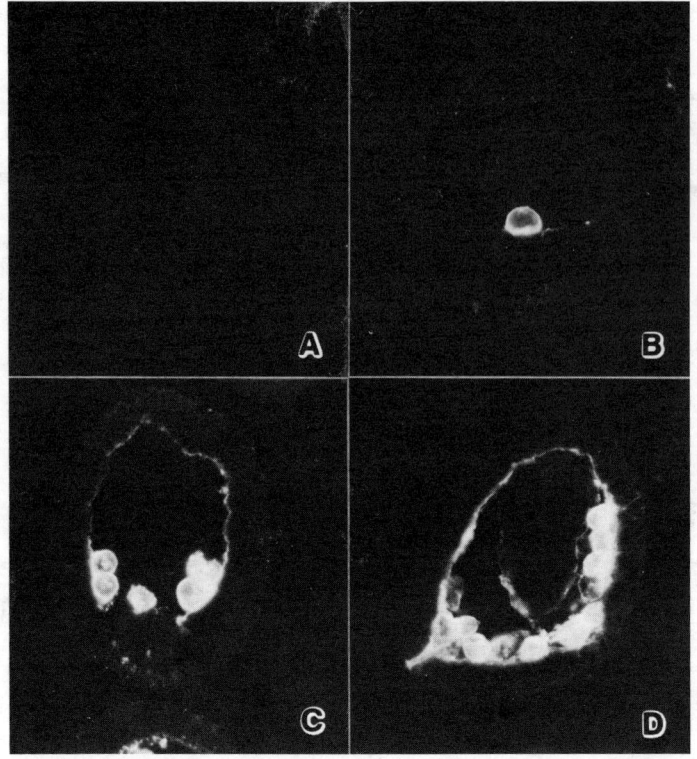

Fig. 4. Mesodermal antigen, AA1g8, that appears at the mesenchyme blastula stage. The antigen is not present at the blastula stage (a). It first appears on the first of the primary mesenchyme cells to complete delamination (b). As each primary mesenchyme cell completes its migration through the basal lamina, it becomes positive for AA1g8 (c). The antigen is deposited along the basal lamina from the vegetal to the animal pole. When spicules appear, two projections of the antigen pass through the ectoderm wall immediately beneath the area of spicule origin (d).

in which one or more cells have completed the delamination process and others are still in the process. Only the cells that have completed delamination express AA1g8. Each of the cells expressing AA1g8 has an intracellular spot of stain, suggesting a site in the synthetic pathway at which antigenicity is first attained. The antigen covers the primary mesenchyme cells and then is deposited along the blastocoelic surface, first at the vegetal pole and then toward the animal pole (Fig. 4b & c). At the prism stage, the antigen penetrates the ectodermal covering immediately beneath the site of spicule synthesis (Fig. 4d). The antigen also covers the spicule envelope (Fink & McClay, 1985). The specificity of AA1g8 has been explored in different species to determine whether it recognizes primary mesenchyme cells in a spectrum of echinoderm species. The species examined were Lytechinus variegatus, Tripneustes esculentus, Strongylocentrotus purpuratus, Eucidaris tribuloides, and Dendraster excentricus. The antigen was present and specific for primary mesenchyme cells in all of these species except Eucidaris, which does not have primary mesenchyme cells. Spicules in Eucidaris originate from secondary mesenchyme cells, and these cells also did not express the AA1g8 antigen. In one case there was a difference in the time of expression; in Dendraster primary mesenchyme cells released the antigen but not until the mid-gastrula stage.

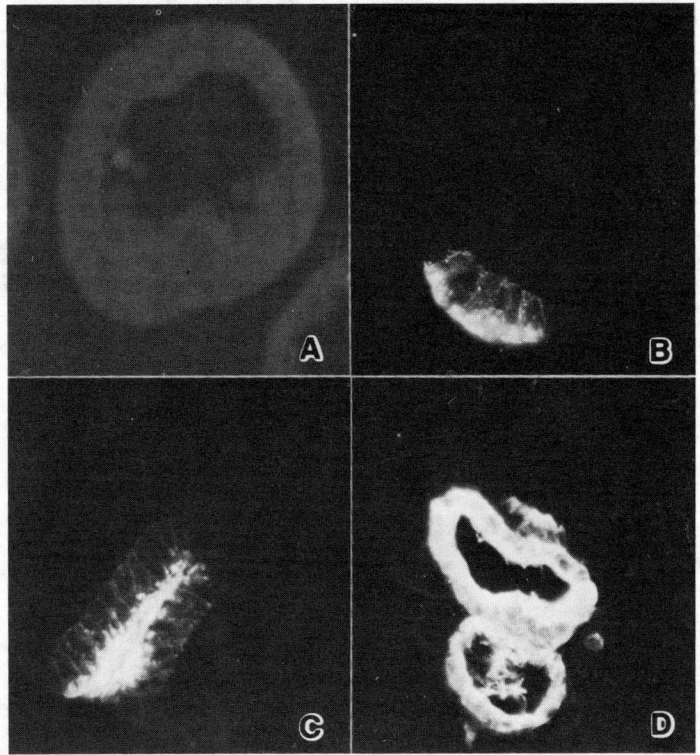

Fig. 5. An endodermal antigen that appears early in the gastrula stage. At the mesenchyme blastula stage (a) the antigen is not present. The antigen, De25c7, first appears in and on cells of the vegetal plate shortly before the first evidence of archenteron invagination (b). At the midgastrula stage (c) the antigen is on the apical surface of each presumptive endodermal cell with the exception of cells in the esophageal region. Finally, in the pluteus stage the antigen is present in the midgut and hindgut but not in the foregut (d). The boundary between cells that express the antigen and those that do not is striking both at the midgastrula stage where there is no apparent anatomical marker delineating a boundary and in the pluteus where the boundary is defined by the constriction between the foregut and the midgut.

Endoderm Antigen

An antigen that is specific for endoderm was observed to appear on cells of the vegetal plate at the beginning of invagination of the archenteron (Fig. 5a & b). The antigen is present on the apical surface of the cells and less so on the basal surface (Fig. 5b & c). At the mid-gastrula stage the antigen is found on cells of the lower 2/3 of the archenteron. Secondary mesenchyme cells and anterior endodermal cells do not express the antigen.

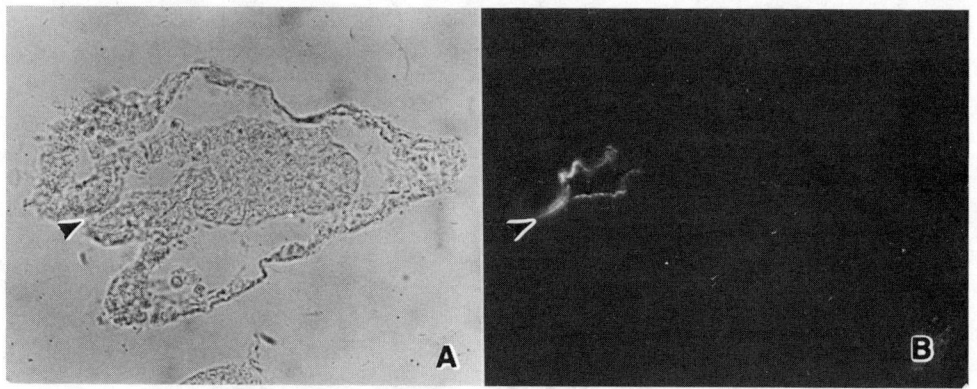

Fig. 6. Antigen restricted to the stomodael opening. At the pluteus stage 'Lips' is detected only on cells lining the stomodaeum. (a) is a bright field of the fluorescent section in (b).

Antigens that Assume Precise Spatial Localizations Without Regard to Germ Layer Boundaries

Many of the monoclonal antibodies identify antigens that appear in a very confined area of the embryo. Two antigens, in particular, serve as examples for a highly restricted distribution. The first, shown in Figure 6, is an antigen that is restricted to the stomodael opening. The cells that are positive for this antigen are both of ectodermal and of endodermal origin.

Figure 7 shows the developmental history of another antigen with a novel distribution. This antigen is secreted at the blastula stage and as a signal of that secretion there is an intracellular spot of antigen within each blastomere. The antigen eventually covers the surface of the gastrula in a gradient of antigen, increasing in concentration from the animal to the vegetal pole (Fig. 7b). The antigen persists

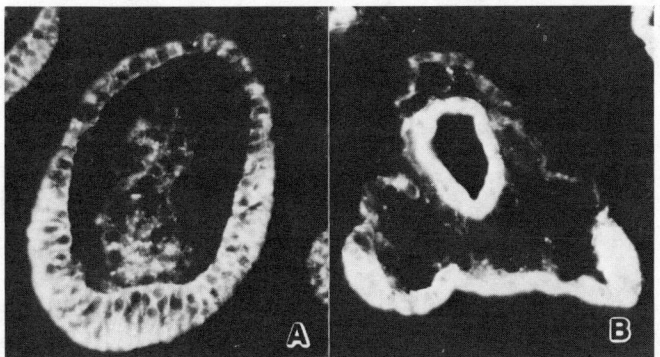

Fig. 7. Antigen restricted to the vegetal half of the embryo. At the blastula stage 'cheeks' is detected as a 'gradient' increasing toward the vegetal pole (a). At the pluteus stage this antigen is restricted to the hindgut and the ectoderm surrounding the anus on the ventral side of the embryo.

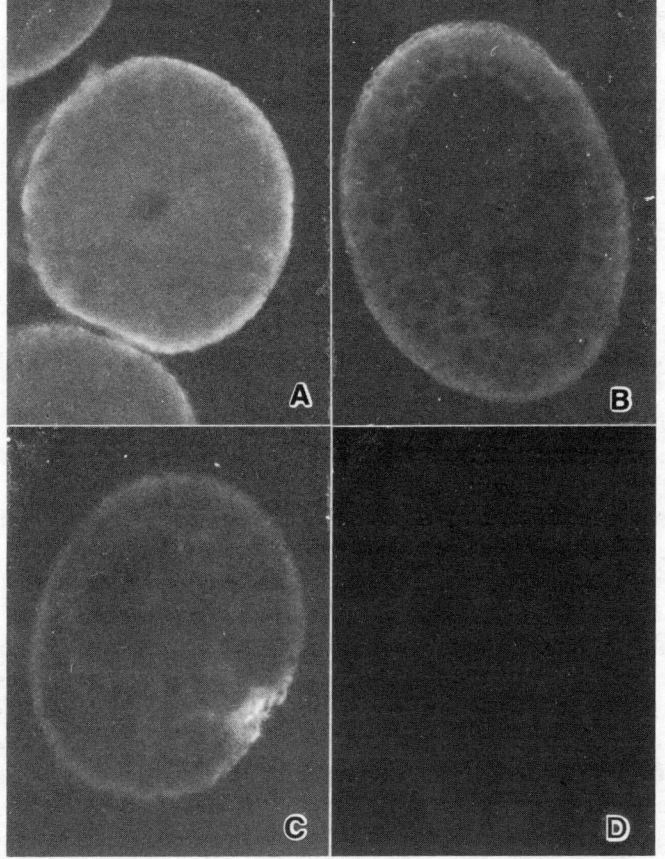

Fig. 8. An antigen that appears early in development and then declines as the pluteus stage approaches. (a) egg, (b) blastula, (c) gastrula, (d) pluteus.

through development on the surface of the ectoderm surrounding the original blastopore and on the surface of the hindgut (Fig. 7b). Thus the ectoderm and endoderm that were originally at the vegetal pole retain the antigen and it is lost from other tissues.

Antigens that are Present Early in Development but Disappear at Gastrulation

Since most of the cell fusions were performed using mice immunized with gastrula-stage material, we did not expect to find antigens that were present during early development and absent later. One fusion was performed using blastula cell membranes, and an antigen was found that recognized blastula cell membranes but disappeared thereafter. In Figure 8 this antigen is present in the egg, but disappears by the late gastrula stage. The immunogen in this case was a butanol extract of membranes that has been shown to promote adhesion of blastula cells (Noll et al., 1979).

DISCUSSION

The goal of this paper has been to document the variety of cell-surface antigenic patterns associated with gastrulation in the sea urchin. Several types of patterns were observed with the more than 100 antigens used. First, the separation of cells into three germ layers is accompanied by a simultaneous expression of antigens on each of the three new germ layers. The function of the antigens is yet to be learned, though other studies in our laboratory suggest that some of these antigens are active in guiding morphogenetic movements (Fink & McClay, 1985). Second, a number of the germ layer-specific antigens were traced back to the egg. These became specific by a process of compartmentalization; antigens disappeared from the surface of two of the germ layers and remained on the surface of the third. The simplest explanation for this behavior is that the antigen persists on the germ layer that continues to synthesize it. Immunoprecipitation of radiolabeled germ layers will be required to determine whether this notion is true. Nevertheless, most of the germ layer-specific molecules were first identified as the three germ layers were organized. This observation at the molecular level complements the observation at the cellular level that the gastrula stage is of fundamental importance to organogenesis.

Some antigens do not fit into the germ layer categories. For example the antibody called 'cheeks' identifies a region that is defined by the hindgut and the ectoderm surrounding the anus on the ventral plate of the pluteus larva (Fig. 7). Thus, when cells differentiate, the boundaries of differential gene expression are not necessarily restricted to germ layer boundaries. Two other antigens, one described by Davidson and Angerer (1984), and the other described as Spec 1 (Carpenter et al., 1984), appear in patterns that are complementary to the cheeks antigen. Spec 1 is found in regions of ectoderm that border the ectodermal region of cheeks antigen expression. The patterns exhibited by these antigens reflect an animal-vegetal organization rather than a germ layer localization. Other patterns appear that are restricted to regions of a germ layer. The lips antigen is an example of this behavior and there are other antigens (Wessel et al., 1984) that also fit this pattern. The lips antigen is on the surface of cells lining the stomodeum only. It is not on the entire surface of these cells, only on the basal surface away from the stomodeal cavity. The lips antigen and the other antigens that assume positions in subregions of germ layers suggest that the terms ectoderm, endoderm, and mesoderm include cells that can be subdivided into dorsal ectoderm, ventral ectoderm, etc. almost from the point of original differentiation of the three germ layers.

The de novo antigens demonstrated several principles that are noteworthy. First, it was striking that the antigens appeared at very precise times in development. The mesodermal antigen appeared precisely at the time at which a primary mesenchyme cell completed its delamination from the blastula wall. Cells in the process of delamination did not express the antigen even when some of their brethren that had completed the process were expressing the antigen. Other mesodermal antigens, in the same family of de novo molecules, were expressed as delamination occurred. Delamination in Lytechinus occurs over a period of about 45 min to 1 h at about 8-9 h into development. This means that there is a coordinated expression of antigens spatially and temporally, to within minutes, on a very small subset of cells. Others using different probes see expression of mesodermal antigens at the same time (Davidson & Angerer, 1984). One might ask what is the trigger for these antigens to be expressed at such a precise time? One hypothesis often expressed is a 'clock' hypothesis (Dan & Ikeda, 1971; Spiegel & Spiegel, 1980). The idea is that some internal chemical cycle provides a clock for the timing of various developmental events. While this may explain the precision of normal differentiation, it is clear that differentiation can operate independently. As an example, the endoderm antigen that appears de novo is sensitive to inhibitors of collagen processing (Wessel et al., 1983). The antigen can be inhibited from its normal appearance for long periods of time; however, when the inhibitor is removed, the antigen appears and development proceeds normally. Clearly, some kind of trigger is necessary before the antigen is expressed. The trigger is often called 'induction' in classical embryology. How then does one reconcile this observation with the clock hypothesis? One possibility is that a clock is reset every time an inductive event is required. Also implied is that each differentiative event initiates the birth of a new timepiece since the two pathways can proceed independently of each other. In the example given above, where the endodermal antigen is inhibited from its appearance, the mesodermal antigens are not inhibited by the same treatments. These cells continue to behave as if development is proceeding normally and even express spicules, their normal function in development. Thus, there does not appear to be an internal clock that runs organogenesis, unless that clock is somehow split each time an inductive event or a differentiative event occurs. There may be a clock as a pacemaker for early cleavage (Horstadius, 1939; Hara et al., 1980), but inductive events disrupt the progression.

Cell rearrangements at gastrulation were described at the outset of this paper as being one of the more dramatic events in the life history of an organism. The purpose of collecting the monoclonal antibodies was to explore the hypothesis that morphogenetic movements use cell-surface cues for cellular rearrangements. Several studies have suggested that such cues are indeed used when primary mesenchyme cells separate from the blastula wall. Fink and McClay (1985) have demonstrated that these cells simultaneously lose an affinity for hyaline (the extraembryonic matrix), they lose an affinity for other cells of the blastula (including other primary mesenchyme cells), and they gain an affinity for the basal lamina, their new substrate. The gain in affinity is specifically for fibronectin. Thus three independent affinity changes contributed to the morphogenetic movements of one cell type. The three changes can be attributed to cell surface changes on the surface of the primary mesenchyme cells since the experiment was conducted by leaving the substrates (hyaline, other cells or basal lamina) constant and testing presumptive primary mesenchyme cells of different ages. The assay was conducted at temperatures that prohibited motility so that simple affinities were measured. Thus, although motility is required in vivo to move cells to their new locations, these studies show that the cells have surface components that can act as directive guides for that movement.

LITERATURE CITED

Carpenter, C.D., A.M. Bruskin, P.M. Hardin, M.S. Keast, J. Anstrom, A.L. Tyner, B.P. Brandhorst, and W.H. Klein. 1984. Novel proteins belonging to the Troponin C superfamily are encoded by a set of mRNAs in sea urchin embryos. Cell 36: 663-671.

Dan, K. and M. Ikeda. 1971. On the system controlling the time of micromere formation in sea urchin embryos. Dev. Growth Differ. 13: 285-301.

Davidson, E. and R. Angerer. 1984. Molecular indices of cell lineage specification in sea urchin embryos. Science 226: 1153-1160.

Fink, R.D. and D.R. McClay. 1985. Three cell recognition changes accompany the ingression of primary mesenchyme cells. Dev. Biol. 107: 66-74.

Gross, P.R. and G.H. Cousineau. 1964. Macromolecule synthesis and the influence of actinomycin on early development. Exp. Cell Res. 33: 368-379.

Gustafson, T. and L. Wolpert. 1967. Cellular movement and contact in sea urchin morphogenesis. Biol. Rev. 42: 442-498.

Hara, K., P. Tydeman, and M.W. Kirschner. 1980. A cytoplasmic clock with the same period as the division cycle in Xenopus eggs. Proc. Natl. Acad. Sci. USA 77: 462-466.

Horstadius, S. 1939. The mechanisms of sea urchin development, studied by operative methods. Biol. Rev. Camb. Phil. Soc. 14: 132-179.

Kohler, G. and G. Milstein. 1975. Continuous cultures of fused cells secreting antibody of predefined specificity. Nature 256: 495-497.

McClay, D.R. and A.F. Chambers. 1978. Identification of four classes of cell surface antigens appearing at gastrulation in sea urchin embryos. Dev. Biol. 63: 179-186.

McClay, D.R., G.W. Cannon, G.M. Wessel, R.D. Fink, and R.B. Marchase. 1983. Patterns of antigenic expression in early sea urchin development. pp. 157. In: Time,

Space and Pattern in Embryonic Development. W. Jeffries and R.A. Raff (eds.). A.R. Liss, N.Y.

McClay, D.R. and G.M. Wessel. 1984. Spatial and temporal appearance and redistribution of cell surface antigens during sea urchin development. pp. 165-184. In: Molecular Biology of Development. E.H. Davidson and R.A. Firtel (eds.). A.R. Liss, N.Y.

Noll, H., V. Matranga, D. Cascino, and L. Vittorelli. 1979. Reconstitution of membranes and embryonic development in dissociated blastula cells of the sea urchin by reinsertion of aggregation-promoting membrane proteins extracted with butanol. Proc. Natl. Acad. Sci., USA 76: 288-292.

Ozaki, H. 1982. Vitellogenesis in the sand dollar Dendraster excentricus. Cell Diff. 11: 315-318.

Spiegel, E. and M. Spiegel. 1980. The internal clock of reaggregating embryonic sea urchin cells. J. Exptl. Zool. 213: 271-281.

Wessel, G.M., R.B. Marchase, and D.R. McClay. 1983. Sequential expression of two germ-layer specific antigens in the sea urchin embryo is coupled to development of the extracellular matrix. J. Cell Biol. 97: 68a.

Wessel, G.M., R.B. Marchase, and D.R. McClay. 1984. Ontogeny of the basal lamina in the sea urchin embryo. Dev. Biol. 103: 235-245.

A SHORT REVIEW OF GERM CELL DETERMINATION IN *DROSOPHILA MELANOGASTER*
Robert E. Boswell

ABSTRACT

The germ plasm of Drosophila melanogaster provides one with a unique opportunity to study the mechanisms involved in cellular determination during early embryogenesis. A variety of experimental procedures have been used to establish a correlation between a localized cytoplasmic component and germ cell determination. Furthermore, a number of recessive maternal effect mutations have been obtained that disrupt germ cell formation. Here, I briefly discuss the experimental evidence indicating that there is a localization of cytoplasmic components required for germ cell formation, and then I describe the phenotype of a new recessive maternal effect mutation.

The properties of mutations of the recessive maternal effect gene tudor (tud) indicate that the gene product of the tudor locus is required for the proper determination of germ cells and for the normal pattern of segmentation in Drosophila melanogaster. In particular, the tud^+ gene product appears to be a necessary component for the proper assembly of the germ plasm.

INTRODUCTION

It is a fundamental concept in the developmental biology of invertebrates that the fate of embryonic cells is governed by cytoplasmic determinants or morphogens localized in the ooplasm (cf. Davidson, 1976). In Drosophila, as in a number of insect orders, a distinctively staining cytoplasmic region of the egg is present at fertilization and subsequently becomes incorporated into the primordial germ cells (cf. Hegner, 1914; Counce, 1973; Mahowald, 1977; Boswell & Mahowald, 1984). A variety of experimental procedures have been used to show that the posterior pole region is essential for germ cell determination (see, for reviews Counce, 1973; Boswell & Mahowald, 1984). Typically, these experiments have involved the selective destruction (Hegner, 1909, 1914; Geigy, 1931) or dispersal (Geyer-Duszynska, 1959; Jazdowska-Zagrodzinska, 1966) of this cytoplasmic region resulting in a failure in pole cell formation (the primordial germ cells). Moreover, it has been possible to restore the ability to produce germ cells by transplanting germ plasm from untreated embryos to embryos in which the germ plasm has been experimentally disrupted (Okada et al., 1974; Warn, 1975).

Pole cells were first traced from their site of origin into the embryonic gonads of insects by Leuckart (1865), Mestchinkoff (1866), Balbiani (1882), and Noack (1901). In Drosophila (Huettner, 1923) and in other insects the germ cells are often the first cells to differentiate (see, for reviews Counce, 1973; Beams & Kessel, 1974; Eddy, 1975; Niewkoop & Sutasurya, 1981). This early differentiation of the germ cells and their formation at topographically defined regions associated with cytoplasmic organelles known as polar granules has led to extensive work on germ cell determination with the intent of defining the requisite factors for germ cell determination.

Experimental and Cytological Studies of the Germ Plasm

Hegner (1908, 1914) was able to destroy the ability of insect embryos to form germ cells by cauterizing the posterior pole plasm prior to the migration of nuclei into the posterior pole. This result indicated to Hegner that pole plasm with its associated polar granule material was required for germ cell determination. Although he and others (see, for review Boswell & Mahowald, 1984) have since attempted to obtain direct evidence for the involvement of polar granules or some polar granule-associated component in germ cell determination, no direct evidence for this or other proposals has been forthcoming.

More recently, experimental embryologists have used a combination of ultraviolet irradiation (UV) and cytoplasmic transplantation to obtain evidence for the the localization of germ cell determinants in Drosophila. UV irradiation of the posterior pole of newly fertilized Drosophila eggs was found by Geigy (1931) to disrupt the ability of embryos to form germ cells. These results have subsequently been confirmed by various researchers (cf. Boswell & Mahowald, 1984). Okada et al. (1974) and Warn (1975) have extended our understanding of the germ plasm by restoring the ability of irradiated embryos to form pole cells by injecting posterior cytoplasm from an unirradiated embryo. They found that only posterior pole plasm was capable of reversing the damage induced by UV irradiation.

Definitive evidence for the localization of cytoplasmic information requisite for the formation of germ cells was provided by Illmensee and Mahowald (1974, 1976). The posterior polar plasm, containing polar granules, was transplanted to the anterior or mid-ventral regions of a recipient pre-blastoderm embryo. The posterior polar plasm, transplanted to an ectopic site within a recipient embryo, led to the formation of histologically discernible pole cells at these ectopic sites. When these putative germ cells were transplanted to the posterior tip of a blastoderm embryo, these cells were able to migrate into the embryonic gonads and it was possible to show that these cells functioned as germ cells. These results have recently been repeated by Niki (pers. comm.).

In Drosophila polar granules consist of electron-dense organelles without limiting membranes (Mahowald, 1962). Because polar granules do exhibit species-specific morphological characteristics in Drosophila (Mahowald, 1968), I will limit my discussion to the analysis of the germ plasm in Drosophila melanogaster. Polar granules can first be identified during the mid-vitellogenic stages of oogenesis (stages 9 and 10 as defined by King, 1970) (Mahowald, 1962). During stages 10 through 12 polar granules increase in size and become associated with one another and mitochondria (Mahowald, 1962). Cytochemical analysis of the germ plasm indicates that the polar granules in mature oocytes consist of protein and RNA (Nicklas, 1959; Mahowald, 1971a; Counce, 1973). Fertilization of the egg leads to a dissociation of the polar granules from the mitochondria and it is then possible, by transmission electron microscopy, to observe polysome-like structures associated with the polar granules (Mahowald, 1968). However, once pole cells form, the polar granules coalesce, and it is no longer possible to observe polysomes or cytochemically detectable RNA (Mahowald, 1971a). In other insects with well defined germ plasms similar fine structure changes in the polar granules are observed

(Alléaumé, 1971; Schwalm et al., 1971; Mahowald, 1975). Although the true function of these structural changes is not known, it has been proposed that these changes are intimately involved with germ cell determination (Mahowald, 1968, 1971a, 1975; Alléaumé, 1971; Schwalm et al., 1971).

Mahowald and his co-workers have attempted to isolate polar granules and their associated RNA using biochemical techniques (Allis et al., 1977; Mahowald, 1977; Mahowald et al., 1979). However, to date, it has not been possible to isolate the RNA associated with the polar granules (Mahowald et al., 1979). Although it has been possible to isolate a 95,000 dalton protein that co-purifies with pole cells and polar granules, it has not been possible to raise antibodies to this protein (Mahowald, pers. comm.).

Genetic and Development Analysis of the Grandchildless-like Mutant, Tudor

An alternative approach to the analysis of germ cell determination, and one that should ultimately provide direct evidence as to the role of particular ooplasmic components in germ cell determination, is a combined genetic, developmental and molecular study of the mechanisms involved in germ cell determination. Previously, mutations that disrupt the formation of germ cells have been identified (Boswell & Mahowald, 1984). These mutations are strict recessive maternal effect mutations, i.e., mutations in which the phenotype of the progeny is dependent solely on the genotype of the mother. These mutants have been termed grandchildless (gs) or grandchildless-like mutations because the progeny derived from the homozygous mutant females are sterile. This sterility has been shown to be due to a failure of pole cell formation (cf. Boswell & Mahowald, 1984). These mutants and their phenotypic characteristics are listed in Table 1. These mutations are not fully penetrant for the grandchildless phenotype and have been recently reviewed (cf. Boswell & Mahowald, 1984). Therefore, these mutations will not be discussed.

Recently, Wieschaus and Nusslein-Volhard (pers. comm.) have isolated a new grandchildless mutant, designated tudor, which has been characterized by Boswell and Mahowald (unpubl. obs.). The tudor mutant is fully penetrant for the grandchildless phenotype, and the sterility is due to a failure in pole cell formation. Like other maternal effect mutations, tudor is pleiotropic, and approximately 40% of the embryos from the homozygous gs females die during embryogenesis and exhibit segmentation-pattern abnormalities. The remaining 60% of the embryos are phenotypically normal, but they are sterile. The germ plasm of six different alleles has been analyzed, and it is found that different alleles contain different amounts of assembled polar granule material. The severity of the segmentation-pattern abnormalities observed in the lethal embryos and the sterility due to a failure in pole cell formation correlate with the amount of assembled polar granule material in the germ plasm. For example, one allele of tudor is not grandchildless as a homozygote, but it is grandchildless when heterozygous with other tudor alleles. This leaky allele expresses few, if any, of the segmentation-pattern abnormalities observed in embryos derived from homozygous females of the other tudor alleles. Therefore, mutations at the tudor locus disrupt the assembly of the germ plasm in a quantitative fashion. Independently of how one defines the role of polar granules in the determination of pole cells, it is clear that mutants at the tudor locus disrupt the normal assembly of the germ plasm which, in turn, results in the failure to properly localize the germ cell determinants to the posterior pole plasm.

Table 1. Grandchildless mutations in Drosophila.

Name	Location	Penetrance (% agametic)	Embryonic Development
gs (subobscura)	autosomal	90 - 100%	Fragile blastoderm; very little lethality (Mahowald et al., 1979)
gs^{87}	1 - 20	ts 17% at 15°C 60 at 28.5°C	ts 38% lethal at 15°C 83% lethal at 28.5°C (Thierry-Mieg, 1976)
par	3B1	sterile at 29°C 40% at 23°C	60% embryonic lethality at 23°C (Thierry-Mieg, 1982)
gs(2)M	between S & Sp	ts 40% agametic at 28.5°C	40 - 60% late embryo lethals at 28.5°C (Mariol, 1981)
gs(1)N126	1 - 33.8	ts 25% no pole cells 18 - 50% no pole cells (Niki & Okada, 1981)	
gs(1)N441	1 - 39.6	ts 25°C 70% no pole cells 18°C 20% no pole cells (Niki & Okada, 1981)	
ag	1 - 20	semi-dominant ts 50% agametic at 25°C	no embryonic defects (Engstrom et al., 1982)

OUTLOOK

The excellent genetics and cytogenetics available in Drosophila melanogaster allow one to induce mutations in particular developmental processes and then to isolate the gene of interest at the molecular level. I have briefly described the evidence for the localization of informational molecules required for the determination of germ cells. The genetic and developmental analysis of a fully penetrant grandchildless mutation, tudor, has been presented. Mutations exhibiting a fully penetrant grandchildless phenotype are being selected and characterized. These mutations should provide one with significant insights into the molecular mechanisms involved in the localization of determinants to particular ooplasmic regions and to the molecular mechanisms by which these determinants specify cell fates during early embryogenesis.

LITERATURE CITED

Alléaumé, N. 1971. Contribution à l ánalyse expérimentale des facteurs de la détermination et de la différenciation des ébauches dans le germe des Diptères supérieurs (Calliphora erythrocephala Meig. et Drosophila melanogaster Meig.). Thèse, Université de Bordeaux.I.

Allis, C.D., G.L. Waring, and A.P. Mahowald. 1977. Mass isolation of pole cells from Drosophila melanogaster. Dev. Biol. 56: 372-381.

Balbiani, E.G. 1882. Sur la signification des cellules polaires des insectes. C.R. Acad. Sci., Paris 95: 927-929.

Beams, H.W. and R.G. Kessel. 1974. The problem of germ cell determinants. Int. Rev. Cytol. 39: 413-479.

Boswell, R.E. and A.P. Mahowald. 1985. Cytoplasmic determinants in embryogenesis. pp. 387-405. In: Comprehensive Insect Physiology, Biochemistry and Pharmacology, Vol. I. G.A. Kerkut and L.I. Gilbert (eds.). Pergamon Press, Elmsford.

Counce, S.J. 1973. The casual analysis of insect embryogenesis. pp. 1-156. In: Developmental Systems: Insects, Vol. 2. S.J. Counce and C.H. Waddington (eds.) Academic Press, N.Y.

Davidson, E. 1976. Gene Activity in Early Development. Academic Press, N.Y.

Eddy, E.M. 1975. Germ plasm and the differentiation of the germ-line. Int. Rev. Cytol. 43: 229-280.

Engstrom, L., J.H. Caulton, E.M. Underwood, and A.P. Mahowald. 1982. Developmental lesions in the agametic mutant of Drosophila melanogaster. Dev. Biol. 91: 163-170.
Fielding, C.J. 1967. Developmental genetics of the mutant grandchildless of Drosophila subobscura. J. Embryol. Exp. Morph. 17: 375-384.
Geigy, R. 1931. Action de l'ultra-violet sur le pole germinal dans l'oeuf de Drosophila melanogaster (castration et mutabilité). Rev. Suisse Zool. 38: 187-288.
Geyer-Duszynska, I. 1959. Experimental research on chromosome elimination in Cecidomyidae (Diptera). J. Exp. Zool. 141: 391-448.
Hegner, R.W. 1908. Effects of removing the germ-cell determinants from eggs of some chrysomelid beetles. Preliminary Report. Biol. Bull. 16: 19-26.
Hegner, R.W. 1909. The origin and early history of the germ cells of some chrysomelid beetles. J. Morphol. 20: 231-296.
Hegner, R.W. 1914. Studies on germ cells. I. The history of the germ cells in insects with special reference to the "Keimbahn" determinants. J. Morphol. 25: 375-509.
Huettner, A.F. 1923. The origin of germ cells in Drosophila melanogaster. J. Morph. 37: 385-423.
Jazdowska-Zagrodzinska, B. 1966. Experimental studies on the role of 'polar granules' in the segregation of pole cells in Drosophila melanogaster. J. Embryol. Exp. Morphol. 16: 391-399.
King, R.C. 1970. Ovarian Development in Drosophila melanogaster. Academic Press, N.Y.
Leuckart, R. 1865. Die ungeschlechtliche Fortpflanzung der ceidomyien larven. Arch. Naturgesch. 32: 286-303.
Mahowald, A.P. 1962. Fine structure of pole cells and polar granules in Drosophila melanogaster. J. Exp. Zool. 151: 201-215.
Mahowald, A.P. 1968. Polar granules of Drosophila II. Ultrastructural changes during early embryogenesis. J. Exp. Zool. 167: 237-262.
Mahowald, A.P. 1971a. Polar granules of Drosophila IV. The loss of RNA from polar granules during early stages of embryogenesis. J. Exp. Zool. 176: 329-344.
Mahowald, A.P. 1971b. Origin and continuity of polar granules. pp. 158-169. In: Results and Problems in Cell Differentiation, Vol. 2. J. Rienert and H. Usprung (eds.). Springer Verlag, Berlin.

Mahowald, A.P. 1975. Ultrastructural changes in the germ plasm during the life cycle of Miastor (Ceidomyidae, Diptera). Wilhelm Roux Arch. 176: 223-240.

Mahowald, A.P. 1977. The germ plasm of Drosophila: An experimental system for the analysis of determination. Amer. Zool. 17: 551-563.

Mahowald, A.P., J.H. Caulton, and W.J. Gehring. 1978. Ultrastructural studies of oocytes and embryos derived from female flies carrying the grandchildless mutation in Drosophila subobscura. Dev. Biol. 69: 118-132.

Mahowald, A.P., C.D. Allis, K.M. Karrer, E.M. Underwood, and G.L. Waring. 1979. Germ plasm and pole cells of Drosophila. pp. 127-146 In: Determinants of Spatial Organization. S. Subtelny and I.R. Konigsberg (eds.). Academic Press, London.

Mahowald, A.P. and R.E. Boswell. 1983. Germ plasm and germ cell development in invertebrates. pp. 1-17. In: Current Problems in Germ Cells. A. McLaren and C.C. Wylie (eds.). Cambridge University Press, Cambridge.

Mahowald, A.P., K. Illmensee, and F.R. Truner. 1976. Interspecific transplantation of polar plasm between Drosophila embryos. J. Cell Biol. 70: 358-373.

Mariol, M.-C. 1981. Genetic and developmental studies of a new grandchildless mutant of Drosophila melanogaster. Mol. Gen. Genet. 181: 505-511.

Mestchinkoff, E. 1866. Embryologische studien an insekten. Z. wiss. Zool. 16: 389-500.

Nicklas, R.B. 1959. An experimental and descriptive study of chromosome elimination in Miastor sp. (Cecidomydae, Diptera). Chromosoma 10: 301-336.

Nieuwkoop, P.D. and L.A. Sutasurya. 1981. Primordial Germ Cells in the Invertebrates. Cambridge University Press, N.Y.

Niki, Y. and M. Okada. 1981. Isolation and characterization of grandchildless-like mutants in Drosophila melanogaster. Wilhelm Roux Archiv. 190: 1-10.

Noack, W. 1901. Beiträge zur Entwicklungsgeschichte der Muschiden. Z. wiss. Zool. 70: 1-57.

Okada, M., I.A. Kleinman, and H.A. Schneiderman. 1974. Restoration of fertility in sterilized Drosophila eggs by transplantation of polar cytoplasm. Dev. Biol. 37: 43-54.

Poulson, D.F. and D.F. Waterhouse. 1960. Experimental studies on pole cells and midgut differentiation in Diptera. Aust. J. Biol. Sci. 13: 541-567.

Schwalm, F.E., R. Simpson, and H.A. Bender. 1971. Early development of the kelp fly, Coelopa frigida (Diptera) ultrastructural changes within the polar granules during pole cell formation. Wilhelm Roux Archiv. 166: 205-218.

Smith, L.D. and M.A. Williams. 1975. Germinal plasm and determination of the primordial germ cells. pp. 3-24. In: The Developmental Biology of Reproduction. C.L. Markert and J. Papaconstantinou (eds.). Academic Press, N.Y.

Warn, R. 1975. Restoration of the capacity to form pole cells in UV irradiated Drosophila embryos. J. Embryol. Exp. Morph. 33: 1003-1011.

THE MANY MOTORS OF MORPHOGENESIS: THE ROLE OF MUSCLES, CILIA, AND MICROFILAMENTS IN THE METAMORPHOSIS OF MARINE BRYOZOANS

Christopher G. Reed

ABSTRACT

The cataclysmic metamorphoses of marine invertebrates are presented as opportune systems to investigate the mechanistic bases of rapid morphogenetic movements. Ultrastructural and experimental evidence is analyzed in an attempt to elucidate mechanisms of rapid morphogenetic movements during the metamorphosis of marine bryozoans. A spectrum of hitherto undescribed morphogenetic movements is found to depend upon a variety of mechanisms, from muscle and microfilament contractility to ciliary motility. These studies demonstrate that, while the motors of morphogenesis ultimately are resolvable as cytoskeletal elements, they may assume novel associations and interactions in the metamorphoses of marine invertebrates. The diversity of larval form and the wide range of metamorphic patterns found within the phylum Bryozoa alone indicates that many variations in the mediation of rapid morphogenetic movements still await discovery.

INTRODUCTION

Morphogenesis, or the generation of form, is an integral aspect of development at the cellular, tissue, organ, and organismal levels of organization. Although it is a developmental phenomenon that transcends stages of development as well as levels of organization, evidence is accumulating that indicates that morphogenesis has a common mechanistic basis. Whether it is the movement of individual cells or sheets of cells, the morphogenetic movements that shape the embryo depend upon the integrity and function of intrinsic populations of cytoskeletal elements (such as microfilaments and microtubules) and on the selective, differential adhesion of the cells to each other. Whereas the mechanisms that shape the embryo have been awarded a great deal of investigative attention, the cataclysmic metamorphosis characteristic of many sessile marine invertebrates is another developmental phase in the ontogeny of organisms with indirect development that affords an opportunity to explore a wide range of novel morphogenetic movements of unknown mechanistic bases.

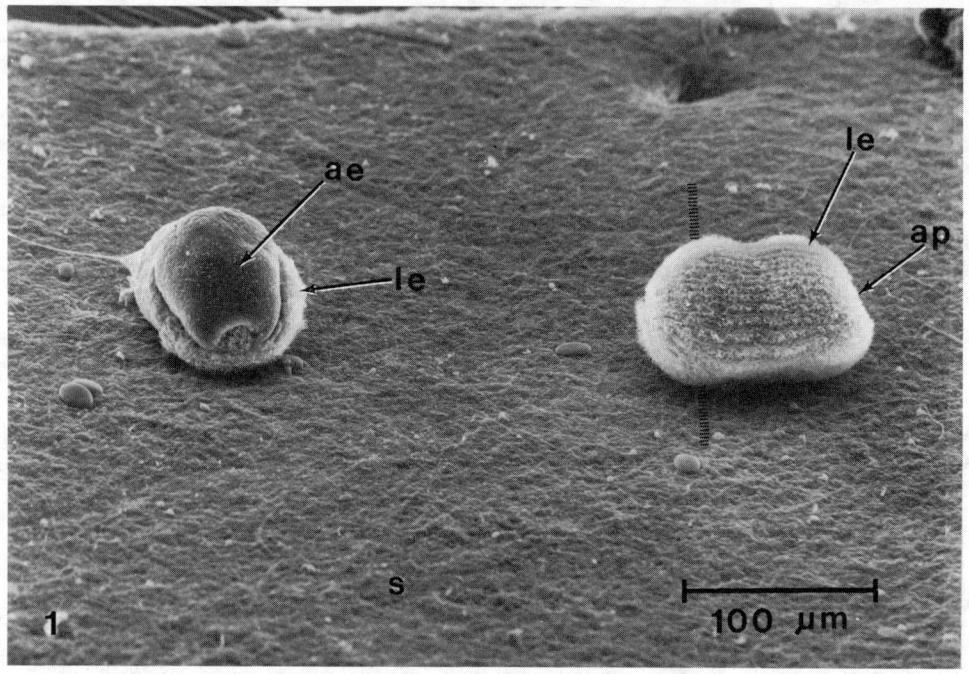

Fig. 1. Scanning electron micrograph (SEM) of the larvae of the marine bryozoan Bowerbankia gracilis before (right) and during metamorphosis (left). The larva exploring the substratum (s) on the right (lateral view) is covered by the densely ciliated larval epithelium (le). The cilia constitute the larval locomotory organ and propel the larva apical pole (ap) foremost. The dashed line indicates the plane of section illustrated in Figure 3. Within 3 min of its initial attachment to the substratum, the metamorphosing larva (left) has almost completely internalized the ciliated larval epithelium and covered itself with the presumptive adult epithelium (ae). x 320

Many benthic marine invertebrates possess free-swimming larvae that settle after a variable planktonic period and metamorphose into the adult form (see Chia & Rice, 1978, for review). Metamorphosis is a critical period in the life cycle of the organism; the larva must rapidly transform itself from a mobile, free-swimming organism specialized for a pelagic existence into a sedentary or sessile adult adapted to the benthic biotope (Fig. 1). Larvae may be considered composite organisms composed of both transitory larval tissues and rudimentary or precociously developed adult tissues that will become functional only after metamorphosis. In addition, certain tissues or organs may function only during metamorphosis to effect the rapid shape changes and tissue rearrangements that are characteristic of this stage of development (Fig. 1).

Metamorphosis, therefore, is typically characterized by a period of rapid morphogenesis, followed or accompanied by the inactivation and degeneration of larval tissues and the simultaneous differentiation and/or activation of adult rudiments and preformed tissues. The wide range of metamorphic patterns, even within a single phylum, and the rapidity with which this transformative morphogenesis is effected, indicate that novel mechanisms and tissue interactions may underly the rapid morphogenetic movements that drive metamorphosis. Recent experimental and ultrastructural evidence on the metamorphosis of marine bryozoans supports this contention.

Bryozoans are colonial organisms that, with few exceptions, brood their embryos and release them as lecithotrophic larvae in response to light. The non-feeding lecithotrophic larvae have a short free-swimming existence (between 1-12 h for most species), and are competent to metamorphose soon after release. These aspects of the reproductive and developmental biology of bryozoans make them particularly convenient subjects for studies of metamorphosis. Sexually reproductive colonies may be collected and maintained in marine aquaria and induced, on a daily basis, to release thousands of larvae at the same stage of development. Because the free-swimming larvae of most bryozoans are positively phototactic during the initial part of their natatory period, they may be easily collected from the lighted side of the aquarium. Moreover, the larvae may be induced to settle and metamorphose readily under laboratory conditions. Finally, metamorphosis can be resolved into a specific sequence of rapid morphogenetic movements, each individually amenable to experimental and ultrastructural analysis (see Reed, in press a).

Despite the facility with which this system can be manipulated, and the potential rewards inherent in such investigations, few studies have been conducted on the mechanisms of rapid morphogenetic movements during metamorphosis. In this paper, recent experimental and ultrastructural evidence is analyzed in an attempt to elucidate the mechanisms that underlie the morphogenetic movements that effect attachment of the larva to the substratum and the internalization of the larval ciliated epithelium. A spectrum of hitherto undescribed morphogenetic movements is found to rely upon a range of mechanisms, from muscle- and microfilament-mediated contractions to ciliary motility. The elucidation of these mechanisms, while adding to the reservoir of known forces that drive morphogenesis, also demonstrates the need to consider each morphogenetic movement within the overall context of the developing system.

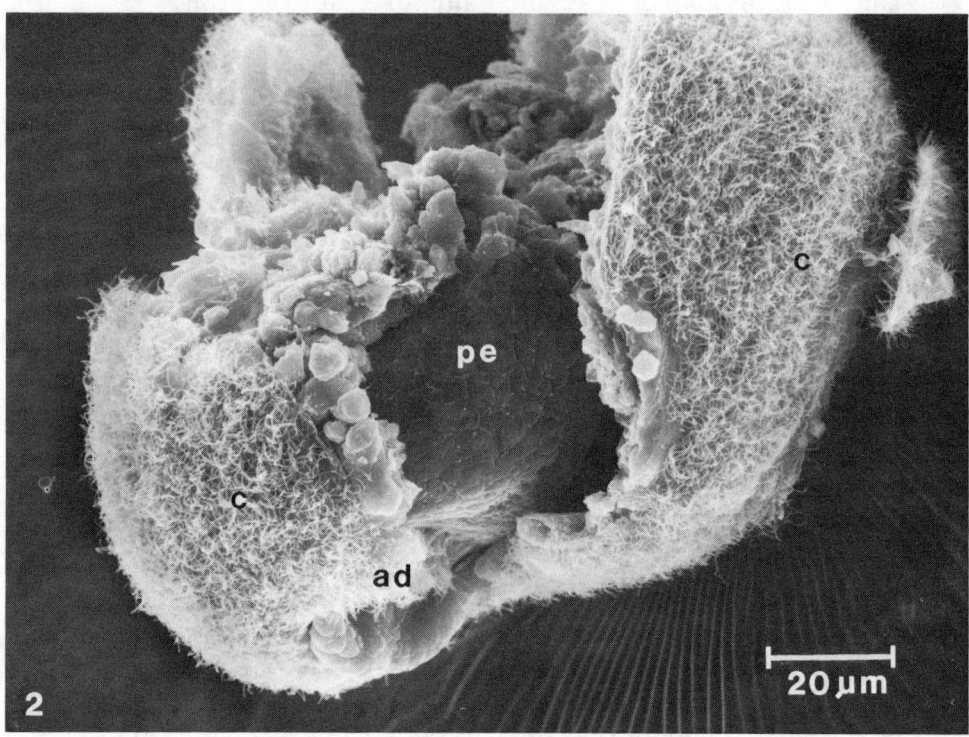

Fig. 2. Bowerbankia. SEM of a larva broken open to reveal the infolded, preformed adult epithelium, or pallial epithelium (pe) around the apical disc (ad). The ciliated larval locomotory organ or corona (c) forms most of the larval surface. x 1100.

The Role of Muscles in Morphogenesis

The larva of the ctenostome bryozoan Bowerbankia gracilis is barrel-shaped and covered almost entirely by a densely ciliated epithelium that constitutes the locomotory organ (called the corona; Fig. 1; Reed & Cloney, 1982a) (Fig. 2). At the apical end of the larva, however, there is an extensive circular infolding of epithelium around a ciliated sensory patch called the apical disc. The nonciliated infolded epithelium constitutes the presumptive adult epithelium, or pallial epithelium, which at the proper moment during metamorphosis will be exposed and become functional (Figs. 1 & 2). At the opposite end of the larva is a smaller infolding of glandular epithelium that forms the adhesive organ that attaches the larva to the substratum at the onset of metamorphosis. Beneath the infolded glandular epithelium lies a hemispherical layer of undifferentiated cells that form part of the rudiment of the adult digestive tract (Fig. 3). The layer of undifferentiated cells is embraced on its convex surface by a network of muscle fibers called the rete muscularis (Figs. 3 & 4).

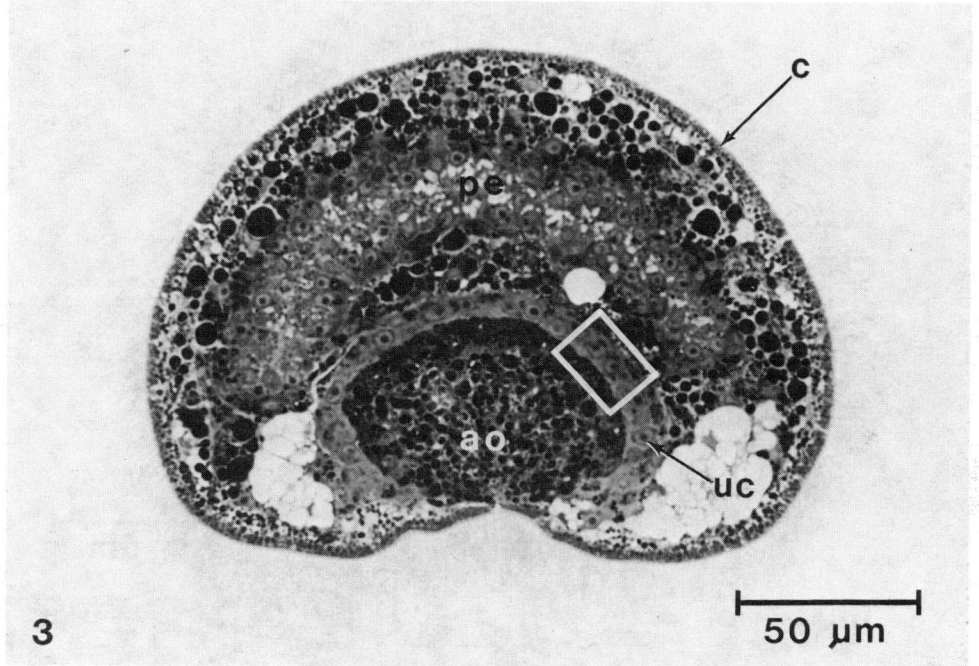

Fig. 3. <u>Bowerbankia</u>. Photomicrograph of a transverse section through a larva at the level indicated in Figure 1. The infolded glandular epithelium that forms the attachment organ (ao) is embraced by a cup-shaped layer of undifferentiated cells (uc). The rectangle indicates the area illustrated in Figure 5. x 560 (after Reed & Cloney, 1982a).

Attachment is effected by the sudden eversion of the infolded glandular epithelium against the substratum and the secretion of the adhesive (Reed & Cloney, 1982b). As the glandular epithelium is everted, the underlying hemispherical layer of undifferentiated cells flattens and doubles in thickness (Fig. 4). Ultrastructural evidence reveals that the individual cells of the layer change shape from short columnar cells (8 µm tall by 4 µm wide) to flask-shaped cells (16 to 20 µm tall) as the layer flattens (Figs. 5 & 6). The large, bulbous ends of the flask-shaped cells bulge into the basal surface of the infolded glandular epithelium, forcing it out against the substratum (Reed, In press b) (Fig. 5).

The force for this morphogenetic movement does not appear to be generated within the undifferentiated cells; the constricted ends of the flask-shaped cells lack the microfilaments associated with intrinsic cell shape deformations, and exposure to concentrations of cytochalasin B ranging from 0.25 to 5.0 µg/ml does not affect the change in cell shape or

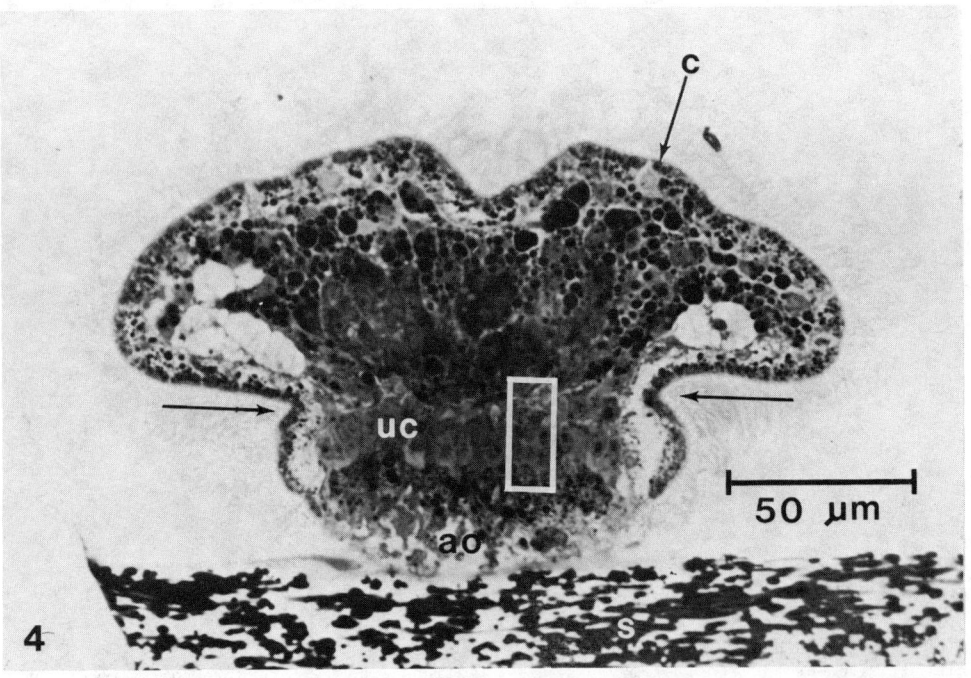

Fig. 4. <u>Bowerbankia</u>. Photomicrograph of a transverse section through a metamorphosing larva during attachment (at the same level as in Fig. 3). The larva is constricted (arrows) in the plane of the flattened layer of undifferentiated cells (uc), and the glandular epithelium of the attachment organ (ao) is pressed against the substratum (s). The rectangle indicates the area illustrated in Figure 6. x 460 (Figs. 4-7 after Reed, in press b).

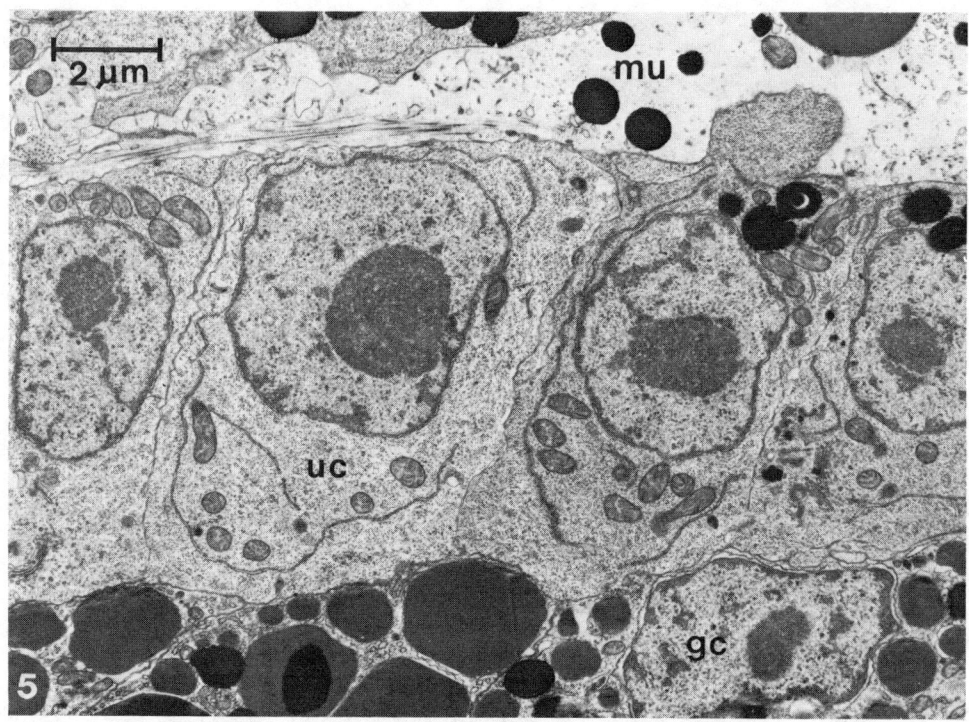

Fig. 5. Bowerbankia. Transmission electron micrograph (TEM) of the layer of undifferentiated cells (uc) before eversion of the glandular cells (gc) of the attachment organ (area indicated by the rectangle in Fig. 3). The undifferentiated cells are embraced by the muscle fibers (mu) of the rete muscularis. x 7870.

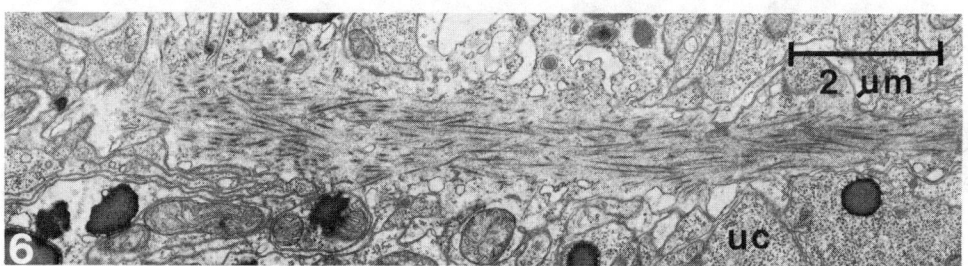

Fig. 6. Bowerbankia. TEM of a contracted muscle fiber of the rete muscularis during eversion of the glandular epithelium. The lateral sarcolemma is folded and the myofilament field consists of an irregular array of thick and thin myofilaments. x 11,630.

eversion of the glandular epithelium. Rather, the force appears to be generated extrinsically by the contraction of the muscle fibers that embrace the hemispherical layer of undifferentiated cells. As the layer flattens and the

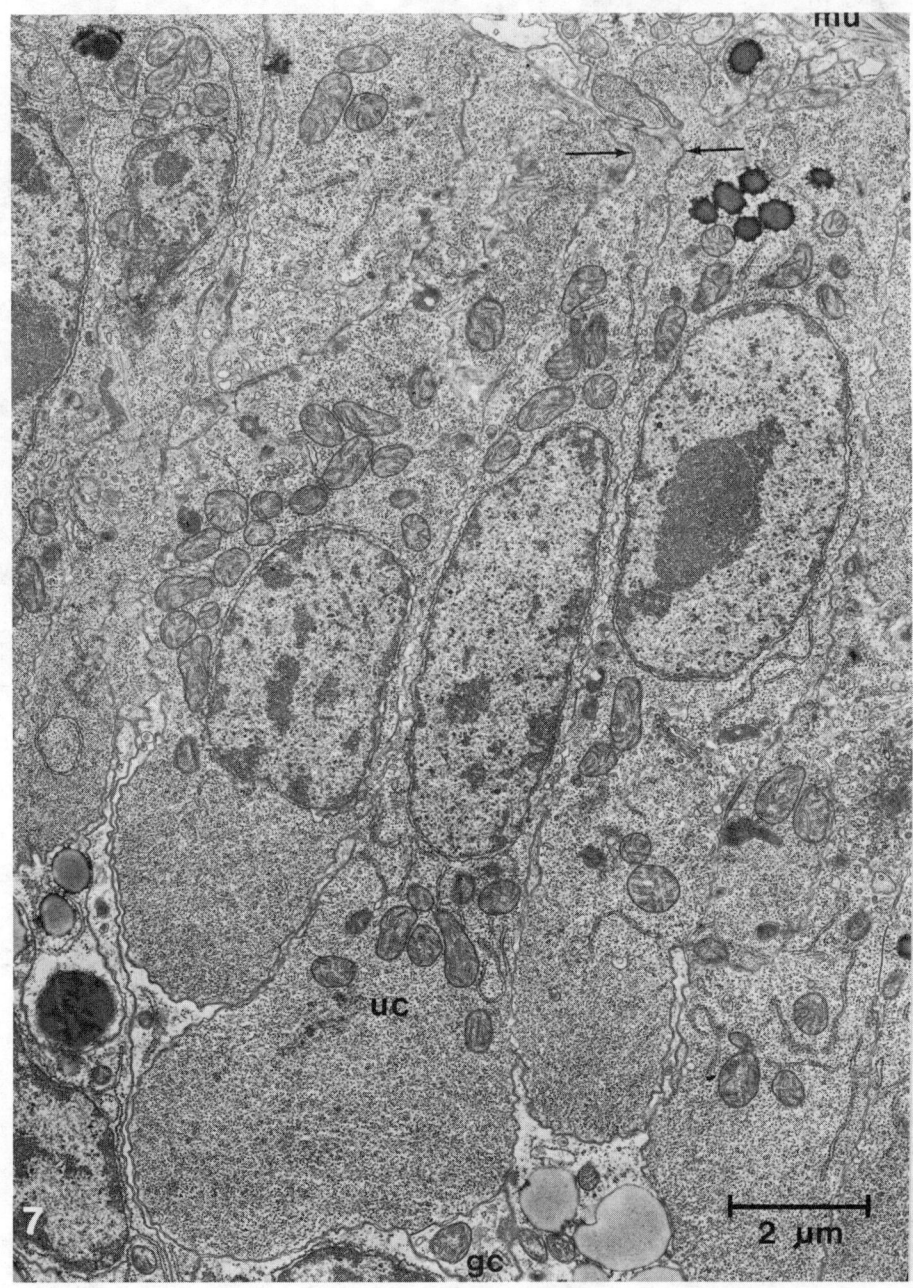

Fig. 7. Bowerbankia. TEM of the layer of undifferentiated cells during eversion of the glandular cells (gc) of the attachment organ (area indicated by the rectangle in Fig. 4). The undifferentiated cells have become flask-shaped with their narrow ends (arrows) next to the muscle fibers (mu) of the rete muscularis. The bulbous ends of the cells bulge into the glandular cells of the attachment organ. x 11,080.

glandular epithelium is everted, the subjacent muscle fibers of the rete muscularis shorten and become packed with an irregular array of thick and thin myofilaments (Fig. 7). The lateral sarcolemmata of the muscle fibers also become highly folded, indicating that the muscle fibers are in a contracted state. This is not a condition stimulated by fixation, because the muscle fibers are not contracted in free-swimming larvae fixed at the same time after release. In addition, the contraction of the rete muscularis and the coincident eversion of the adhesive organ can be inhibited by exposure to isotonic $MgCl_2$, which is believed to interfere with the innervation of the muscle fibers. These results indicate that attachment to the substratum by eversion of the glandular adhesive organ in Bowerbankia gracilis is effected by a muscle-mediated morphogenetic movement. The sudden change in shape of the undifferentiated cells is believed to transmit the extrinsic force of contraction of the muscle fibers to the infolded glandular epithelium, causing it to evert (Figs. 6 & 7).

The Role of Cilia in Morphogenesis

At the same time that eversion of the adhesive organ is occurring, the cilia of the larval epithelium abruptly reverse their direction of beat (Reed & Cloney, 1980, 1982b). The immediate effect of this action is that the adhesive secreted by the everted glandular epithelium is wafted up over the metamorphosing larva, encasing it in a tent throughout the subsequent morphogenetic movements (Figs. 8 & 9). In the next minute, the larva essentially turns itself inside out, internalizing the ciliated larval epithelium while at the same time exposing the incipient adult epithelium (Figs. 10 & 11). This dramatic reorganization of the larva takes only 60 sec at 12.5°C (Figs. 8-11).

What generates the forces that drive this dramatic reorganization of the larva? The rapidity with which the ciliated larval epithelium is involuted implicates muscle contraction, a mechanism that already has been demonstrated for earlier morphogenetic movements. Initial observations appeared to support this hypothesis. In addition to the rete muscularis, the larva has an equatorial muscle ring and an anterior axial muscle that runs the length of the larva. At the beginning of involution of the larval epithelium, the anterior side of the larva shortens to about 65% of its original length. Observations made by light microscopy and transmission electron microscopy indicate that the anterior

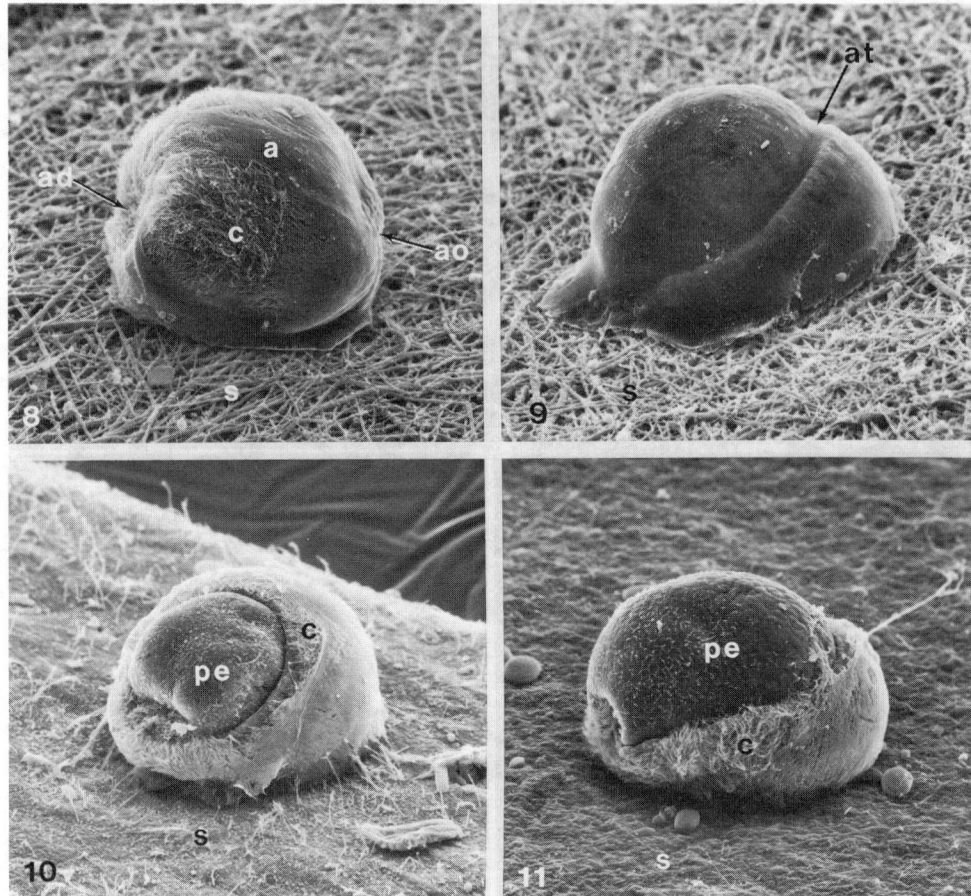

Fig. 8. Bowerbankia. SEM of a larva (lateral view) at the onset of metamorphosis. The reversed beating of the coronal cilia (c) wafts the adhesive (a) secreted by the attachment organ (ao) up over the metamorphosing larva toward the apical disc (ad). x 470 (Figs. 8-11 after Reed & Cloney, 1982b).

Fig. 9. Bowerbankia. SEM of a metamorphosing larva (lateral view) within the adhesive tent (at) during coronal involution. x 460.

Fig. 10. Bowerbankia. SEM of metamorphosing larva (apico-lateral view) during early coronal involution with the adhesive tent partially removed. x 470.

Fig. 11. Bowerbankia. SEM of a metamorphosing larva (lateral view) near the end of coronal involution with the adhesive tent removed. Most of the pallial epithelium (pe) is exposed. x 470.

axial muscle undergoes a corresponding contraction. The shortened anterior side of the larva, however, subsequently undergoes a rotation of about 90° as the ciliated larval epithelium involutes during the last 40 sec of this stage of metamorphosis. Examination of thick and thin serial sections has not revealed any other musculature in a position that could produce the force for rotation of the anterior side of the larva and the complete involution of the larval epithelium. Moreover, involution is not inhibited when the metamorphosing larva is exposed to isotonic $MgCl_2$ during the last 40 sec of this morphogenetic movement. These results indicate that, although muscle contraction may be involved in the initiation of coronal involution, it is not responsible for the completion of this morphogenetic movement.

What are possible alternatives if the driving force for involution of the larval epithelium is not generated extrinsically by muscle contractions? One possibility is a temporally synchronized assembly, alignment, and contraction of microfilaments, either in the corona or in adjacent tissues. The corona, however, does not change length during involution, and no microfilaments have been observed by transmission electron microscopy. In addition, cytochalasin B applied in the range of concentrations described previously does not affect the involution of the ciliated larval epithelium. Microfilaments, then, do not appear to be involved in coronal involution.

Another possibility is that the propulsive force for involution of the ciliated larval epithelium is generated by the cilia themselves. Although cilia are ubiquitous organelles of motility, they have never been shown to mediate morphogenesis. There are several lines of evidence, however, to indicate that ciliary motility may be a viable mechanism for this morphogenetic movement (Reed & Cloney, in prep.).

First, the effective strokes of the cilia after reversal are in the appropriate direction to move the corona toward the substratum and up into the interior of the metamorphosing larva. Second, the cilia continue to beat apically in the reverse direction throughout this morphogenetic movement, and may be seen beating within the metamorphosing larva for a short time after their internalization. Finally, the hardened adhesive tent and the substratum provide the required surface for the cilia to beat against to generate the force necessary to move the larval epithelium. The reversed beating of the cilia against the adhesive tent appears to be a plausible mechanism to provide the propulsive force for the completion of coronal involution.

How might this hypothesis be tested? The temporal separation of the postulated muscular and ciliary mechanisms for coronal involution make it possible to test this model experimentally. First it is necessary to demonstrate that the cilia are interacting with the adhesive tent, and that the intact adhesive tent is essential for the successful completion of this movement. This may be demonstrated readily by mechanically removing the adhesive tent (with a fine tungsten needle) from metamorphosing larvae during involution of the ciliated larval epithelium. In every case the larval epithelium stops involuting, although the cilia continue to beat apically, as indicated by the movement of carborundum particles sprinkled on the metamorphosing larva. In addition, metamorphosing larvae may be dislodged from the substratum, although they remain within the adhesive tent. In these cases the larval epithelium does not involute, but the metamorphosing larva rotates posteriorly within the tent, opposite to the direction of beat of the cilia. Moreover, when rotating larvae are removed from their tents, they move posteriorly but stop rotating. These observations indicate that the adhesive tent provides a surface for the cilia to beat against. The ability of the cilia to rotate the entire larva within the adhesive tent indicates that ciliary motility provides a sufficient force to propel the involution of the larval epithelium.

Another way to test the role of ciliary motility in this morphogenetic movement is to inhibit the beat of the cilia during coronal involution. This may be done by hypertonic shock or by exposure to nickel ions. The locomotory cilia of free-swimming larvae, as well as those of metamorphosing larvae, are rendered immotile and easily detachable by exposure to double-strength sea water. When the cilia are immobilized in this fashion during coronal involution, the larval epithelium abruptly ceases its movement toward the substratum. Unfortunately, the cilia are irreversibly damaged, so that this inhibition of coronal involution is not reversible upon transfer of the specimen back to normal sea water. This experiment does demonstrate, however, that the structural and functional integrity of the cilia is necessary for the movement to occur.

Unlike osmotic shock, exposure to nickel ions reversibly inhibits ciliary motility in many organisms, including the larvae of Bowerbankia gracilis. Coronal involution ceases when ciliary motility is altered by exposure to 1% $NiCl_2$ in sea water during metamorphosis. Upon rinsing with fresh sea water, however, normal ciliary motility is restored, along with the resumption of coronal involution. Again, this experimental evidence indicates that the reversed beating of the cilia provides the force for this morphogenetic movement.

Recent investigations on the biochemical composition of cilia and the interaction between the axoneme and the ciliary membrane may allow for a more specific inhibition of ciliary motility to further test this hypothesis. The lipophilic photochemical 4,4'-dithiobisphenylazide has been used effectively to crosslink specific ciliary membrane proteins and inhibit ciliary motility in the ciliate Tetrahymena and in the scallop Aequipecten (Dentler et al., 1980). The specificity of this photochemical for cilia and the ease with which it is handled offer exciting possibilities for its experimental application to metamorphosing larvae.

The hypothesis that the cilia, beating against the adhesive tent, provide the force for involution of the corona can be further tested by controlling the direction of ciliary beat. This may be done by manipulating the concentration of calcium ions in the sea water. It is well documented that ciliary reversal in a variety of organisms is triggered by a sudden influx of extracellular calcium ions upon depolarization of the ciliary membrane (Naitoh, 1968; Eckert, 1972; Naitoh & Kaneko, 1972). The larvae of Bowerbankia may be induced to swim backward by artificially depolarizing the ciliated epithelial cells by exposure to 1% KCl in sea water, although under normal conditions ciliary reversal occurs only at the onset of metamorphosis. In normally metamorphosing larvae, the cilia continue to beat in the reversed direction throughout the period of rapid morphogenetic movements. If the cilia are indeed responsible for coronal involution, the movement of the corona should reverse when the direction of ciliary beat is reversed. This is what happens when metamorphosing larvae are exposed to calcium-free sea water during coronal involution. The coronal cilia reorient and beat in the original direction, as in the free-swimming larva, and the corona stops involuting and begins to move back up over the larva, recovering the exposed adult epithelium. This reversal of coronal involution demonstrates the sufficiency of the cilia to move the larval epithelium. Moreover, the ability to redirect movement of the larval epithelium according to direction of beat of the cilia provides further strong experimental evidence that the cilia are responsible for generating the force that drives this dramatic reorganization of the metamorphosing larva.

In summary, both the empirical and experimental evidence is consistent with the hypothesis that the driving force for coronal involution is generated in part by the reversed beating of the cilia against the adhesive tent. These observations and experimental results indicate that a novel mechanism for morphogenesis has been discovered. Furthermore, the occurrence of ciliary reversal and adhesive tent formation in the metamorphoses of other marine bryozoans indicates that this mechanism of morphogenesis may be widespread within the phylum (Reed, 1981; Reed & Woollacott, 1982).

The Role of Microfilaments in Morphogenesis

In Bowerbankia, the infolded pallial epithelium at the apical pole of the larva forms the adult epithelium and is exposed as the ciliated larval epithelium involutes (Hondt, 1977a; Reed & Cloney, 1982b). In some species, however, the adult epithelium originates entirely from the opposite end of the larva. These species have a unique pattern of metamorphosis, replete with a novel repertoire of morphogenetic movements that clothe the metamorphosing larva with the adult epithelium. This pattern of metamorphosis is exemplified by the cheilostome superfamily Cellularioidea, of which Bugula neritina is a member (Woollacott & Zimmer, 1971, 1978; Lyke et al., 1983).

The larval morphology and the metamorphosis of Bugula neritina differs radically from that of Bowerbankia. The apical infolded pallial epithelium forms only a shallow trough around the apical disc, instead of the extensive invagination that occurs in Bowerbankia. Correspondingly, the glandular invagination at the opposite pole in Bugula is much larger, forming not only the attachment organ, but the presumptive adult epithelium as well. At the onset of metamorphosis, the adult epithelium is everted against the substratum, forming a circular disc composed of the adult epithelium folded upon itself around the base of the metamorphosing larva (Fig. 12). The ciliated larval epithelium subsequently is internalized above the everted adult epithelium. At the end of coronal involution, the apical hemisphere of the metamorphosing larva is covered transitorily by the squamous pallial epithelium, which contacts and adheres to the margin of the basal disc of adult epithelium (Fig. 13). Then ensues a remarkable morphogenetic movement unique to this particular group of bryozoans: the pallial epithelium recedes toward the apical pole as the presumptive adult epithelium slowly rises over the apical hemisphere and completely covers the metamorphosing larva (Figs. 12-15).

MORPHOGENESIS OF BRYOZOANS   211

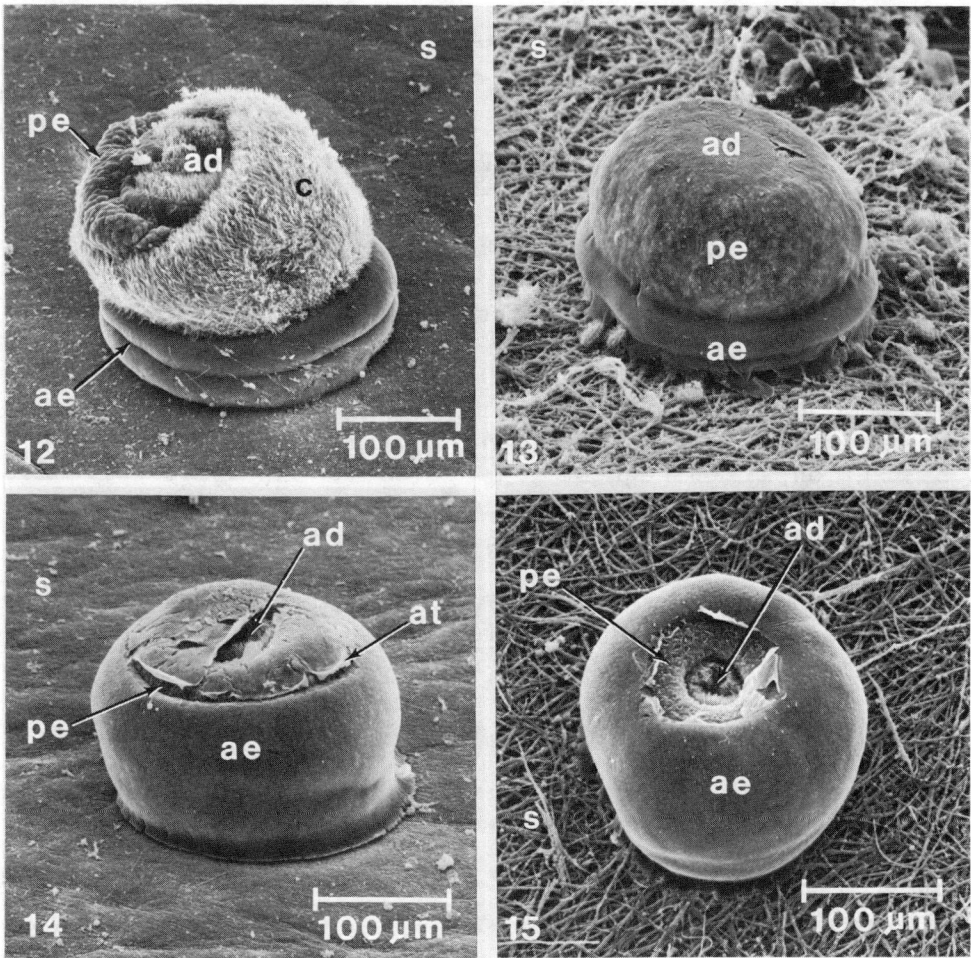

Fig. 12.   SEM of a metamorphosing larva (lateral view) of the bryozoan Bugula neritina at the beginning of coronal involution with the adhesive tent removed. The adult epithelium (ae) is temporarily folded upon itself at the base of the metamorphosing larva. The pallial epithelium (pe) is being exposed as the corona (c) involutes above the adult epithelium. x 200 (Reed & Woollacott, 1982).

Fig. 13.   Bugula. SEM of a metamorphosing larva within the adhesive tent at the end of coronal involution. The apical hemisphere is covered by the pallial epithelium (pe), which contacts the adult epithelium (ae) to form an equatorial suture line. x 220 (Figs. 13-17 after Reed & Woollacott, 1983).

Fig. 14.   Bugula. SEM of a metamorphosing larva during elevation of the adult epithelium (ae). The pallial epithelium (pe), partially covered by the adhesive tent, is receding toward the apical disc (ad). x 220.

Fig. 15.   Bugula. SEM of a metamorphosing larva nearly covered by the adult epithelium (ae). The pallial epithelium (pe) has receded to form a narrow annulus around the apical disc (ad). x 220.

What is the mechanistic basis for this dramatic morphogenetic movement? Woollacott and Zimmer (1971) have hypothesized that it is mediated by microfilaments in the pallial epithelium. During elevation of the adult epithelium, which takes about 2 min at 20°C, the receding pallial epithelium thickens and develops numerous apical folds (Fig. 16). Examination by transmission electron microscopy reveals that there is an accumulation of microfilaments 5.5 nm in diameter in an apical band 0.3 µm thick in the thickening pallial cells during this movement (Reed & Woollacott, 1983; Figs. 17 & 18). No microfilaments have been found in this position immediately before or after this morphogenetic movement. These observations support the hypothesis that a temporal assembly of microfilaments is responsible for the shape changes observed in the pallial epithelium; the force of contraction is subsequently conveyed to the adult epithelium via the nascent zone of adhesion between the two epithelia (Figs. 16-18).

According to this hypothesis, the mechanism that generates the force for elevation of the adult epithelium resides in the apical hemisphere. This model lends itself to experimental manipulations designed to inhibit the elevation of the adult epithelium by either interfering with the contact and/or adhesion of the pallial epithelium to the adult epithelium, or by interfering with the contraction of the pallial epithelium. The first approach may be readily achieved by mechanically disrupting the attachment of the pallial epithelium to the adult epithelium with a fine tungsten needle. In this event the pallial epithelium recedes as in normally metamorphosing larvae, but the adult epithelium does not rise. Instead, the involuted corona is re-exposed as the pallial epithelium retracts, indicating that the force for elevation of the adult epithelium does, indeed, reside in the apical hemisphere, in accordance with the model.

Results of attempts to inhibit contraction of the pallial cells with cytochalasin B, however, are more equivocal. When metamorphosing larvae are exposed to 0.5 - 1.0 µg/ml cytochalasin B (CB) during contraction of the pallial epithelium, the adult epithelium continues to rise, as in normal metamorphosis (Reed & Woollacott, 1983). This range of concentration of CB was used because it effectively inhibits cytokinesis in bryozoan embryos (pers. obs.). However, the population of microfilaments in the pallial epithelium may be more resistant to CB, or adequate penetration by the drug may not be achieved in the short duration

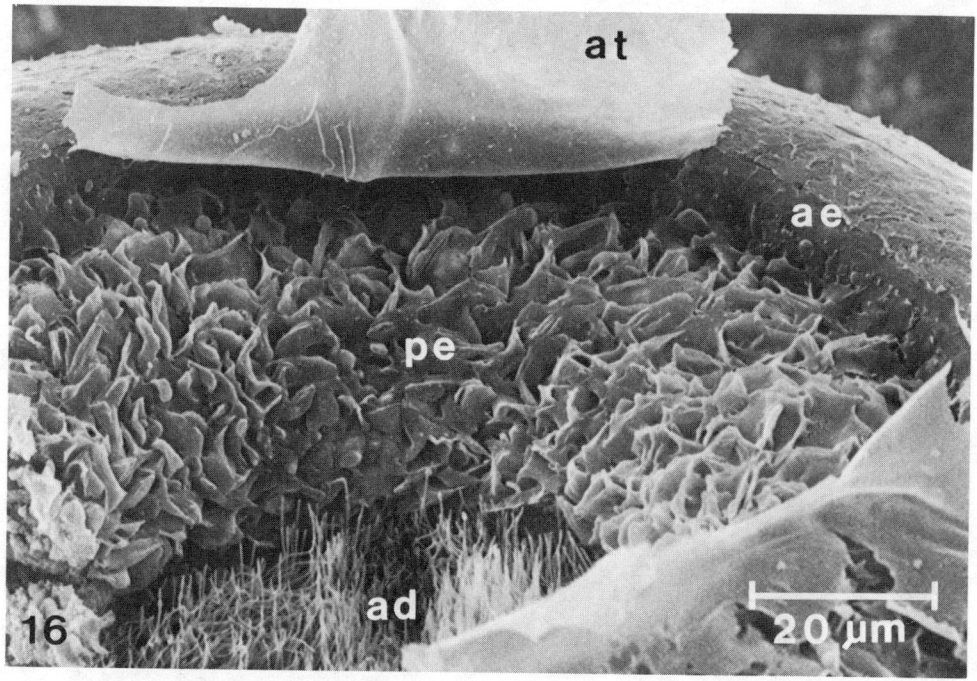

Fig. 16. <u>Bugula</u>. SEM of the contracted pallial epithelium (pe) at the stage of metamorphosis shown in Figure 15. The apices of the pallial cells form numerous folds. x 1200.

of the morphogenetic movement. To test these possibilities, we alternately increased the concentration of CB or lengthened the exposure period. The contraction of the pallial epithelium, and the coincident elevation of the adult epithelium, were arrested in the presence of 2.0 µg/ml CB applied at the onset of the morphogenetic movement. The inhibition was not reversible, however, so that the possibility of a cytotoxic effect at this concentration cannot be ruled out. To increase the incubation period, we exposed free-swimming larvae to lower concentrations of CB and induced them to settle and metamorphose. The results were not expected. In the presence of 0.25 to 1.0 µg/ml CB, the larvae settled and metamorphosed normally up to the end of coronal involution. In every case (N = 167) the adult epithelium failed to rise over the apical hemisphere, although metamorphosis was normal in all controls with the appropriate concentrations of dimethyl sulfoxide. These are the results expected if we have successfully inhibited contraction of the pallial epithelium with CB. Contrary to this interpretation, however, the ciliated larval epithelium subsequently was re-exposed after a delay beyond the normal timing for elevation

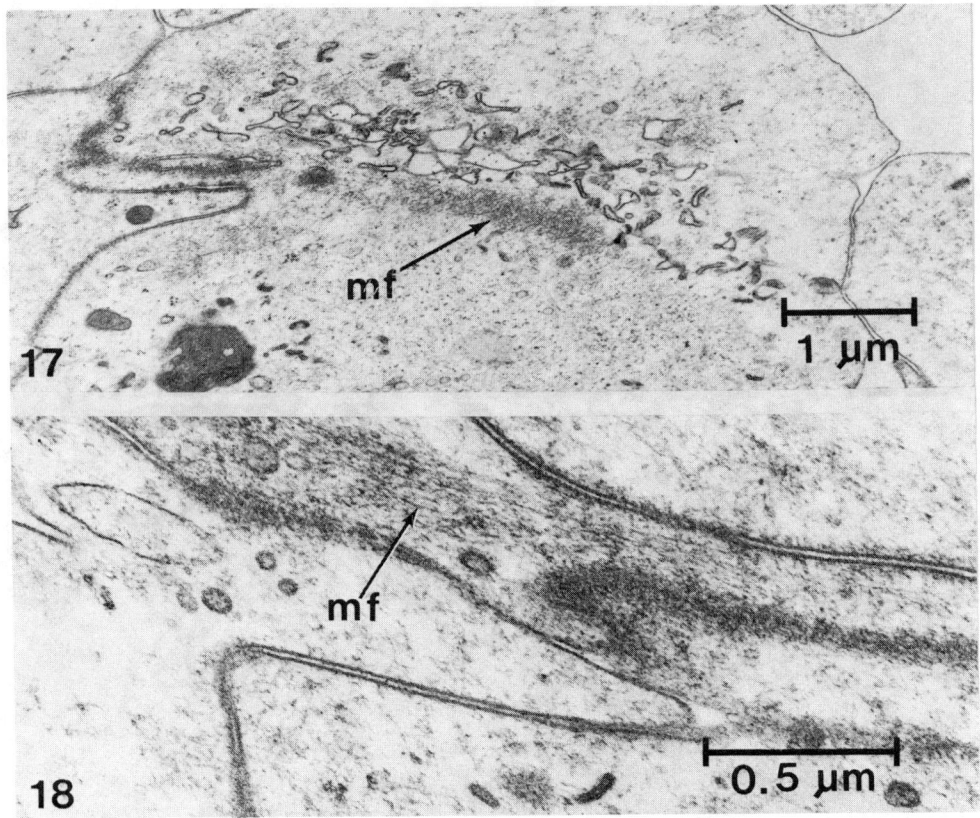

Fig. 17. <u>Bugula</u>. TEM of the apical band of microfilaments (mf) and vesicles present in the contracting pallial cells during elevation of the adult epithelium. x 18,100.

Fig. 18. <u>Bugula</u>. Higher magnification TEM of the apical band of microfilaments in the contracting pallial cells. x 53,625.

of the adult epithelium. Two conclusions may be drawn from this observation: 1) the adhesion of the pallial epithelium to the reflected adult epithelium does not occur in the presence of CB, and 2) a force-generating mechanism remains unaffected in the apical hemisphere in the presence of CB.

How does cytochalasin B affect the adhesion of the pallial epithelium to the adult epithelium? A reevaluation of normal metamorphosis reveals that contact between these two epithelia is ensured by two closely coordinated morphogenetic movements. At the end of coronal involution, the margin of the pallial epithelium constricts, followed by a compression of the apical hemisphere. Together, these morphogenetic movements establish the contact between the pallial epithelium and the adult epithelium that is essential for the successful completion of metamorphosis. The compression of the apical hemisphere is correlated with the contraction of an axial muscle that extends from the apical disc to the attachment organ. The constriction of the margin of the pallial epithelium, on the other hand, appears to be a microfilament-mediated morphogenetic movement. During the constriction, the pallial cells in this region contain a dense array of microfilaments 5.5 nm in diameter. Moreover, the microfilaments disappear, and constriction of the pallial margin is inhibited, in the presence of 0.25 to 1.0 µg/ml CB. As a consequence, the pallial epithelium never establishes contact with the adult epithelium. In our attempt to test the role of microfilaments in one morphogenetic movement, we have discovered their role in an antecedent movement.

What generates the force for the re-exposure of the corona in the presence of CB? Either the microfilaments in the pallial epithelium are not affected at the concentrations used, or there is another force-generating mechanism in the apical hemisphere that is insensitive to CB. Examination by transmission electron microscopy reveals that the microfilaments have disappeared, appearing to falsify the first supposition. The axial muscle, however, is strongly contracted, indicating that re-exposure of the corona in the presence of CB may be muscle-mediated. Indeed, the timing of this movement coincides with the contraction of the axial muscle to retract the apical disc in normal metamorphosis.

These observations indicate that the sequence of morphogenetic movements is closely coordinated to effect the proper association of tissues necessary to complete metamorphosis. The experimental modification of one morphogenetic movement, therefore, may have profound epigenetic consequences on metamorphosis, and should be considered only in light of the entire sequence of morphogenetic movements.

Since the initial work of Cloney (1966) and Schroeder (1968, 1969, 1970), the contractile role of microfilaments in cell and tissue motility in developmental processes has been documented many times. With the exception of the ascidians, however, there are few examples of microfilament-mediated rapid morphogenetic movements during metamorphosis. The results of these experiments on Bugula reveal that two rapid morphogenetic movements (constriction of the margin of the pallial epithelium, and contraction of the pallial epithelium during the subsequent elevation of the adult epithelium) are caused by two distinct populations of microfilaments that assemble in different regions of the pallial epithelium at specific times during metamorphosis. This is the first documentation of microfilament-mediated morphogenetic movements in the phylum Bryozoa.

In summary, these studies reveal that the complex series of morphogenetic movements that effect metamorphosis in marine bryozoans is mediated by a variety of mechanisms, from muscle and microfilament contractility to ciliary motility. While this research supports the basic tenet that the motors of morphogenesis are ultimately resolvable as cytoskeletal elements, it illustrates that they may assume novel associations and interactions in the cataclysmic metamorphoses of marine invertebrates. Indeed, even within the Bryozoa, the morphological diversity of larvae (Hondt, 1977b; Zimmer & Woollacott, 1977a; Reed & Cloney, 1982a) and the wide range of metamorphic patterns (Zimmer & Woollacott, 1977b; Reed & Cloney, 1982b) indicate that many variations in the mediation of rapid morphogenetic movements still await discovery.

The studies discussed in this paper utilize both empirical and experimental evidence to elucidate mechanisms of morphogenesis. This approach was a central theme in the research of Ernest Everett Just, who is being honored by this symposium for his scientific achievements in the field of marine invertebrate embryology. Just used descriptive and experimental techniques as complementary approaches to investigate various developmental phenomena, while always maintaining a clear appreciation for the need to interpret his results within the overall context of the developing system. On the strength of this holistic approach, Just bridged the naturalist-experimentalist dichotomy of the time. In The Biology of the Cell Surface (1939), Just wrote, "Both description and experiment are utilized by all the natural sciences--the extent to which each is used being determined by the state of the matter investigated. Just as the complex organization of living matter demands the large employment of description, so it prescribes experimental method." The work of E. E. Just, like his life, may be most distinguished by the voyage, rather than the destination.

ACKNOWLEDGMENTS

Portions of this research were conducted at the Friday Harbor Laboratories, the Museum of Comparative Zoology, the Smithsonian Institution (Fort Pierce Bureau), and the Department of Biological Sciences at Dartmouth College. I am indebted to Drs. Thomas E. Schroeder, A. O. D. Willows (Friday Harbor Laboratories), Robert M. Woollacott (Museum of Comparative

Zoology), and Mary E. Rice (Smithsonian Institution) for providing the facilities and atmosphere conducive to the completion of these studies. Special appreciation is extended to Dr. Richard A. Cloney (Department of Zoology, University of Washington), who, in his advisorial capacity, was instrumental in the initiation and completion of much of this research. I would especially like to acknowledge the counsel and guidance of the late Dr. Robert L. Fernald, former Director of the Friday Harbor Laboratories (1959-1974) and Professor Emeritus of the Department of Zoology at the University of Washington, who initiated, and continues to inspire, my interest in the development of marine invertebrates.

I am grateful to Marjorie Audette and Kendra Ballou (Dartmouth College) for their help in preparing this manuscript. This research has been supported by United States Public Health Training Grant HD-00266 and HD-07183 from the NICHD to the University of Washington, by ONR Contract N00014-78-C-0064 with Harvard University, and by Biomedical Research Support Grant RR-07056-11 to Dartmouth College from the Biomedical Support Branch, Division of Research Facilities and Resources, National Institute of Health.

LITERATURE CITED

Chia, F.S. and M.E. Rice. 1978. Settlement and Metamorphosis of Marine Invertebrate Larvae. Elsevier/North-Holland Biomedical Press, N.Y.

Cloney, R.A. 1966. Cytoplasmic filaments and cell movements: Epidermal cells during ascidian metamorphosis. J. Ultrastruct. Res. 14: 300-328.

Dentler, W.L., M. M. Pratt, and R.E. Stephens. 1980. Microtubule-membrane interactions in cilia. II. Photochemical cross-linking of bridge structures and the identification of a membrane-associated dynein-like ATPase. J. Cell Biol. 84: 381-403.

Eckert, R. 1972. Bioelectric control of ciliary activity. Science 176: 473-481.

Hondt, J.L. d' 1977a. Structure larvaire et histogenese postlarvaire chez Bowerbankia imbricata (Adams, 1798) Bryozoaire Ctenostome (Vesicularines). Arch. Zool. Exp. Gen. 118: 211-243.

Hondt, J.L. d' 1977b. Valeur systèmatique de la structure larvaire et des particularitès de la morphogenèse postlarvaire chex les bryozoaires Gymnolaemates. Gegenbaurs Morph. Jahrb. 123: 463-483.

Just, Ernest Everett. 1939. The Biology of the Cell Surface. P. Blakiston's Son and Co., Inc., Philadelphia.

Lyke, E.B., C.G. Reed, and R.M. Woollacott. 1983. Origin of the cystid epidermis during the metamorphosis of three species of gymnolaemate bryozoans. Zoomorph. 102: 99-110.

Naitoh, Y. 1968. Ionic control of the reversal response of cilia in Paramecium caudatum. A calcium hypothesis. J. Gen. Physiol. 51: 85-103.

Naitoh, Y. and H. Kaneko. 1972. Reactivated triton-extracted models of Paramecium. Modification of ciliary movement by calcium ions. Science 176: 523-524.

Reed, C.G. 1981. The role of ciliary reversal in the settlement of the marine bryozoan, Bugula neritina. J. Cell Biol. 91: 45a.

Reed, C.G. In Press a. The reproductive and developmental biology of marine bryozoans. In: Handbook of Marine Invertebrate Embryology and Larval Biology. M. Strathmann and R. L. Fernald (eds.).

Reed, C.G. In Press b. Larval attachment by eversion of the internal sac in the marine bryozoan Bowerbankia gracilis (Ctenostomata: Vesicularioidea): A muscle-mediated morphogenetic movement. Acta Zoologica.

Reed, C.G. and R.A. Cloney. 1980. The role of ciliary motility in the metamorphosis of a bryozoan. Amer. Zool. 20: 953.

Reed, C.G. and R.A. Cloney. 1982a. The larval morphology of the marine bryozoan Bowerbankia gracilis (Ctenostomata: Vesicularioidea). Zoomorph. 100: 23-54.

Reed, C.G. and R.A. Cloney. 1982b. Settlement and metamorphosis of the marine bryozoan Bowerbankia gracilis (Ctenostomata: Vesicularioidea). Zoomorph. 101: 103-132.

Reed, C.G. and R.M. Woollacott. 1982. Mechanisms of rapid morphogenetic movements in the metamorphosis of the bryozoan Bugula neritina (Cheilostomata, Cellularioidea). I. Attachment to the substratum. J. Morph. 172: 335-348.

Reed, C.G. and R.M. Woollacott. 1983. Mechanisms of rapid morphogenetic movements in the metamorphosis of the bryozoan Bugula neritina (Cheilostomata, Cellularioidea). II. The role of dynamic assemblages of microfilaments in the pallial epithelium. J. Morph. 177: 127-143.

Schroeder, T.E. 1968. Cytokinesis: Filaments in the cleavage furrow. Exp. Cell Res. 53: 272-276.
Schroeder, T.E. 1969. The role of the 'contractile ring' filaments in dividing Arbacia eggs. Biol. Bull. 137: 413-414.
Schroeder, T.E. 1970. The contractile ring. I. Fine structure of dividing mammalian (HeLa) cells and the effects of cytochalasin B. Z. Zellforsch. Mikrosk. Anat. 109: 431-449.
Woollacott, R.M. and R.L. Zimmer. 1971. Attachment and metamorphosis of the cheilo-ctenostome bryozoan, Bugula neritina (Linne). J. Morph. 134: 351-382.
Woollacott, R.M. and R.L. Zimmer. 1978. Metamorphosis of cellularioid bryozoans. pp. 49-63. In: Settlement and Metamorphosis of Marine Invertebrate Larvae. F.-S. Chia and M.E. Rice (eds.). Elsevier/North-Holland Biomedical Press, N.Y.
Zimmer, R.L. and R.M. Woollacott. 1977a. Structure and classification of gymnolaemate larvae. pp. 57-89 In: Biology of Bryozoans. R.M. Woollacott and R.L. Zimmer (eds.). Academic Press, N.Y.
Zimmer, R.L. and R.M. Woollacott. 1977b. Metamorphosis, ancestrulae, and coloniality in bryozoan life cycles. pp. 91-148 In: Biology of Bryozoans. R.M. Woollacott, and R.L. Zimmer (eds.). Academic Press, N.Y.

TISSUE ARCHITECTURE AND HYDROID MORPHOGENESIS: THE ROLE OF LOCOMOTORY TRACTION IN SHAPING THE TISSUE

Richard D. Campbell

ABSTRACT

This paper examines the tissue architecture of the polyp generation of hydrozoan coelenterates, with the aim of deducing what cellular behaviors control the form of the animals. From the study of morphogenesis in other animals, the following two generalizations may be drawn: (1) tissues that are undergoing morphogenesis are usually under tension, and (2) cell rearrangements (and thus tissue morphogenesis) are caused by pulling forces exerted by the cells. Morphogenesis of polyps probably accords with these two generalizations. Tension is set up by hydrostatic fluid in the gastric cavity. The epithelial cells probably pull on the tissue through locomotory traction of the muscle processes, and this causes cell rearrangement. The muscle processes are aligned axially in the ectoderm and circumferentially in the endoderm, so that these two layers apply morphogenetic forces in complementary directions.

INTRODUCTION

Polyps of hydrozoan coelenterates have been a favorite object of developmental research because of their simplicity and amenability to experimental study. Most of the work that has been done relates to problems of patterning, notably the spacing and ordering of the major body parts (hypostome, tentacles, buds, and so forth) along the length of the polyp. In this paper I will explore the problem of the mechanical causes of morphogenesis. This approach to development has been very fruitful in other developing systems, and we have a reasonably concrete understanding of what cellular behaviors are responsible for tissue foldings and rearrangements in many kinds of embryos.

Mechanics of Morphogenesis in Animals

Morphogenesis, the mechanical shaping of tissues, involves the rearrangement of cells. Tissues have viscous and elastic properties, and forces must be applied in order to rearrange the cells and change the shape of a tissue. Pioneers in the study of cell behavior, such as Holtfreter (1943, 1944) and Weiss (1961), led to the concept that cells, through their behavior, provide the forces that cause morphogenesis.

This modern viewpoint became firmly established when Gustafson and Wolpert (1967) showed that most of the shaping of the gastrulating sea urchin embryo could be seen as the logical outcome of simple cellular behaviors such as migration and filopodial attachment and contraction. In the last two decades, numerous cases of morphogenesis in animal tissue have similarly been found to be the result of the behaviors of the component cells. The types of behavior underlying morphogenesis are few and widespread, and we can make the following important generalizations on the basis of studies of many animal tissues (Cloney, 1966; Baker & Schroeder, 1967; Burnside, 1971; Spooner & Wessells, 1972; Jacobson & Gordon, 1976):

A. _Tissues undergoing morphogenesis are under tension_. This generalized tension is usually set up by a hydrostatic cavity (such as the blastocoele), but sometimes by skeletal elements or by cell behavior.

B. _Morphogenetic forces are applied by the cells themselves and are tensile rather than compressive: cells pull but do not push_. These morphogenetic forces are superimposed on the general tensile stress carried by the tissue.

C. _It is usually possible to deduce the pattern of morphogenetic forces by simple inspection of a tissue_. Tensile stress aligns whatever tissue elements set up and transmit the stress. Long, smooth contours in a tissue that is otherwise irregular are suggestive of tensile stresses. Thus one can see morphogenetic forces, using cellular or subcellular alignments as a guide.

D. _Most morphogenetic forces in animal tissues are set up by a limited number of cell behaviors_. The major morphogenetic behaviors are pulling by means of filopodia, locomotive hauling by cells at the edge of a tissue, contraction of an apical band of microfilaments, and perhaps cell adhesion. All of these mechanisms except for adhesion involve force generation by contraction of microfilaments. These are reviewed by Trinkaus (1984).

A new insight into the mechanical underpinning of morphogenesis has been recently offered by Stopak and Harris (1982) and Oster et al. (1983). The overriding behavior of cells, at least as seen in culture, is that of locomotion. A cell is propelled forward by a pseudopod that pulls the substratum towards it. This force has been designated as locomotory traction by Stopak and Harris (1982). Since cells frequently tend to be stretched in several directions simultaneously by their locomotory pseudopodia located around the margin of the cells, locomotory traction tends to pull the substratum centripitally toward a cell, from two or more directions. Stopak and Harris (1982) demonstrate and review the elegant experiments that allow the tractional forces to be visualized. Oster et al. (1983) examine the implications for such tractional forces on morphogenesis of patterns, finding that numerous known patterns can arise automatically from the dynamics of traction in a population of cells. The concept of locomotory traction suggests how the collected behavior of individual cells can result in large-scale tissue deformations and movements.

Various lines of inquiry have been made into the cellular basis of morphogenesis in coelenterates, but generally these have not been closely correlated with tissue architecture. Gierer (1977) and Graf and Gierer (1980) have proposed that shaping is due to differential expansion (or contraction) of the inner and outer surfaces of the tissue. Numerous suggestions have been put forth that ordered cell division causes shaping (reviewed in Berrill, 1961; see also Webster, 1971). The most analytical studies have involved morphogenesis of the stolon tip (reviewed in Campbell, 1974) and of hydranth development in Calyptoblastic hydroids by Beloussov and his collaborators (see Beloussov & Dorfman, 1974). Beloussov has paid great attention to the relation between cell shape and behavior and tissue morphogenesis in these organs.

However, in view of the great popularity of studying patterning in hydroid polyps, it is disappointing that so little is known about the mechanical aspects of morphogenesis. In addition, most of the existing work looks at polyps as 1-dimensional tissues, without regard to the radial organization. Ignoring the radial organization is unfortunate since the phylum Coelenterata is characterized by its radial organization. It is likely, as discussed below, that morphogenesis in the two dimensions results from an interplay of circumferential and longitudinal components.

Structure of Coelenterate Polyps

Coelenterate polyps are roughly cylindrical with a whorl of tentacles surrounding one end (Fig. 1). There may be another whorl of tentacles lower on the column and sometimes there are eversions in the form of young polyps or medusae protruding from the middle region. In a few polyps, such as hydra, gonads may be formed on the side of the column. The top of the column, above the tentacles, is called the hypostome because the mouth forms at its tip. The base of a polyp may be continuous with a colonial stolon or it may end in a blind, sticky adhesive disk as it does in hydra.

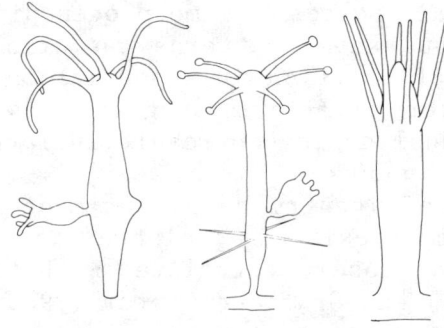

Fig. 1. Diagram of three hydroid polyps: Hydractinia, Cladonema, and Hydra (left to right).

The body wall of polyps is trilaminar: it consists of two epithelia, ectoderm and endoderm, resting back-to-back on a common basement membrane called the mesolamella (Fig. 2). The cavity of the cylinder is filled with a fluid and is called the gastric cavity. This tissue construction is continuous everywhere over the animal, extending out along the tentacles and polyp buds. Each species or type of polyp has its characteristic proportions and size. Morphogenesis of polyps thus consists partly of shaping the trilaminar cylinder into the proper proportions and extensions.

This paper draws the general picture of polyp structure,

but there are many exceptional cases (which here will be ignored), and of course much of the data has been collected from only a few species. The entire group of Calyptoblastic hydrozoans, however, is quite different in construction and their morphogenesis occurs by quite different means to those discussed here (see Beloussov & Dorfman, 1974).

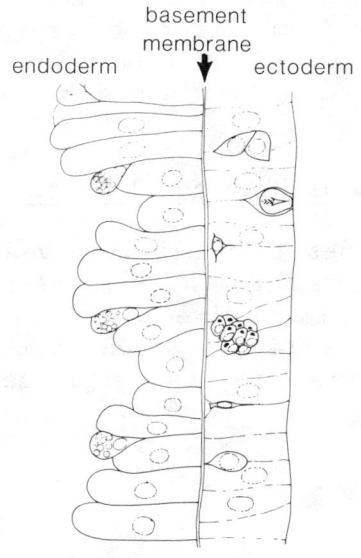

Fig. 2. General tissue plan of hydroid polyp, consisting of ectoderm and endoderm resting on a common basement membrane called the mesolamella. The polyp consists of a shell of this tissue and is filled centrally with gastric fluid.

Epithelial Cells Drive Morphogenesis in Polyps

Coelenterate polyps, at least as exemplified by hydra, contain three independent lineages of cells: ectodermal epithelial cells, endodermal epithelial cells, and interstitial cells. The population of interstitial cells contains an undifferentiated stem cell (frequently called the Interstitial or I-cell) and numerous specialized derivatives including nerve cells, gland cells, stinging cells (nematocytes) and gametes. It is possible to remove permanently the entire lineage of interstitial cells from hydra while maintaining the viability of the animal (Sugiyama & Fujisawa, 1978; Campbell, 1979). Such animals are called "epithelial" polyps because they contain only epithelial cells. Epithelial polyps have nearly normal morphogenetic activities, such as regeneration, indicating that epithelial cells carry out the processes of morphogenesis. Epithelial hydra that are experimentally repopulated with the interstitial cell lineage from another strain of hydra retain the shape of the hydra that donated the epithelial cells, rather than the shape of the hydra that donated the interstitial cells, again showing that the epithelial cells mediate morphogenesis.

Recently it has been demonstrated that both ectodermal and endodermal epithelial cells are jointly involved in morphogenesis. Chimeras have been constructed by recombining ectoderm and endoderm from different strains of hydra (Wanek, 1983). The morphology of the body column, tentacles, hypostome and stalk have determinants in both ectoderm and endoderm.

Ectoderm and Endoderm as Orthogonal Tissues

The two cell types that are involved in morphogenesis, ectodermal and endodermal epithelial cells, are anisotropic. The epithelia they form are directional in their organization and developmental properties. Endoderm is organized circumferentially while ectoderm is organized longitudinally.

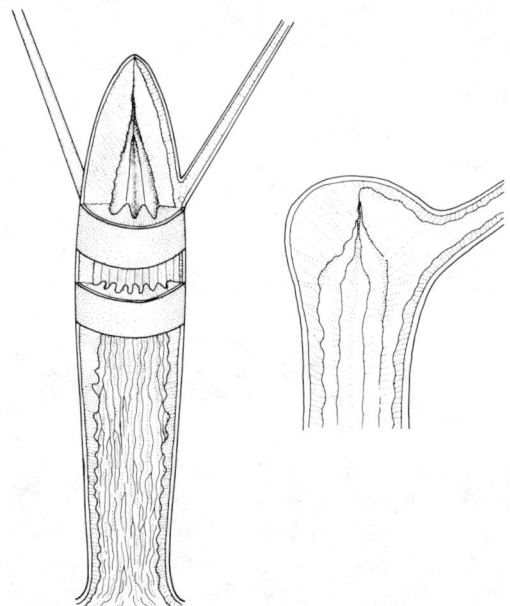

Fig. 3. Diagram of the three-dimensional structure of polyps of Hydractinia (left) and Hydra (right; only the top portion is shown). The endoderm is ridged conspicuously near the top of the column, above the tentacles. The ridges, or taeniolae, alternate with the tentacles. Beneath the tentacles the taeniolae gradually branch and lose their distinctiveness.

Endoderm. The endoderm of polyps exhibits a circumferential organization. The endoderm at the top of the polyp is organized as regular ridges and intervening troughs (see Figs. 3 & 4). The ridges (termed taeniolae by Hamann, 1882) give the cross section of the polyp a star-shaped lumen. Mitosis is localized in the intertaeniolate positions

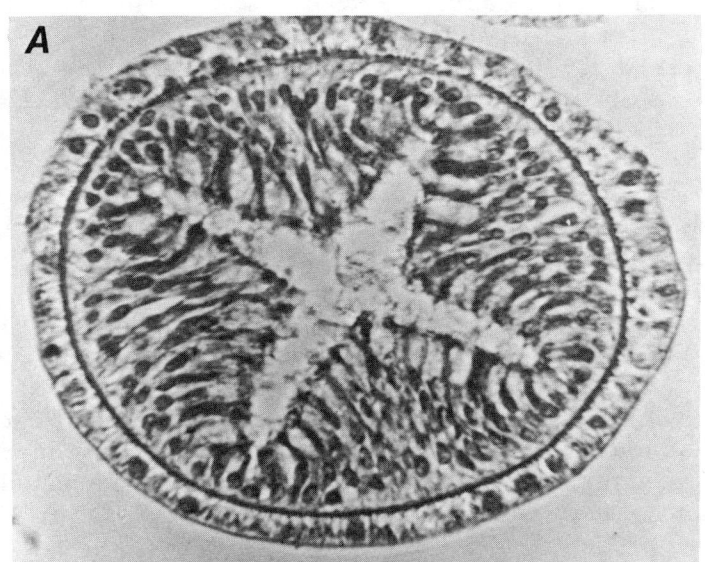

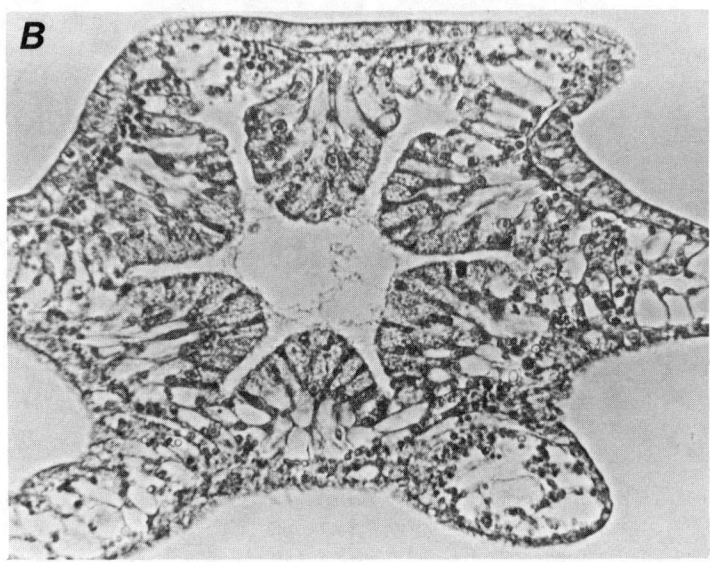

Fig. 4. Cross sections of Hydractinia (A) and Hydra (B) showing the organization of the endoderm. The section of Hydractinia is through the hypostome, that is, above the tentacles. The section of Hydra is through the level where tentacles insert into the column; the tentacles insert between adjacent taeniolae.

in some species (Campbell, 1967; Braverman, 1968). The ridges rather than the tentacles appear to represent the <u>fundamental</u> radial organization of the endoderm. The troughs arise before the tentacles during regeneration, and even in species which do not possess tentacles, the endoderm is still organized into radial taeniolae.

That the polyp's endoderm is organized radially is consistent with the structure of other coelenterate tissue as well. The stems of tubularian hydroids have the endoderm organized into 2-20 radial domains that divide the cavity into longitudinal canals. The endoderm of the polyp stage of Scyphozoan jellyfish is organized as a radially folded structure. The endoderm of Anthozoan polyps (including the sea anemones) is conspicuously organized as radially arranged mesenteries whose circumferential ordering is one of the primary characteristics of these animals. In embryos of Siphonophores, endodermal tissue around the float is often organized into numerous radial components. The radial canals of medusae are another instance of radial organization of endoderm.

The phylum Cnidaria is characterized as one of radially-symmetric animals. Some of the structures that manifest this radial symmetry, such as the tentacles, are composed of both ectoderm and endoderm. But when radial symmetry is expressed by a single tissue, this tissue is endoderm.

<u>Ectoderm</u>. The ectoderm of polyps demonstrates a tendency toward longitudinal organization. Those polyps that have gonads as ectodermal structures bear them at a particular axial level, scattered around the circumference of the column. The complex population of interstitial cells shows axial partitioning with respect to where different types of cells are located, where they differentiate and to where the mature cells migrate (Bode & David, 1978). Nematocytes migrate in the ectoderm, strongly oriented along the axis (Campbell & Marcum, 1980). Mature nematocytes become stationed at particular axial levels, scattered around the circumference (Weill, 1934). The first morphogenetic event in many coelenterate embryos is an invagination of an apical zone of ectoderm. There are virtually no instances of circumferential ordering in the ectoderm alone. The endoderm also displays some longitudinal organization; the most conspicuous example is that the radial organization of the endoderm is itself graded in intensity down the column (see Fig. 3, left), as are the muscular processes of the endodermal epithelial cells (described below).

Thus, the overall development and morphology of polyps suggest that ectoderm and endoderm play complementary roles in morphogenesis. The ectoderm is concerned with longitudinal organization and the endoderm is concerned primarily with a radial organization.

The basal surfaces of both the ectodermal and endodermal epithelial cells are specialized and complicated through the formation of extensions that resemble filopodia or lamellipodia. Figures 5 and 7 show the appearance of the muscle processes of ectodermal cells of hydra. The muscles processes are precisely oriented: in the ectoderm they are all aligned longitudinally along the polyp's axis; in the endoderm they are all aligned circumferentially. Their orientations thus coincide with the tendency for ectoderm and endoderm to demonstrate the longitudinal and circumferential organization, respectively. This raises the suspicion that these processes apply the morphogenetic forces. These cellular extensions are called "muscle processes" because they have the organization and behavior of smooth muscles. When the ectodermal processes contract, the polyp shortens; when the endodermal processes contract, the fluid-filled polyp decreases in girth and hence elongates. Although these processes serve as muscles for the polyp, they probably also have another function: that of migratory pseudopodia providing morphogenetic forces through locomotory traction.

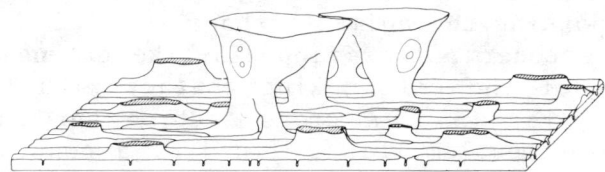

Fig. 5. Organization of the ectodermal epithelial cells of hydra. The cell bodies are raised on legs that rest on the mesolamella (stippled). From these legs, long muscle processes extend along the mesolamella which is completely covered by the processes.

Evidence That Epithelial Cell Processes Apply Morphogenetic Forces

Morphogenetic forces stretch and therefore align those components of the tissue that create and transmit them. The forces presumably could be transmitted from cell to cell through regions of specialized cell junctions. Morphogenetic forces, in animal tissues, are set up by parts of cells that are rich in microfilaments. These are all features that can be seen in histological sections. By looking at the tissue architecture of polyps one can conclude that morphogenetic forces are applied in the vicinity of the mesolamella.

The ectodermal and endodermal epithelial cells are fixed into their tissues by means of septate and gap junctions. These junctions are found in the apical and basal regions of the epithelia, with the lower-middle lateral regions being free of junctions. Hence, morphogenetic forces must be transmitted through the tissue at one of three surfaces in the polyp: apical surface of the ectoderm, apical surface of the endoderm, or in the vicinity of the mesolamella. Of these three surfaces, only the tissue in the vicinity of the mesolamella is smooth. The apical surfaces of both the ectoderm and endoderm are generally wrinkled, bunched, or fluted and cannot be carrying morphogenetic forces.

There are other reasons supporting the conclusion that morphogenetic stresses are applied in the vicinity of the mesolamella. First, the epithelial cells are very vacuolate, and the basal cytoplasm is the densest cytoplasm in the cell. The cytoplasm of both epithelia here is rich in microfilaments, but there are few microfilaments in other regions of the cells. Second, it is frequently the case in morphogenetic systems that force is applied near the mesolamella. Finally, the ectoderm may be disrupted by the formation of gonads, but these do not influence the shape of the polyp. In the vicinity of the gonads, only the basal part of the ectoderm retains its normal configuration, and hence must be the part supporting the animal's shape.

In the ectoderm, the filopodia-like extensions are the only portion of the cell making contact with the basement membrane (see Fig. 5). A cross section through the polyp shows that the mesolamella is lined by a continuous mat of these ectodermal cell processes (Fig. 6). In the endoderm, the "muscle processes" are usually flattened extensions of the basal margin of the cell. The processes are quite short and most of the contractile filaments lie along the basal surface of the cell itself which, in contrast to the ectodermal epithelial cell, is directly abutted against the mesolamella.

If the microfilament-laden cytoplasm is the morphogenetically active portion of the cells, then the morphogenetic machinery for the polyp consists of the trilaminar juxtaposition of ectodermal muscle processes, mesolamella, and endodermal contractile cytoplasm (Fig. 6). The microfilaments all lie within the plane of this trilaminar structure, and those of the ectoderm (axial) are perpendicular to those of the endoderm (circumferential).

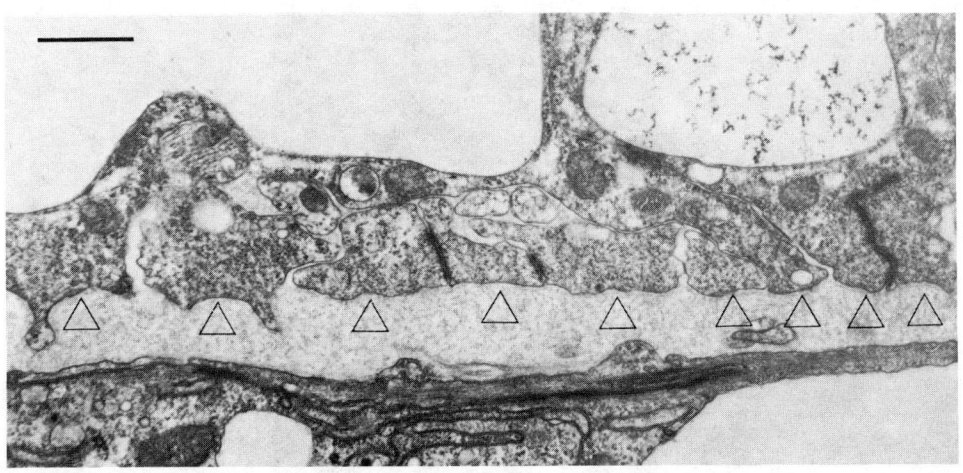

Fig. 6. Cross section through the column of a hydra showing the trilaminar region that is postulated to generate morphogenetic forces. This consists of the layer of ectodermal muscle processes (shown in cross section each marked by the top of a triangle); the mesolamella (light area, containing the triangles) and the endodermal muscle processes (below the mesolamella and cut along the direction of the microfilaments). Bar = 1 μm.

Another reason for suspecting that the muscle processes are primarily mediators of morphogenesis is that they have peculiar dispositions in all the body regions where major morphogenetic shaping is occurring (such as the budding region where tissue is being incorporated into a bud; the base of the tentacles where tissue is being incorporated into tentacles; and the bottom of the peduncle where cells are transforming into the basal disk). These dispositions are shown in Figure 7. In the other regions of the column (such as the central region of the column), where the only morphogenetic event seems to be a gradual proportioning, there are only shallowly graded alterations in muscle processes.

Thus, one would conclude from the structure of a polyp that morphogenetic forces are applied along the mesolamella. The cellular structures located here, the muscle processes of the epithelial cells, have the appearance of locomotory organelles and represent the major accumulation of microfilaments in the polyp. The nature of the processes suggests that they exert morphogenetic forces by means of locomotory traction. Hence in the ectoderm the forces result in an axial tension while in the endoderm they result in circumferential tension.

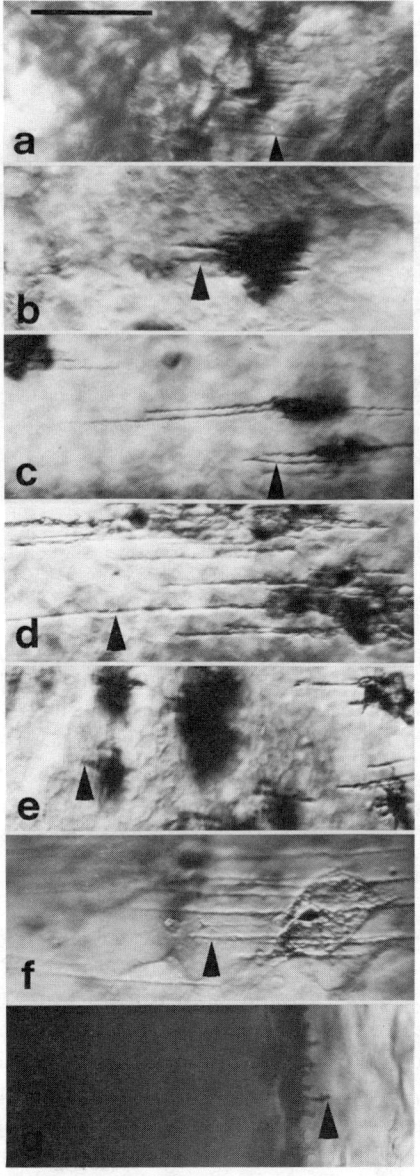

Fig. 7. Ectodermal muscle processes, seen in surface view of whole mounts of hydra in which individual or small groups of cells were stained by the method of Locke and Huie (1981). In all photographs the top of the hydra is oriented towards the right. Each photograph represents a different region of the column. Arrowheads mark representative muscle processes. Bar = 50 μm. a: hypostome (only the distally directed processes are visible), b: vicinity of the base of the tentacles, c: upper column region, d: lower column region, e: budding zone, a quarter of the circumference away from a bud, f: peduncle, g: edge of basal disk (black). Only the distally directed processes are visible here; the processes directed toward the center of the disk are withdrawn into the cell.

Integration of Ectodermal and Endodermal Traction

The postulated traction resulting from ectodermal and endodermal cell processes should lead to a shrinkage of the entire polyp tissue because it pulls all parts of the tissue toward one another. Polyps do not normally shrink, for this "tractional compression" is balanced by distension due to gastric pressure. But when gastric fluid pressure is experimentally relieved, polyps do shrink (Wanek et al., 1980). Shrinkage occurs gradually over the course of several days. When gastric pressure is allowed to build up again, the shrunken polyps re-extend, a process that takes one or two days. The shrinkage of polyps seems to involve only a compression of the tissue rather than a decrease in cell number, and in fact cell proliferation continues. The mesolamella becomes many times thicker during shrinkage and the cells become more crowded by becoming more columnar.

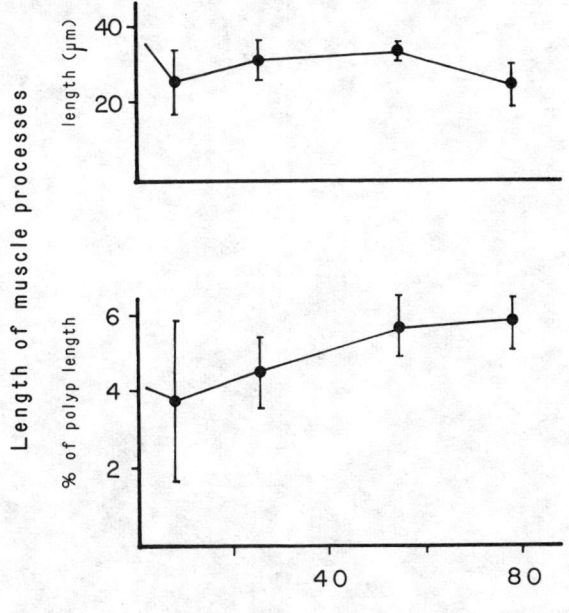

Fig. 8. Lengths of ectodermal muscle processes in the column of epithelial hydra in which the gastric fluid was not allowed to build up pressure. The upper curve shows the absolute lengths of the muscle processes, measured from the center of the cell in either direction. The lower curve shows the lengths of the processes as a percentage of the lengths of the whole polyps. Data were gathered separately for each hydra (n = 29) by measuring 8-10 processes per polyp. Error bars show standard deviations of the average process length for each polyp. Polyps were treated as described in Wanek et al. (1980) and measurements were made on cells stained by the method of Locke and Huie (1981).

While animals deprived of gastric pressure shrink, the lengths of the muscle processes remain more or less constant (Fig. 8, top). Hence the processes are elongating through the tissue relative to neighboring cells, increasing their

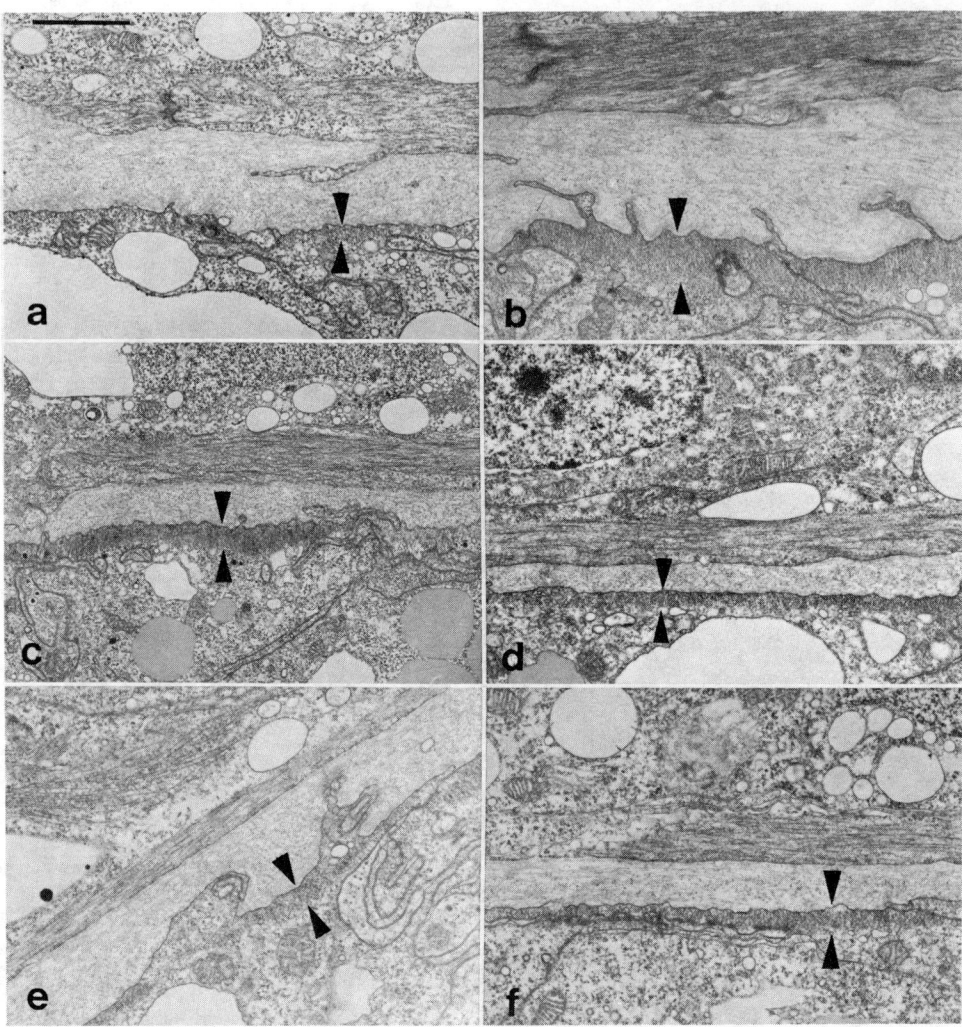

Fig. 9. Appearance of ectodermal (at the top of each panel) and endodermal (bottom) processes and the mesolamella (center) in the following regions of the hydra polyp: a: hypostome, intertaeniolate radius; b: hypostome, taeniolate radius; c: upper column; d: lower column; e: budding zone (across the column from a bud); f: peduncle. All micrographs were sectioned parallel to the polyps' axis, so the ectodermal processes are cut longitudinally and the endodermal processes are cut in cross section. The endodermal microfilamentous region is visible as a dense cytoplasmic region just below the mesolamella (opposed arrow-heads show the extent of the microfilamentous region in the endoderm). Bar = 2 μm (1.2 μm for panel c).

span through the tissue by 50% over a period of three days (Fig. 8, bottom). In light of the work of Stopak and Harris (1982), it is reasonable to suppose that this extension is providing the tractional force for shrinkage of the polyp.

Now one must consider the simplest aspect of morphogenesis of the polyp, that of maintaining the proportions of the body column. A tractional model of morphogenesis would hold that greater circumferential traction would promote a narrowing of the girth of the column, while a greater longitudinal traction would lead to a shortening of the column. The relative balance of these two tractional activities would determine the proportions of the column. The column is most slender just under the tentacles, and it gradually increases in circumference toward the budding region. The peduncle also is slender. The most slender region is, of course, the hypostome. The endodermal processes vary markedly in their development among these regions, from prominent in the hypostome to weak in the lower column (see Fig. 9; see also Mueller, 1950). The ectodermal muscle processes are not notably different in these various regions of the column, although they are somewhat shorter in the hypostome and slightly longer in the lower column.

If the general extent of microfilament development is related to the tractional force that the processes exert, the proportioning of the column would be explained as a balance between ectodermal and endodermal traction.

Morphogenesis in other regions of the body is obviously more complex, but the role of the muscle processes is in evidence. It is possible that the hypostome is constricted to a dome because of the extensive development of the endodermal, circumferential processes there. (Another site of notable constriction occurs as a bud separates from the parent; this is accompanied by a ring of several very prominent, circularly oriented, ectodermal muscle processes.) The taeniolate organization of the endoderm, and hence the radial organization of the tentacles, might well be a consequence of instability of the tissue homogeneity prompted by strong locomotory traction in the hypostome (as analyzed in the general case, Oster et al., 1983). Finally, it is worth noting that at the base of buds and tentacles, where tissue is transiting from the column organization to that of the extremity, the processes of the ectoderm are short and contorted and the mesolamella is frequently thrown into a thick compressed mass.

CONCLUSIONS

The tissue architecture of the coelenterate polyp has been analyzed by microscopists for 100 years. In the last decade we have learned enough about morphogenesis in other animals to be able to "read" morphogenetic mechanisms from histological appearance. Examined in this light, polyps seem to shape themselves by means of processes extending out from the epithelial cells along the basement membrane. While these processes serve physiologically as muscles, in the context of morphogenesis they presumably act by providing locomotory tractional forces. Ectodermal and endodermal processes are orthogonal to each other, and thus along with the mesolamella they form a trilaminar structure that displays both circumferential and longitudinal tractional compression of the polyp. Compression is balanced by distension due to hydrostatic pressure of the gastric fluid. The relative tractional strengths of ectoderm and endoderm in any region should thus influence the proportion (length vs. width) of that region. Proportioning of the graded body column is consistent with the graded extents to which the muscle processes are developed, although since traction would be set up by the _tips_ of processes rather than by the muscular portions, this consistency might be fortuitous. Other portions of the polyp where morphogenetic activities are more extreme also display major changes in the disposition of the muscle processes.

LITERATURE CITED

Baker, P.C. and T.E. Schroeder. 1967. Cytoplasmic filaments and morphogenetic movements in the amphibian neural tube. Develop. Biol. 15: 432-450.
Beloussov, L.V. and J.G. Dorfman. 1974. On the mechanics of growth and morphogenesis in hydroid polyps. Amer. Zool. 14: 719-734.
Berrill, N.J. 1961. Growth, Development and Pattern. Freeman, San Francisco. 555 pp.
Bode, H.R. and C.N. David. 1978. Regulation of a multipotent stem, the interstial cell of hydra. Progr. Biophys. Mol. Biol. 33: 189-206.
Braverman, M.H. 1968. Studies on hydroid differentiation III. The replacement of hypostomal gland cells of _Podocoryne carnea_. J. Morphol. 126: 95-106.

Burnside, B. 1971. Microtubules and microfilaments in newt neurulation. Develop. Biol. 26: 416-441.

Campbell, R.D. 1967. Cell proliferation and morphological patterns in the hydroids Tubularia and Hydractinia. J. Embryol. Exp. Morphol. 17: 607-616.

Campbell, R.D. 1974. Development. pp. 179-210. In: Coelenterate Biology. L. Muscatine and H.M. Lenhoff (eds.). Academic Press, N.Y.

Campbell, R.D. 1979. Development of hydra lacking interstitial and nerve cells ("epithelial hydra"). Symp. Soc. Develop. Biol. 37: 267-293.

Campbell, R.D. and B.A. Marcum. 1980. Nematocyte migration in hydra: Evidence for contact guidance in vivo. J. Cell Sci. 41: 33-51.

Cloney, R.A. 1966. Cytoplasmic filaments and cell movements: Epidermal cells during ascidian metamorphosis. Ultrastruct. Res. 14: 300-328.

Gierer, A. 1977. Biological features and physical concepts of pattern formation exemplified by hydra. Curr. Top. Develop. Biol. 11: 17-59.

Graf, L. and A. Gierer. 1980. Size, shape and orientation of cells in budding hydra and regulation of regeneration in cell aggregates. Roux´ Arch. Develop. Biol. 188: 141-151.

Gustafson, T. and L. Wolpert. 1967. Cellular movement and contact in sea urchin morphogenesis. Biol. Rev. 42: 442-498.

Hamann, O. 1882. Der Organismus der Hydroidpolypen. Jen. Z. Naturwiss. 15: 473-544.

Holtfreter, J. 1943. A study of the mechanics of gastrulation. Part I. J. Exp. Zool. 94: 261-318.

Holtfreter, J. 1944. A study of the mechanics of gastrulation. Part II. J. Exp. Zool. 95: 171-212.

Jacobson, A.G. and R. Gordon. 1976. Changes in the shape of the developing vertebrate nervous system analyzed experimentally, mathematically and by computer simulation. J. Exp. Zool. 197: 191-246.

Locke, M. and P. Huie. 1981. Epidermal feet in insect morphogenesis. Nature 293: 733-735.

Mueller, J. 1950. Some observations on the structure of hydra, with particular reference to the muscular system. Trans. Amer. Microscop. Soc. 69: 133-147.

Oster, G.F., J.D. Murray, and A.K. Harris. 1983. Mechanical aspects of mesenchymal morphogenesis. J. Embryol. Exp. Morphol. 78: 83-125.

Spooner, B.S. and N.K. Wessells. 1972. Analysis of salivary gland morphogenesis: Role of cytoplasmic microfilaments and microtubules. Develop. Biol. 27: 38-54.

Stopak, D. and A.K. Harris. 1982. Connective tissue morphogenesis by fibroblast traction. I. Tissue culture observations. Develop. Biol. 90: 383-398.

Sugiyama, T. and T. Fujisawa. 1978. Genetic analysis of developmental mechanisms in hydra. V. Cell lineage and development of chimera hydra. J. Cell Sci. 32: 215-232.

Trinkaus, J.P. 1984. Cells Into Organs: The Forces That Shape The Embryo, 2nd edition. Prentice-Hall, Englewood Cliffs, NJ. 543 pp.

Wanek, N. 1983. Roles of ectodermal and endodermal epithelial cells in hydra morphogenesis: Analysis of chimeras. J. Exp. Zool., 225: 89-97.

Wanek, N., B.A. Marcum, H-T. Lee, M. Chow, and R.D. Campbell. 1980. Effect of hydrostatic pressure on morphogenesis in nerve-free hydra. J. Exp. Zool. 211: 275-280.

Webster, G. 1971. Morphogenesis and pattern formation in hydroids. Biol. Rev. 46: 1-46.

Weill, R. 1934. Contribution à l´étude des cnidaires et de leurs nématocystes. Trav. Sta. Zool. Wimereux 10 and 11: 1-700.

Weiss, P. 1961. Guiding principles in cell locomotion and cell aggregation. Exp. Cell Res. Suppl. 8: 260-281.

ISOLATION AND CHARACTERIZATION OF BLASTODERM-SPECIFIC GENES OF *DROSOPHILA MELANOGASTER*

Judith A. Lengyel, Margaret Roark,
Kritaya Kongsuwan, Paul A. Mahoney,
Paul D. Boyer, and John R. Merriam

ABSTRACT

A dramatic transition occurs at the blastoderm stage of embryogenesis of Drosophila melanogaster. Nuclear multiplication slows and becomes asynchronous, transcription is activated, cells form, and the determination of much of the basic pattern of the larva and adult occur. We review the evidence that there should be only a small number of genes which play a major role in these events and which are expressed differentially or specifically at this stage. Using recombinant DNA techniques, we have isolated genes newly activated at the blastoderm stage. We have succeeded in identifying three highly blastoderm-specific genes, each of which maps to a single locus in the salivary gland polytene chromosomes. Molecular analyses of these cloned loci indicate that all are complex, encoding multiple, and in some cases overlapping transcripts. Genetic analysis of a small chromosomal region (12-13 polytene bands) containing one of the blastoderm-specific genes has revealed the presence of a previously undescribed gene controlling embryonic head formation. This is consistent with our hypothesis that blastoderm-specific genes are involved in events unique to early embryogenesis. Further genetic and biochemical analyses of these cloned genes should provide insight into events specific to the blastoderm stage.

INTRODUCTION

The blastoderm stage of embryogenesis in Drosophila is crucial in setting the stage for later events of morphogenesis and cytodifferentiation. Only a small number of genes are likely to be involved in controlling events which occur at blastoderm; these genes must provide information that is qualitatively different from that of the maternal program.

Molecular and Cellular Events During Early Drosophila Embryogenesis

The morphological events that occur throughout Drosophila embryogenesis have been reviewed (Fullilove & Jacobson, 1978). We describe here the molecular and cellular changes that occur between fertilization and the onset of gastrulation.

Immediately following fusion of the male and female pronuclei (15 min after oviposition at 25°C), there are nine synchronous nuclear divisions (Sonnenblick, 1950) (Fig. 1a). This is the syncytial cleavage stage (0.3-1.6 h), as there is no subdivision of the embryo into cells at this time. During this period, the nuclei divide every 9-10 min and there is little or no nuclear transcription (McKnight & Miller, 1976; Zalokar, 1976; Anderson & Lengyel, 1979). The rate of nuclear multiplication during this stage is the most rapid for any eukaryote, with an S phase of only 3-4 min (Rabinowitz, 1941). By the end of the syncytial cleavage stage, most of the nuclei have migrated from the interior of the egg to the surface of the embryo. Development of the blastoderm embryo continues in two stages, the syncytial and the cellular blastoderm.

A small number of nuclei at the posterior pole of the embryo, now termed a syncytial blastoderm, are enclosed by cell membranes to form the pole cells, which are the germ cell precursors. The remaining nuclei at the surface of the embryo undergo four more rapid and synchronous divisions, with the cycle time gradually increasing to 20 min (Zalokar & Erk, 1976; Warn & Magrath, 1982; Foe & Alberts, 1983). During the syncytial blastoderm stage (1.6-2.5 h) (Fig. 1b), prior to cellularization of the entire embryo, nuclear transcription is activated (McKnight & Miller, 1976; Zalokar, 1976). Histone mRNAs are among the first specific zygotic transcripts that can be detected at this time (Anderson & Lengyel, 1980).

The syncytial blastoderm stage is followed by a period of cellularization (2.5-3.5 h) during which membranes from the egg surface move downward into the embryo and surround individual nuclei (Turner & Mahowald, 1976) (Fig. 1c). The movements of gastrulation in Drosophila begin at 3.5 h, marking the end of the cellular blastoderm stage. During the formation of the cellular blastoderm embryo there is a lull in nuclear replication and DNA synthesis (Madhavan & Schneiderman, 1977; Anderson & Lengyel, 1981). The first true cell cycle is not completed until at least one hour after the beginning of the cellular blastoderm stage (Foe & Alberts, 1983).

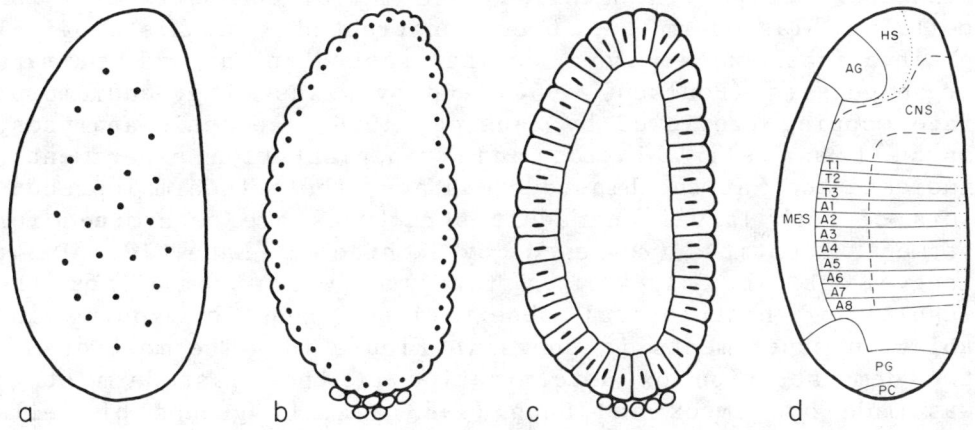

Fig. 1. Early embryonic development in Drosophila. (a) Syncytial cleavage stage (0-1.5 h); nuclei divide in center of embryo. (b) Syncytial blastoderm (1.5-2.5 h); nuclei have migrated to surface of embryo and undergo four more divisions. Note pole cells (primordial germ cells) at posterior. (c) Cellular blastoderm (2.5-3.5 h); membranes form around the surface nuclei. (d) Fate map of cells on surface of cellular blastoderm (redrawn from Anderson & Nusslein-Volhard, 1984). MES = mesoderm; CNS = central nervous system; HS = head skeleton; AG = anterior gut; PG = posterior gut; PC = pole cells; T1, T2, T3, A1, A2, A3, A4, A5, A6, A7 and A8 = ectoderm of thoracic segments one through three and abdominal segments one through eight.

Nuclear transcription at the cellular blastoderm is activated several-fold over its level at the syncytial blastoderm stage, to its highest rate throughout embryogenesis (McKnight & Miller, 1976; Anderson & Lengyel, 1979, 1981). Transcription of ribosomal RNA genes first occurs at the cellular blastoderm stage (McKnight & Miller, 1976). We will refer to the period of 2.5-3.5 h, between the temporary cessation of DNA replication at the end of the syncytial blastoderm and the onset of the movements of gastrulation, as the cellular blastoderm stage.

While rates of DNA and RNA synthesis change dramatically during the period from fertilization to cellularization, protein synthesis and the fraction of total mRNA being translated (i.e., on polysomes) do not change appreciably (Goldstein & Snyder, 1973; Zalokar, 1976; Lovett & Goldstein, 1977; Santon & Pellegrini, 1981; Anderson & Lengyel, 1983). Significant changes in translation efficiency occur for some mRNAs; however, there is a three-fold increase in the fraction of histone mRNA on polysomes during the first four hours of embryogenesis (Anderson & Lengyel, 1983).

Determination of cells to give rise to the ectodermal component of larval and adult segments occurs at the cellular blastoderm stage. A detailed fate map of the surface of the cellular blastoderm has been constructed from histological observations on embryos during gastrulation and primary organogenesis (Poulsen, 1950) and by means of gynandromorph fate mapping (reviewed by Janning, 1978). Genetic analyses, as well as cell ablation and transplantation experiments, indicate that at the blastoderm stage, the ectodermal precursors of both larval and adult structures are determined for segment identity (reviewed by Konrad & Mahowald, 1984; Lengyel et al., 1984). A fate map which summarizes the results of histological observations, genetic mapping and ablation experiments is shown in Figure 1d. The most dramatic demonstration of determination at the blastoderm stage was made by Simcox and Sang (1983): small groups of cells transplanted into ectopic locations differentiate into adult structures consistent with the original position of the cells in the blastoderm fate map of the donor embryo.

It is likely that the blastoderm transition in Drosophila represents a phenomenon in early embryogenesis which is of widespread and general importance. A similar "midblastula transition" has been described in amphibians. In this case, the transition includes a desynchronization and slowing of cell divisions, an activation of transcription, and a change in cell adhesiveness and motility, presumably in preparation for gastrulation (reviewed by Gerhart, 1980).

Maternal Contribution to Embryogenesis

The fact that most known mutations affecting nuclear multiplication and cellularization are maternal effect mutations (Bakken, 1973; Rice & Garen, 1975; Zalokar et al., 1975; Thierry-Mieg, 1976; Swanson & Poodry, 1980) and that little transcription occurs prior to these events, suggests that these events are controlled largely by the maternal genome [a few zygotic lethal mutants, however, also affect these processes (Wright, 1970)]. The phenotype of maternal-effect mutants indicates that the maternal genome also directs the establishment, during oogenesis, of the anterior-posterior and dorsal-ventral axes of the embryo (Nusslein-Vohard, 1979; Nusslein-Volhard et al., 1983; Anderson & Nusslein-Volhard, 1984).

These genetic observations are consistent with biochemical analyses indicating that there is a large store of maternal mRNA of high sequence complexity in the egg (Lovett & Goldstein, 1977; Anderson & Lengyel, 1979; Hough-Evans et al., 1980). Most of the mass of maternal mRNA is still on polysomes in the embryo by the end of the cellular blastoderm stage (Anderson & Lengyel, 1979). Furthermore, the great majority of maternal mRNA sequence complexity is still present in the embryo at the beginning of gastrulation (Arthur et al., 1979).

Zygotic Contribution to Early Embryogenesis

As new embryonic transcripts constitute 14% of the mass of mRNA on polysomes by the start of gastrulation (Anderson & Lengyel, 1981), they clearly have the potential to affect events during early embryogenesis. While most mRNA sequences present at the blastoderm stage are the same as those of maternal mRNA, hybridization studies indicate that a small, but significant, number of new sequences appears by the end of the cellular blastoderm stage (Arthur et al., 1979; Goldstein & Arthur, 1979). Since mRNAs that are present prior to the blastoderm stage are maternal, we hypothesize that the small number of genes newly expressed at the blastoderm stage are likely to be involved in the critical events occurring at this time that are not controlled by the maternal genome: determination and the beginning of morphogenetic movements.

There is strong genetic evidence that early transcription of a number of zygotic genes is required at the blastoderm stage, particularly in establishing the pattern of the Drosophila embryo. A group of 15 embryonic lethal complementation groups affecting segment determination (which is believed to occur at the blastoderm stage) has been isolated. Three of these mutants show an abnormal phenotype shortly after the onset of gastrulation (Nusslein-Volhard & Wieschaus, 1980); the genes defined by these embryonic lethal mutations must therefore be transcribed during the blastoderm stage.

Molecular analyses of a number of embryonic lethal loci also provide evidence for activation of new gene functions at the blastoderm stage. Messenger RNAs specific to, or newly appearing at, the blastoderm stage have been detected by hybridization to cloned DNA of the fushi tarazu and Krüppel genes and the bithorax complex (Scott et al., 1983; Preiss et al., 1985; R. Saint & D. Hogness, pers. comm.). All of these genes are involved in controlling embryonic pattern formation, affecting either segment determination or segment identity.

Additional biochemical evidence for the expression of a small number of new genes at the blastoderm stage comes from the analysis of the pattern of protein synthesis. Of 300-400 resolvable proteins, only 11 appear at the blastoderm stage (Gutzeit, 1980; Sakoyama & Okubo, 1981; Savoini et al., 1981; Trumbly & Jarry, 1984), and of these, only 3 are blastoderm-specific.

CONCLUSION

At the cellular blastoderm stage of embryogenesis in Drosophila, events occur which are crucial to later development: a transition to a slower and asynchronous mode of nuclear multiplication, dramatic activation of transcription, and cellularization. Of greatest interest is the determination of cells of the blastoderm to give rise to a significant portion of the basic body plan of the embryo. Clearly a more complete understanding of Drosophila embryogenesis depends on the characterization of the genes involved in processes that occur at this stage. Genetic and biochemical evidence indicate that only a minority of the total genes active during embryogenesis are involved in controlling events at the blastoderm stage and that genes specific to the zygotic program may make a qualitatively different contribution to early embryogenesis than genes active during oogenesis.

The analysis of genes involved in developmental processes has traditionally been carried out by identifying mutant genes which interrupt the process. It is difficult on the basis of purely genetic criteria, however, to identify genes whose activity is restricted to the blastoderm stage, since all of the phenotypes which would arise are not obvious a priori. An alternative to the classical genetic approach, which we describe here, is the identification, by molecular techniques, of genes specifically expressed at the blastoderm stage. The test of our hypothesis, that blastoderm-specific genes are actually required for events which occur at this stage, will be the isolation and characterization of mutations in the cloned genes.

We show here that it is possible, by molecular cloning, to identify genes whose expression is specific to the blastoderm stage, and by genetic techniques, to initiate studies on the requirement for such genes in embryogenesis. We expect the analysis of such cloned genes to provide information about (1) the number and nature of functions which are required for the events specific to the blastoderm stage, and (2) whether the functions newly activated at the blastoderm stage are qualitatively different from those provided by the maternal program.

EXPERIMENTAL

Isolation of Blastoderm-Specific Clones

We screened a Drosophila genomic DNA library (Yen et al., 1979), using as probe ^{32}P-kinased poly A$^+$ polysomal RNA from cellular blastoderm stage embryos, in the presence of an excess of unlabeled poly A$^+$ RNA from early cleavage-stage embryos as competitor (Mangiarotti et al., 1981). Plaques giving a positive signal were plaque purified and rescreened twice, using ^{32}P-cDNA to cellular blastoderm RNA as a probe. In this way, we identified 16 putative blastoderm-differential phages.

RNA gel blots of poly A$^+$ RNA from early cleavage-, cellular blastoderm-, and late gastrula-stage embryos were probed with ^{32}P-DNA from each blastoderm-differential phage. On the basis of this characterization we determined that the DNA inserts of four of the phage belong to two repeated DNA families (one of which is B104; Scherer et al., 1982), those of three of the phage encode RNAs differentially expressed at the blastoderm stage (but expressed at a higher level later in development), those of two of the phage are the same and encode a blastoderm-differential RNA maximally expressed at the grastrula stage, and those of four of the phage do not encode differentially expressed RNAs (Roark et al., in press). Only three phage carried DNA inserts encoding blastoderm-specific transcripts. We selected these phage and their encoded RNAs for more detailed analysis.

Developmental Analysis of Blastoderm-Specific RNAs

The changing expression during development of the poly A^+ RNAs encoded by the DNA inserts of three phage is shown in the RNA gel blot autoradiograms of Figure 2. The RNAs in the figure display the stage-specific expression pattern that we predicted should exist for a small number of mRNAs, based on the uniqueness of the events which occur at the blastoderm stage. The DNA insert of each phage encodes at least one poly A^+ RNA which is not present in early cleavage-stage embryos (i.e., is not maternal), and which reaches a significant concentration by the cellular blastoderm stage. Other stages probed included additional embryonic stages, as well as first instar larvae, and early and late pupae (Roark et al., In press). None of the blastoderm-specific RNAs reappears at any of the stages probed (Roark et al., in press).

The DNA insert of phage IB142 encodes a highly blastoderm-specific RNA which is 1.6 kb in length (Fig. 2a). This RNA is moderately abundant, constituting 0.07% of the total poly A^+ RNA at the cellular blastoderm stage [determined by slot blot hybridization (Roark et al., in press)]. This RNA is not expressed significantly at any other stage in development. The genomic insert of phage IB142 contains two homologous (or identical) RNA coding regions which appear to be arranged in a 3'-3' orientation (E. Gustavson et al., unpubl.). Thus the hybridization signal in Figure 2a may represent two mRNAs.

The DNA insert of phage IB150 encodes one blastoderm-specific RNA, 2.7 kb in length, as well as three larger RNAs of 3.0, 4.4 and 4.5 kb (Fig. 2b). The blastoderm-specific RNA is the most abundant of the IB150 RNAs and constitutes 0.03% of the poly A^+ RNA at the cellular blastoderm stage. The larger RNAs, although not blastoderm-specific, are maximally expressed during the first eight hours of embryogenesis. Hybridization between blastoderm poly A^+ RNA and cDNA clones homologous to the phage IB150 insert suggests that the 2.7 kb blastoderm-specific RNA shares portions of its coding regions with the 3.0 and the 4.5 kb RNAs, while the 4.4 kb RNA is encoded by an adjacent genomic DNA fragment (P.D. Boyer, unpubl. results). Each of these four RNAs exhibit unique developmental expression profiles, thus (although this remains to be confirmed) it is unlikely that the larger RNAs represent precursors to the blastoderm-specific RNA.

BLASTODERM-SPECIFIC GENES 247

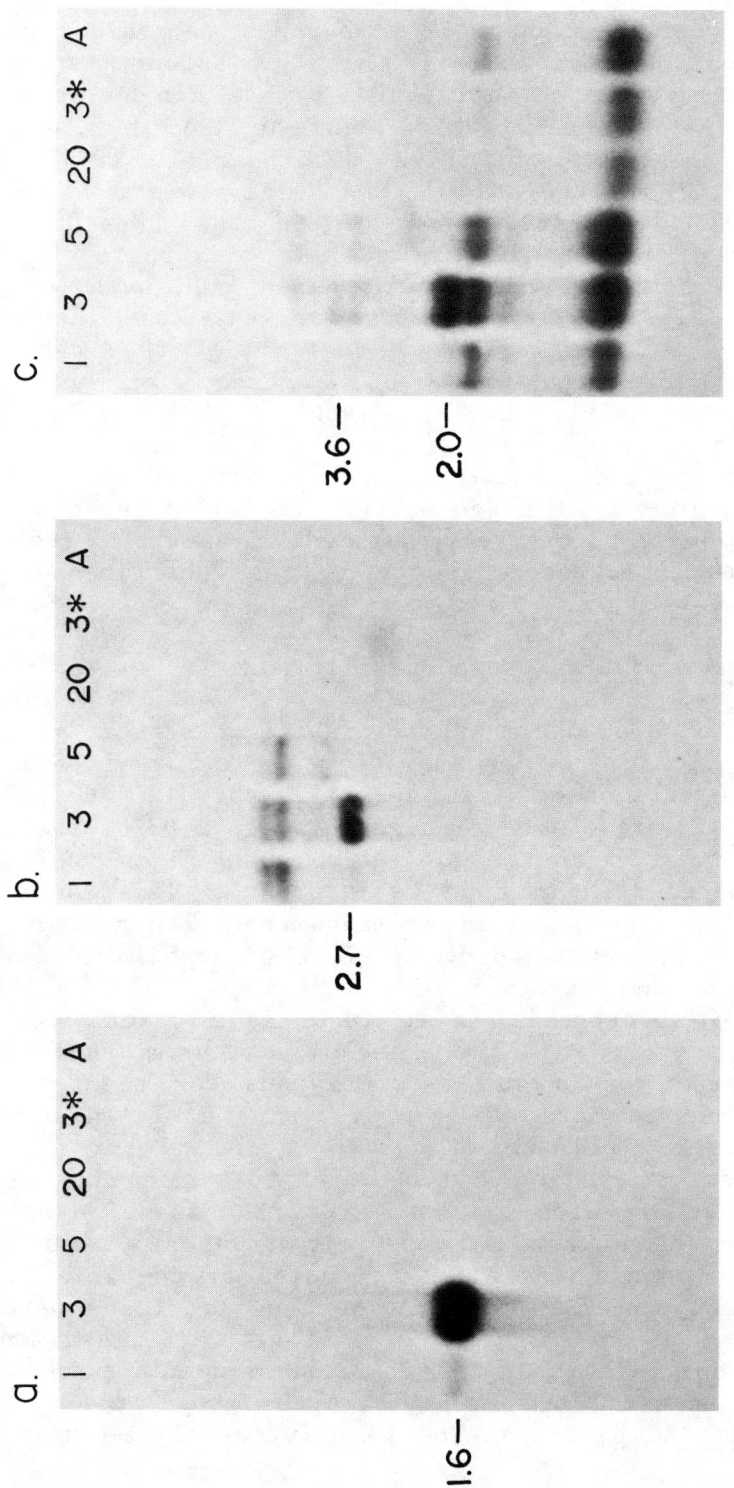

Fig. 2. RNA gel blot analysis of stage specific mRNAs. Staged poly A$^+$ RNAs were denatured with glyoxal, electrophoresed in 1.1% agarose gels and blotted to nitrocellulose. Each lane contains 1 μg of poly A$^+$ RNA from the following time points post-fertilization: 1 = 0.5-1.5 h (syncytial cleavage); 3 = 2.5-3.5 h (cellular blastoderm); 5 = 5-6 h (late gastrula); 20 = 19.5-20.5 h (late embryo); 3* = 4 days (late third instar larva); A = adult. Molecular size markers were Hind III-digested λ-DNA denatured with glyoxal, and Drosophila poly A$^-$ RNA, run in parallel in the same gel. The blots were hybridized with the following phage DNAs, ^{32}P-labeled by nick-translation: (a) IB142; (b) IB150; and (c) IB153. The blastoderm-specific mRNAs are indicated by arrows and their sizes given in kilobases.

The DNA insert of phage IB153 encodes a complex set of RNAs (Fig. 2c). Two of these are highly blastoderm-specific in their expression: one, a relatively more abundant mRNA of 2.0 kb length, the other a less abundant RNA of 3.6 kb in length. The most abundant blastoderm-specific IB153 mRNA constitutes 0.07% of the poly A^+ RNA at blastoderm. Similar to what was found for IB150, most of the other RNAs detected by the labeled IB153 probe are maximally expressed in the early embryo; however, one is expressed continuously. By restriction analysis we have determined that the insert of IB153 is identical over a 17.4 kb region with the insert of phage C25, isolated previously in a screen for ribosomal protein genes (Vaslet et al., 1980). The continuously expressed 0.6 kb RNA thus encodes a ribosomal protein, r-pro49 (Vaslet et al., 1980; Vincent et al., 1984). Within 5 kb, the DNA insert of phage C25 (phage IB153) encodes five overlapping and alternatively spliced transcripts, two of which are the blastoderm-specific RNAs described above (Vincent et al., 1984).

Genetic Analyses of Blastoderm-Specific Loci

Each of the blastoderm-specific phage DNAs hybridizes to a unique location in the euchromatin of the larval salivary gland polytene chromosomes. Phage IB142 hybridizes to 75C on the left arm of chromosome 3, phage IB150 to 25D3 on the left arm of chromosome 2, and phage IB153 to 99D4,8 on the right arm of chromosome 3 (Roark et al., in press). This localization to single chromosomal sites means that each gene can be investigated by classical procedures of Drosophila genetics.
A powerful approach to investigate the function of a cloned gene in Drosophila is to identify a deletion which "uncovers" the gene in the homologous chromosome. The phenotype of embryos which are homozygous for the deletion can be examined to more rapidly discern what developmental steps are blocked because of the lack of the normal gene product. Such a deletion can be used to screen for point mutations, some of which should disrupt the function of the gene encoding the blastoderm-specific transcript. Of the three sequences mapped here, deficiencies of the 25D region cannot presently be used in this way because flies heterozygous for such deficiencies are sterile (Velissariou & Ashburner, 1980). The two other blastoderm-specific loci, at 75C and 99D, however, are amenable to this type of analysis. From both in situ hybridization to polytene chromosomes and

restriction fragment length polymorphism analysis, we have mapped (M. Roark, unpubl. data) the blastoderm-specific sequence IB142 into W^{R4} (obtained from W. Seagraves), a small deletion of 75B8 to 75C5-7 (Fig. 3a). This deletion can therefore be used in the types of studies described above. Although the region from 99C to 99E is not viable in a single dose (Lindsley et al., 1972) we have succeeded in obtaining

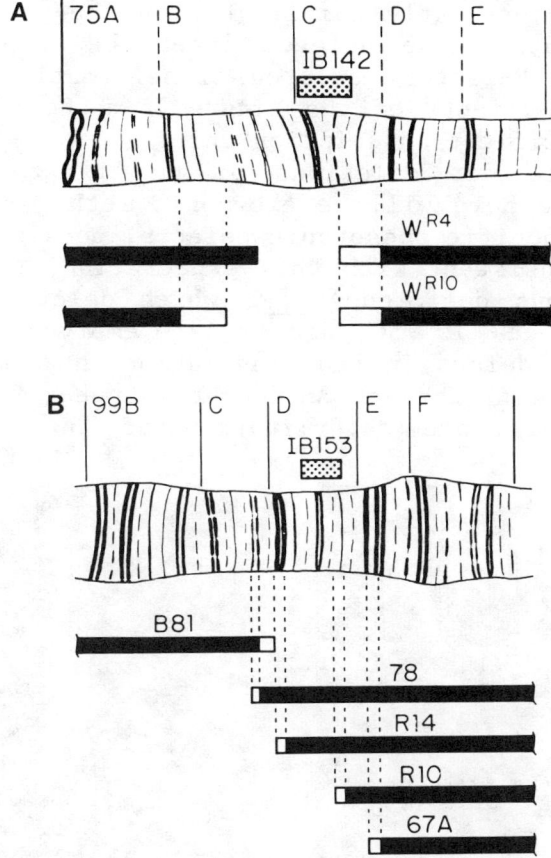

Fig. 3. Deletions available for genetic analysis of the blastoderm-specific loci at 75C and 99D. In both drawings, the heavy black bars represent the DNA which is present in the mutant chromosome. The chromosomal (in situ) localization of each cloned blastoderm-specific sequence is shown as a stippled bar above the chromosome. a. Deletions in the 75BC region. Two deletions, W^{R4} and W^{R10}, isolated by W. Seagraves by reverting the dominant wing mutation Wrinkled, are shown. b. Construction of a 99D deficiency. Region 99B-F of polytene chromosome three, showing the extent of duplications 78, R14, R10, and 67A and the Y;3 translocation B81 used to construct synthetic deficiencies in the 99D region. The chromosomal localization of the IB153 clone is deleted by a combination of B81 and either of the R10 or 67A duplications, since IB153 DNA does not hybridize in situ to either of the duplications R10 or 67A or to the B81 terminal deficiency chromosome shown in the figure (K. Kongsuwan & J.A. Lengyel, unpubl. data).

a smaller deficiency, from 99C8 to 99D8, which uncovers the IB153 DNA and is fertile (although a Minute) as a heterozygote (K. Kongsuwan et al., in prep.). This 99D deficiency is synthetic, constructed by combining the terminal deficiency chromosome B81 (which extends only to chromosome band 99C) with the duplication 67A (which begins in 99D and extends to the tip of the right arm of the third chromosome) (see Fig. 3b). Characterization of the phenotype of the 99D deletion homozygotes is presently complicated by the presence of the Minute gene within the deleted interval. Experiments are underway using P-factor mediated transformation to separate the phenotypic effects of the Minute from those of the IB153 blastoderm-specific transcription unit.

We expect that deletions of the blastoderm-specific genes described here will be embryonic lethal and that they will lead to specific phenotypic defects, possibly in pattern formation. Consistent with this expectation, embryos carrying a homozygous deficiency W^{R4}, which deletes the blastoderm-specifc locus at 75C, die as late embryos. The W^{R4}/W^{R4} embryos show a defect in head involution; however, formation of the segments T1 through A8 does not appear to be affected (see Fig. 4). Since determination of the head segments

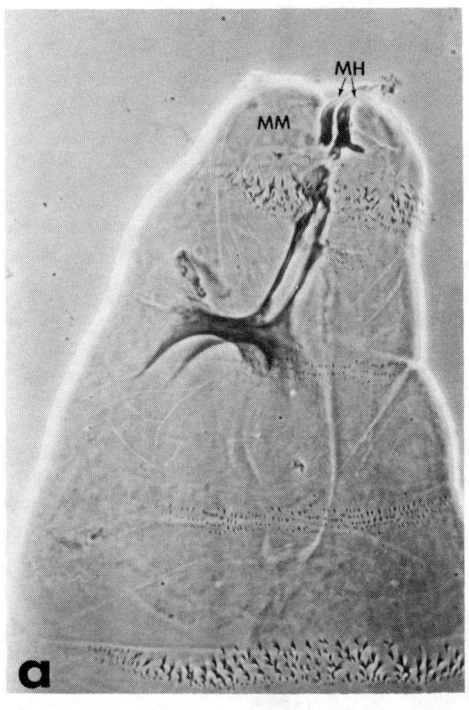
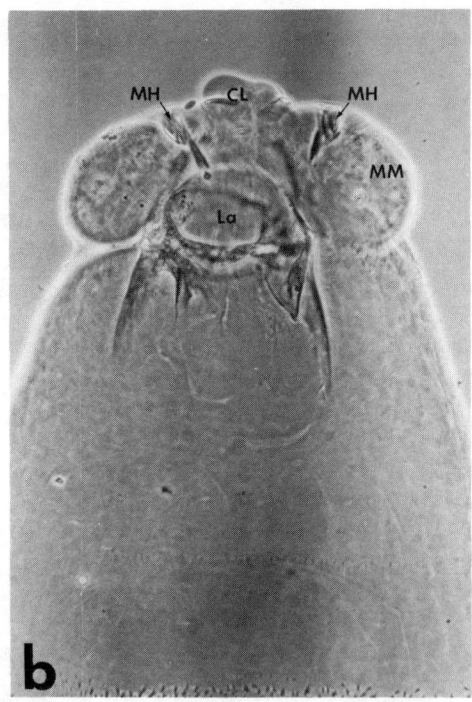

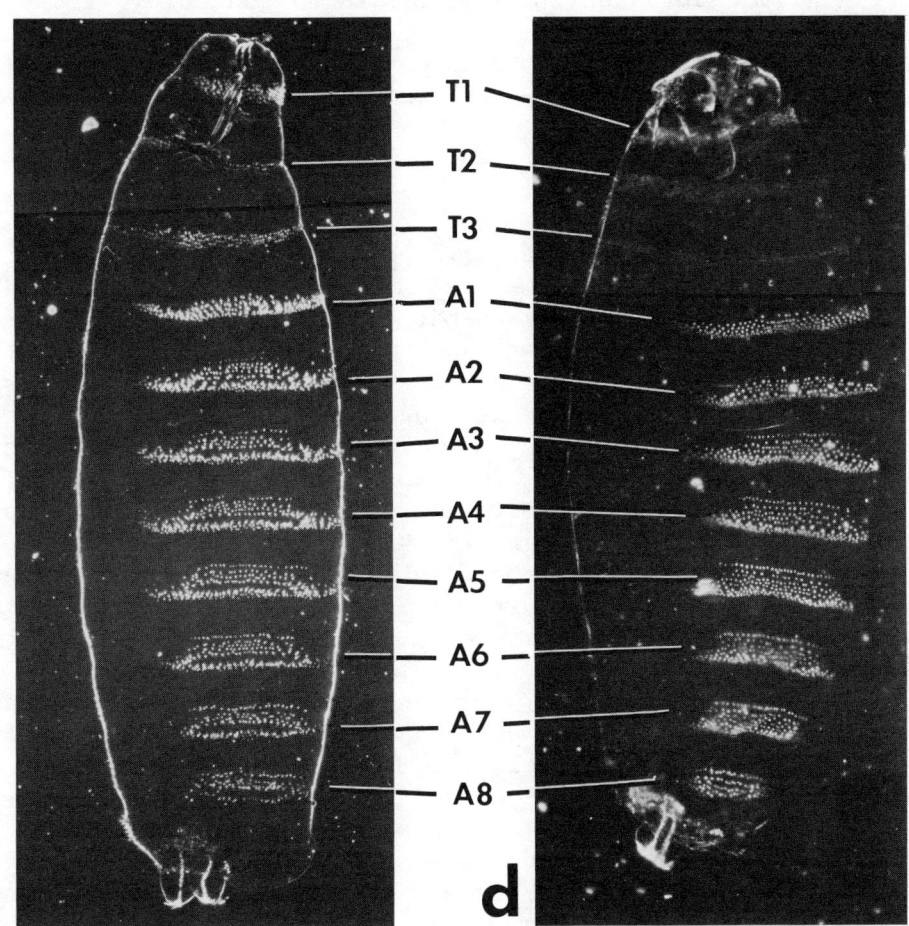

Fig. 4. Embryonic phenotype of the homozygous 75C deficiency. Embryos with homozygous deletions for the 75C region were obtained as one-fourth of the progeny from the W^{R4}/TM3 stock [TM3 is a multiply inverted, "balancer" chromosome (Lindsley & Grell, 1968)]. Phase contrast photographs of the heads of cleared whole mounts of mature embryos show the anterior cuticle of morphologically wild type (TM3/TM3) (a) and mutant (W^{R4}/W^{R4}) (b) embryos. Note the lack of involution of the head in the mutant embryo in (b). The clypeolabral (CL) and labial (La) lobes, which are evident in the mutant head, are completely involuted, and hence not visible, in the wild type. The mandibular/maxillary lobes (MM) and their derivative structures the mouth hooks (MH), widely separated in the mutant, are partially involuted and close together in the wild type. The same mutant head phenotype is seen for embryos of the W^{R4}/W^{R10} genotype, obtained from the cross W^{R4}/TM3 X W^{R10}/TM3 (M. Roark, unpubl. data). This indicates that the head phenotype is due to the absence of genes within the W^{R4} deletion. Dark field photographs of cleared whole mounts of morphologically wild type (TM3/TM3) (c) and mutant (W^{R4}/W^{R10}) (d) embryos show that thoracic (T2-T3) and abdominal (A1-A8) segments are not affected by the homozygous deletion.

occurs at the blastoderm stage (Underwood et al., 1980), it is possible that the head phenotype of the $\overline{WR^4}$ homozygous embryos is due to the absence of the IB142 blastoderm-specific transcription units. It should be possible, by isolating point mutations in the 75C region and by carrying out transformation studies, to determine whether the characteristic defect exhibited by the 75C homozygous-deficient embryos is due to the absence of the IB142 blastoderm-specific sequence.

DISCUSSION

The blastoderm stage of embryogenesis in Drosophila is of particular importance, as it is a time of transitions in DNA replication, RNA synthesis, and cell determination. Yet very few genes are known, on the basis of either genetic or molecular studies, to be expressed primarily at this stage. Our working hypothesis is that there is a small class of genes whose expression is activated and required for events at or immediately following this stage, and that these genes can be identified by molecular hybridization techniques. We have succeeded in identifying three genomic sequences encoding RNAs dramatically limited in their expression to the blastoderm stage. The results of a genetic analysis of one of these loci are consistent with the view that it is required for early embryogenesis.

The blastoderm-specific genes described here are unusual both in their limitation of expression to a stage at which there is no overt differentiation and in their extreme stage specificity. While sequences differentially expressed in Drosophila early embryogenesis have been identified in a number of screens (Scherer et al., 1981; Sina & Pellegrini, 1982), and the genomic DNA of IB153 was isolated serendipitously in a screen for ribosomal protein genes (Vaslet et al., 1980), the group of genes described here are the first to have been isolated in a systematic screen for Drosophila blastoderm-specific genes. A number of genes encoding RNAs specific to the gastrula stage in Xenopus have also been isolated by molecular cloning procedures (Sargent & Dawid, 1983). Thus molecular hybridization screens allow the identification of genes specific to brief embryonic stages, prior to the appearance of differentiated cell types, in both Drosophila and Xenopus.

The goal of our molecular screen was to identify genes (and therefore functions) specifically active at the blastoderm stage, using an approach complementary to that of genetic techniques. None of our blastoderm-specific sequences have been localized to sites of known early embryonic mutants (that are well-localized) nor do they correspond to any loci that are believed, on the basis of genetic analyses, to act at the blastoderm stage (Nusslein-Volhard & Wieschaus, 1980; Jurgens et al., 1984; Nusslein-Volhard et al., 1984). We consider it likely that we have identified new functions required at the blastoderm stage. The technique of genomic screening followed by in situ hybridization should thus provide an approach, complementary to that of classical genetics, for the identification of functions required at the blastoderm.

The DNA inserts of all three phage isolated in this screen for blastoderm-specific sequences are complex, in that each contains multiple coding regions and two possess overlapping coding regions. It may be significant that three loci involved in cell determination and pattern formation, Notch, Antennapedia and Ultrabithorax, also contain DNA segments encoding multiple transcripts (Artavanis-Tsakonas et al., 1983; Garber et al., 1983; Scott et al., 1983; R. Saint et al., pers. comm.).

A number of approaches can be used to investigate the functional significance of the blastoderm-specific genes we have isolated. A genetic analysis will include the isolation of mutations in the genes and characterization of the mutant phenotypes. At the molecular level, the genes can be sequenced and this sequence compared to known protein sequences for homology. The location of cells in which the gene is active can be determined by genetic cell marking techniques, by in situ hybridization to RNA in embryo sections, and by immunological techniques, using antibodies prepared against the gene product (or peptide fragments of it) to stain embryo sections. By investigating the function of the blastoderm-specific genes described here, we should increase our understanding of the unique processes occurring at the blastoderm stage.

ACKNOWLEDGMENTS

This work was supported by grants from the NIH (HD-09948) and NSF (PCM 21830) to J.A. Lengyel, by training grants USPHS GM 07104 to M. Roark and GM07185 to P.D. Boyer and P.A. Mahoney, and by an NSF graduate fellowship to P.A. Mahoney. We thank Ross MacIntyre and Bill Seagraves for providing Drosophila stocks, Vincenzo Pirrotta for providing us with a B104 clone, Michael Rosbash for providing an unpublished restriction map of C25 and Elizabeth Gustavson and Richard Baldarelli for providing us with unpublished data.

LITERATURE CITED

Anderson, K.V. and J.A. Lengyel. 1979. Rates of synthesis of major classes of RNA in Drosophila embryos. Dev. Biol. 70: 217-231.

Anderson, K.V. and J.A. Lengyel. 1980. Changing rates of histone mRNA synthesis and turnover in Drosophila embryos. Cell 21: 717-727.

Anderson, K.V. and J.A. Lengyel. 1981. Changing rates of DNA and RNA synthesis in Drosophila embryos. Dev. Biol. 82: 127-138.

Anderson, K.V. and J.A. Lengyel. 1983. Histone gene expression in Drosophila development: Multiple levels of gene regulation. pp. 135-161. In: Histone Genes and Histone Gene Expression. G. Stein, J. Stein, and W. Marzluff (eds.). John Wiley and Sons, N.Y.

Anderson, K.V. and C. Nusslein-Volhard. 1984. Genetic analysis of dorsal-ventral embryonic pattern in Drosophila. pp. 269-189. In: Pattern Formation. G.M. Malacinski and S.V. Bryant (eds.). MacMillan Publishing Co., N.Y.

Artavanis-Tsakonas, S., M. Muskavitch, and B. Yedvobnick. 1983. Molecular cloning of Notch, a locus affecting neurogenesis in Drosophila melanogaster. Proc. Natl. Acad. Sci., USA 80: 1977-1981.

Arthur, C.G., C.M. Weide, W.S. Vincent, and E.S. Goldstein. 1979. mRNA sequence diversity during early embryogenesis in Drosophila melanogaster. Exp. Cell Res. 121: 87-94.

Bakken, A.H. 1973. A cytological and genetic study of oogenesis in Drosophila melanogaster. Dev. Biol. 33: 100-122.

Foe, V.E. and B.M. Alberts. 1983. Studies of nuclear and cytoplasmic behaviour during the five mitotic cycles that precede gastrulation in Drosophila embryogenesis. J. Cell Sci. 61: 31-70.

Fullilove, S.L. and A.G. Jacobson. 1978. Embryonic development: Descriptive. pp. 105-227. In: The Genetics and Biology of Drosophila, Vol. 2c. M. Ashburner and T.R.F. Wright (eds.). Academic Press, N.Y.

Garber, R., A. Kuroiwa, and W. Gehring. 1983. Genomic and cDNA clones of the homotic locus Antennopedia in Drosophila. EMBO J. 2: 2027-2036.

Gerhart, J.C. 1980. Mechanisms regulating pattern formation in the amphibian egg and early embryo. pp. 133-316. In: Biological Regulation and Development, Vol. 2. R.F. Goldberger (ed.). Plenum Press, N.Y.

Goldstein, E.S. and Arthur, C.G. 1979. Isolation and characterization of cDNA complementary to transient maternal poly(A)$^+$ RNA from the Drosophila oocytes. Biochim. Biophys. Acta 565: 265-274.

Goldstein, E.S. and L.A. Snyder. 1973. Protein synthesis and distribution of ribosomal elements in ovarian oocytes and developmental stages of Drosophila melanogaster. Exp. Cell Res. 81: 47-56.

Gutzeit, H.O. 1980. Expression of the zygotic genome in blastoderm stage embryos of Drosophila: An analysis of a specific protein. Wilhelm Roux's Archiv. 188: 153-156.

Hough-Evans, B.R., M. Jacobs-Lorena, M.R. Cummings, R.J. Britten, and E.H. Davidson. 1980. Complexity of RNA in eggs of Drosophila melanogaster and Musca domestica. Genetics 95: 81-94.

Janning, W. 1978. Gynandromorph fate maps in Drosophila. pp. 1-28. In: Genetic Mosaics and Cell Differentiation. W.J. Gehring (ed.). Springer Verlag, N.Y.

Jurgens, G., H. Kluding, C. Nusslein-Volhard, and E. Wieschaus. 1984. Mutations affecting the pattern of the larval cuticle in Drosophila melanogaster. II. Zygotic loci on the third chromosome. Wilhelm Roux's Arch. 193: 283-295.

Konrad, K.D. and A.P. Mahowald. 1984. Genetic and developmental approach to understanding determination in early development. pp. 167-188. In: Molecular Aspects of Early Development. G.M. Malacinski and W.H. Klein (eds.). Plenum Press, New York.

Lengyel, J.A., S.R. Thomas, P.D. Boyer, F. Salas, T.R. Strecker, I. Lee, M.L. Graham, M. Roark, and E.M. Underwood. 1984. Isolation and characterization of genes differentially expressed in early Drosophila embryogenesis. pp. 219-251. In: Molecular Aspects of Early Development. G.M. Malacinski and W.H. Klein (eds.). Plenum Press, N.Y.

Lindsley, D.L. and E.H. Grell. 1968. Genetic variations of Drosophila melanogaster. Carnegie Institution of Washington, Publication No. 627. 407 pp.

Lindsley, D.L., L. Sandler, B.S. Baker, A.T.C. Carpenter, R.E. Denell, J.C. Hall, P.A. Jacobs, G.L. Gabor Miklos, B.K. Davis, R.C. Gethmann, R.W. Hardy, A. Hessler, S.M. Miller, H. Nozawam, D.M. Parry, and M. Gould-Somero.

1972. Segmental aneuploidy and the genetic gross structure of the Drosophila genome. Genetics 71: 157-184.

Lovett, J.A. and E.S. Goldstein. 1977. The cytoplasmic distribution and characterization of poly(A)$^+$ RNA in oocytes and embryos of Drosophila. Dev. Biol. 61: 70-78.

Madhavan, M.M. and H.A. Schneiderman. 1977. Histological analysis of the dynamics of growth and imaginal discs and histoblast nests during the larval development of Drosophila melanogaster. Wilhelm Roux's Archiv. 183: 269-305.

Mangiarotti, G., S. Chung, C. Zuker, and H. Lodish. 1981. Selection and analysis of cloned developmentally regulated Dictyostelium discoideum genes by hybridization competition. Nucl. Acids Res. 9: 947-963.

McKnight, S.L. and O.L. Miller, Jr. 1976. Ultrastructural patterns of RNA synthesis during early embryogenesis of Drosophila melanogaster. Cell 8: 305-319.

Nusslein-Volhard, C. 1979. Maternal effect mutations that alter the spatial coordinates of the embryo. pp. 185-211. In: Determinants of Spatial Organization. S. Subtelny (ed.). Academic Press, New York.

Nusslein-Volhard, C. and E. Wieschaus. 1980. Mutations affecting segment number and polarity in Drosophila. Nature 287: 795-801.

Nusslein-Volhard, C., E. Wieschaus, and H. Kluding. 1984. Mutations affecting the pattern of the larval cuticle in Drosophila melanogaster. I. Zygotic loci on the second chromosome. Wilhelm Roux's Arch., 193: 267-282.

Nusslein-Volhard, C., E. Wieschaus, and G. Jurgens. In Press. Segmentation in Drosophila - a genetic analysis. Verhandlung der deutsche zoologische Gesellschaft.

Poulsen, D.F. 1950. Histogenesis, organogenesis and differentiation in the embryo of Drosophila melanogaster Meigen. pp. 168-174. In: Biology of Drosophila. M. Demerec (ed.). Hafner Publishing Co., N.Y.

Preiss, A., U.B. Rosenberg, A. Kienlin, E. Seifert, and H. Jäckle. 1985. Molecular genetics of Krüppel, a gene required for segmentation of the Drosophila embryo. Nature 313: 27-32.

Rabinowitz, M. 1941. Studies on the cytology and early embryology of the egg of Drosophila melanogaster. J. Morphol. 69: 1-49.

Rice, T.B. and A. Garen. 1975. Localized defects of blastoderm formation in maternal effect mutants of Drosophila. Dev. Biol. 43: 277-286.

Roark, M., P.A. Mahoney, M.L. Graham, and J.A. Lengyel. In Press. Blastoderm-differential and blastoderm-specific genes of Drosophila melanogaster. Dev. Biol.

Sakoyama, Y. and S. Okubo. 1981. Two-dimensional gel patterns of protein species during development of Drosophila embryos. Dev. Biol. 81: 361-365.

Santon, J.B. and M. Pellegrini. 1981. Rates of ribosomal protein and total protein synthesis during Drosophila early embryogenesis. Dev. Biol. 85: 252-257.

Sargent, T.D. and F.B. Dawid. 1983. Differential gene expression in the gastrula of Xenopus laevis. Science 222: 135-139.

Savoini, A., F. Micali, R. Marzari, F. de Cristini, and G. Graziosi. 1981. Low variability of the protein species synthesized by Drosophila melanogaster embryos. Wilhelm Roux's Arch. 190: 161-167.

Scherer, G., J. Telford, C. Baldani, and V. Pirrotta. 1981. Isolation of cloned genes differentially expressed at early and late stages of Drosophila embryonic development. Dev. Biol. 86: 438-447.

Scherer, G., C. Tschudi, J. Perera, H. Delius, and V. Pirrotta. 1982. B104, a new dispersed repeated gene family in Drosophila melanogaster and its analogies with retroviruses. J. Mol. Biol. 157: 435-452.

Scott, M.P., A.J. Weiner, T.I. Hazelrigg, B.A. Polisky, V. Pirrotta, F. Schalenghe, and T.C. Kaufman. 1983. The molecular organization of the Antennapedia locus of Drosophila. Cell 35: 763-776.

Simcox, A.A. and J.H. Sang. 1983. When does determination occur in Drosophila embryos? Dev. Biol. 97: 212-221.

Sina, B.J. and M. Pellegrini. 1982. Genomic clones coding for some of the initial genes expressed during Drosophila development. Proc. Natl. Acad. Sci., USA. 79: 7351-7355.

Sonnenblick, B.P. 1950. The early embryology of Drosophila melanogaster. pp. 62-167. In: Biology of Drosophila. M. Demerec (ed.). Hafner Publishing Co., N.Y.

Swanson, M.M. and C.A. Poodry. 1980. Pole cell formation in Drosophila melanogaster. Dev. Biol. 75: 419-430.

Thierry-Mieg, D. 1976. Study of a temperature sensitive mutant grandchildless-like in Drosophila melanogaster. J. Microsc. Biol. Cell. 25: 1-6.

Trumbly, R.J. and B. Jarry. 1983. Stage specific protein synthesis during early embryogenesis in Drosophila melanogaster. EMBO J. 2: 1281-1290.

Turner, R.F. and A.P. Mahowald. 1976. Scanning electron microscopy of Drosophila embryogenesis. Dev. Biol. 50: 95-108.

Underwood, E.M., F.R. Turner, and A.P. Mahowald. 1980. Analysis of cell movements and fate mapping during early embryogenesis in Drosophila melanogaster. Dev. Biol. 74: 286-301.

Vaslet, C.A., P. O'Connell, M. Izquierdo, and M. Rosbash. 1980. Isolation and mapping of a cloned ribosomal protein gene of Drosophila melanogaster. Nature 285: 674-676.

Velissariou, V. and N. Ashburner. 1980. The secretory proteins of the larval salivary gland of Drosophila melanogaster. Chromosoma 77: 13-27.

Vincent, A. P. O'Connell, M.R. Gray, and M. Rosbash. 1984. Drosophila maternal and embryo mRNAs transcribed from a single transcription unit use alternate combinations of exons. EMBO J. 3: 1003-1013.

Warn, R.M. and R. Magrath. 1982. Observations by a novel method of surface changes during the syncytial blastoderm stages of the Drosophila embryo. Dev. Biol. 89: 540-548.

Wright, T.R.F. 1970. The genetics of embryogenesis of Drosophila. Advan. Genet. 15: 261-395.

Yen, P., N.D. Hershey, R. Robinson, and N. Davidson. 1979. Sequence organization of Drosophila tRNA genes. pp. 133-142. In: ICN-UCLA Symposium on Eukaryotic Gene Regulation. R. Axel, T. Maniatis, and C.F. Fox (eds.). Academic Press, N.Y.

Zalokar, M. 1976. Autoradiographic study of protein and RNA formation during early development of Drosophila eggs. Dev. Biol. 49: 425-437.

Zalokar, M., C. Audit, and I. Erk. 1975. Developmental defects of female-sterile mutants of Drosophila melanogaster. Dev. Biol. 47: 419-432.

Zalokar, M. and I. Erk. 1976. Division and migration of nuclei during early embryogenesis of Drosophila melanogaster. J. Microsc. Biol. Cell. 25: 97-106.

TRANSLATIONAL CONTROL OF CELL CYCLE-RELATED PROTEINS IN EARLY EMBRYOS

Joan V. Ruderman

ABSTRACT

The mature oocytes of most animals, including the marine clam Spisula, contain a large pool of mRNA. Before fertilization, Spisula oocytes translate a small subset of these mRNAs; within 10 min after fertilization, these mRNAs are released from polysomes, and another completely different subset of pre-existing mRNAs is translationally activated and recruited onto polysomes. This translational switch leads to dramatic changes in the kinds of proteins that are made at fertilization. The activation of certain mRNAs is essential for development: it has long been known that, despite the possession of a large maternal stockpile of structural proteins and enzymes, ongoing protein synthesis is absolutely required for the embryo to proceed throughout first cleavage and beyond. What is responsible for the translational repression of one group of mRNAs and the activation of another group right after fertilization? What roles do these translationally regulated proteins play? Why is it so important for certain mRNAs to be kept inactive until after fertilization? Using cDNA clones complementary to maternal mRNA sequences, we have found that the mRNAs active in the oocyte are poly A$^+$ and lose their poly A tails when they are released from polysomes after fertilization. In contrast, mRNAs that are stored in the oocyte in an inactive form are poly A-deficient and gain a poly A tail when they are activated in stored mRNAs. Two mRNAs, those encoding proteins called A and C, have been sequenced and compared to see if they contain potential translational regulatory sequences.

Both have A,T-rich 5' and 3' noncoding regions. As to the functions of these proteins, some hints are beginning to surface. Protein A (53 kd), also known as cyclin, is synthesized continuously in early embryos but is destroyed at each early cleavage division. Its amino acid sequence suggests that it may be a DNA-binding, membrane-associated protein. Protein C (41 kd) encodes the small subunit of the enzyme ribonucleotide reductase, the first enzyme in the de novo DNA synthesis pathway.

INTRODUCTION

In most species, the developing oocyte synthesizes and accumulates large stockpiles of structural proteins, enzymes and messenger RNAs that are used after fertilization and sustain the rapid cleavage divisions of the early embryo. Examples of such proteins include actin (Mabuchi & Spudich, 1980), myosin (Kane, 1980), DNA polymerase (Racine & Morris, 1978), histones (Poccia et al., 1981), non-histone chromosomal proteins (Kuhn & Wilt, 1981) and tubulins (Raff, 1975). Because it is difficult to obtain populations of unmature, developing oocytes from most animals, very little is known about the synthesis of these proteins during oogenesis. The mature oocyte is usually the first readily available stage. While awaiting fertilization or hormonal activation, the oocyte is relatively quiescent with respect to protein synthesis: only a small portion of its mRNA is actively translated (Davidson, 1976). Very soon after activation many of those mRNAs cease being translated, and other stored mRNAs are activated and recruited for protein synthesis (Rosenthal et al., 1980; Evans et al., 1983). Thus, unlike later stage embryos and somatic cells, where transcription plays the major role in controlling the kinds of proteins that are made, oocytes and early embryos control the pattern of gene expression primarily at the translational level. In this paper, I will discuss three aspects of translational control in embryos of the marine clam Spisula. First, I will review earlier experiments demonstrating that fertilization results in a rapid change in the kinds of proteins made, and that this change is controlled entirely at the translational level. Next, I will discuss evidence for various mechanisms that might be used to inactivate one set of mRNAs and activate another at the time of fertilization. Finally, I will consider the issue of why certain mRNAs, rather than their protein products, should be stored in the oocyte. Why not just store all of the proteins needed for early cleavage and forego protein synthesis altogether?

MATERIALS AND METHODS

Spisula solidissima were collected by the Marine Resources Department at the Marine Biological Laboratory, Woods Hole, MA and kept in running sea water. Oocytes and embryos were prepared and labelled in vivo with ^{35}S-methionine as originally described by Rosenthal et al. (1980). RNA preparation and analysis, and cDNA cloning procedures were all carried out as described by Rosenthal et al. (1983).

Appropriate restriction enzyme fragments of cDNA cloning for proteins A and C were subcloned into M13 phage. DNA sequence analysis was carried out using the dideoxynucleotide chain extension method of Sanger et al. (1977).

RESULTS AND DISCUSSION

Oocytes and Early Embryos Contain the Same Sets of mRNAs but Translate Different Subsets

The mRNA populations of oocytes in two-cell embryos (10 min post-fertilization) were compared by in vitro translation. Phenol-extracted oocyte RNA and two-cell embryo RNA were translated in a mRNA-dependent reticulocyte lysate containing ^{35}S-methionine and the in vitro translation products were compared by gel electrophoresis followed by autoradiography. As shown in Figure 1, the patterns of proteins synthesized in vitro by oocyte (lane a) and embryo (lane b) RNA are indistinguishable. Both contained, for example, the proteins labelled A, B, C, X, Y, and Z. In contrast, comparisons of proteins synthesized in vivo by oocytes (lane c) and two-cell embryos (lane d) showed that different sets of proteins are made before and after fertilization.

As first shown by Rosenthal et al. (1980), this switch in the pattern of protein synthesis in vivo occurs within ten min after fertilization and is regulated exclusively at the translational level. For example, in oocytes the mRNAs encoding proteins A, B and C sediment in the free ribonucleoprotein region of sucrose gradients, whereas these mRNAs are almost entirely associated with polysomes in the two-cell embryo (Fig. 2). Transcriptional contributions to the change in protein synthesis are negligible since the change occurs in the presence of Actinomycin D, an inhibitor of transcription (Rosenthal et al., 1980).

Two kinds of translational changes occur at fertilization. This is most easily seen when radioactively labelled, cloned cDNA probes are used to monitor the association of particular mRNAs with polysomes (Rosenthal et al., 1983).

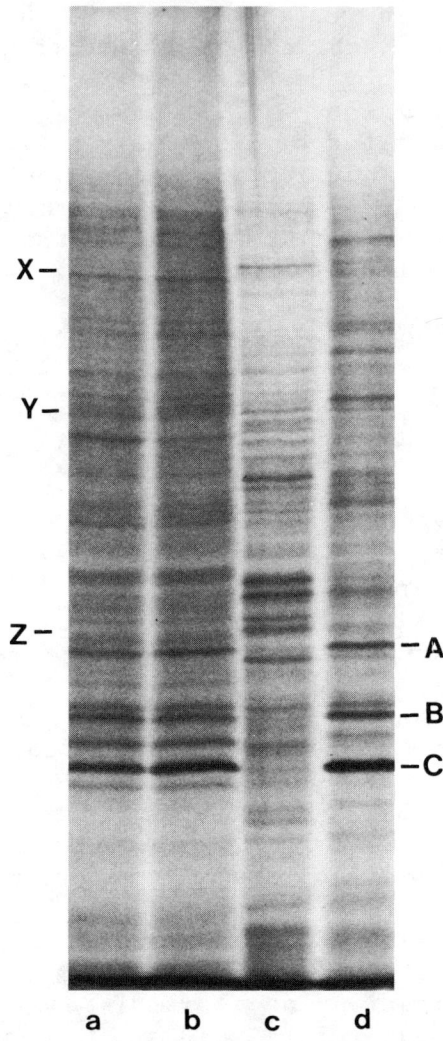

Fig. 1. Comparisons of in vitro translation products of oocyte and embryo RNA with proteins synthesized in vivo. RNAs were phenol-extracted from oocytes and embryos and translated in a reticulocyte lysate containing ^{35}S-methionine. The translation products were electrophoresed on a 11% acrylamide gel. The autoradiogram shows proteins made by oocyte RNA (lane a) and 30-min embryo RNA (lane b). ^{35}S-methionine-labelled proteins made in vivo by oocytes (lane c) and embryos (lane d) were compared on the same gel with the in vitro translation products. (Photograph courtesy of N. Standart.)

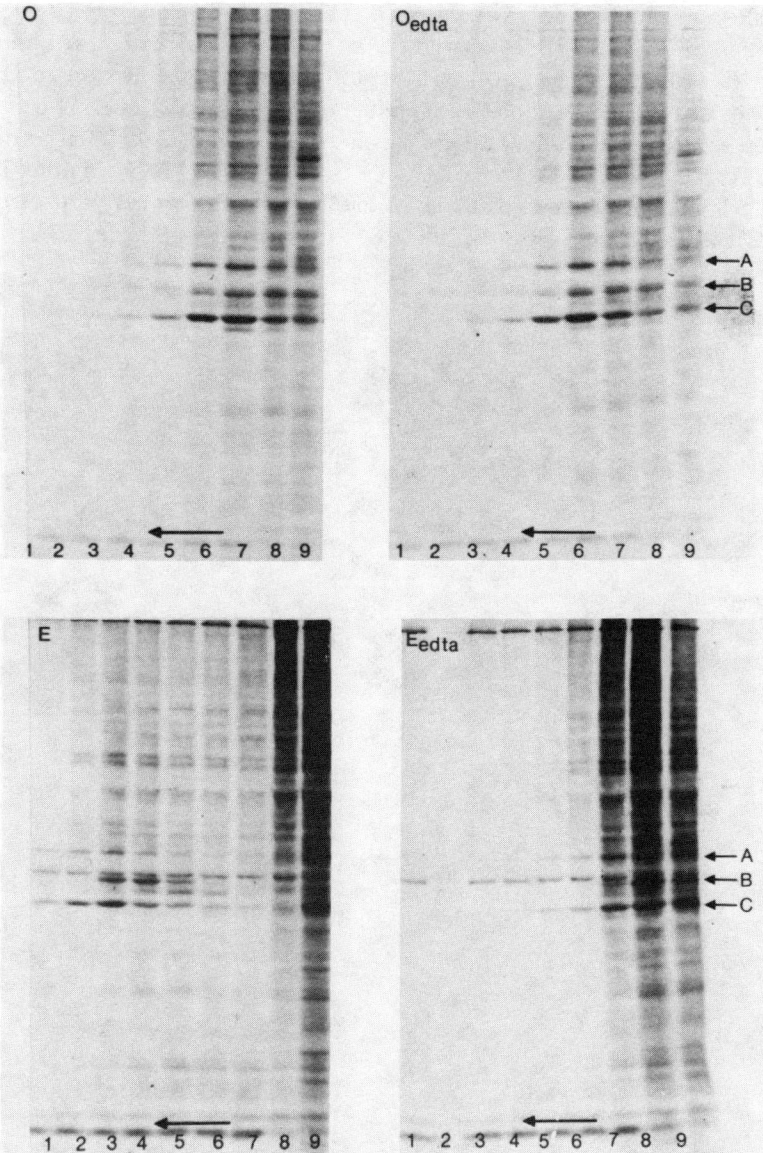

Fig. 2. Comparison of mRNAs associated with polysomes before and after fertilization. Oocytes (O) or embryos (E) were homogenized and a 12,000 g supernatant was sedimented through 15-40% linear sucrose gradients as described by Rosenthal et al. (1983). The direction of sedimentation indicated on the figure is from top (right) to bottom (left). Gradient fractions were collected, RNA was extracted and the RNAs were translated in reticulocyte lysate plus ^{35}S-methionine. O-edta and E-edta panels represent samples which had been treated with EDTA (to dissociate mRNAs from polysomes) prior to centrifugation. The panels show autoradiograms of the lysate translation products. (Photograph courtesy of E. Rosenthal.)

First, mRNAs that are active in the oocyte are released from polysomes within a few minutes after fertilization. For example, a large fraction of α-tubulin mRNA is on polysomes in the oocyte (Fig. 3A). Whereas virtually all of the α-tubulin mRNA in the 30-min embryo sediments in the free RNP region (Fig. 3B). At least six other individually identified mRNAs undergo this same phenomenon of inactivation after fertilization (Rosenthal & Ruderman, in prep.).

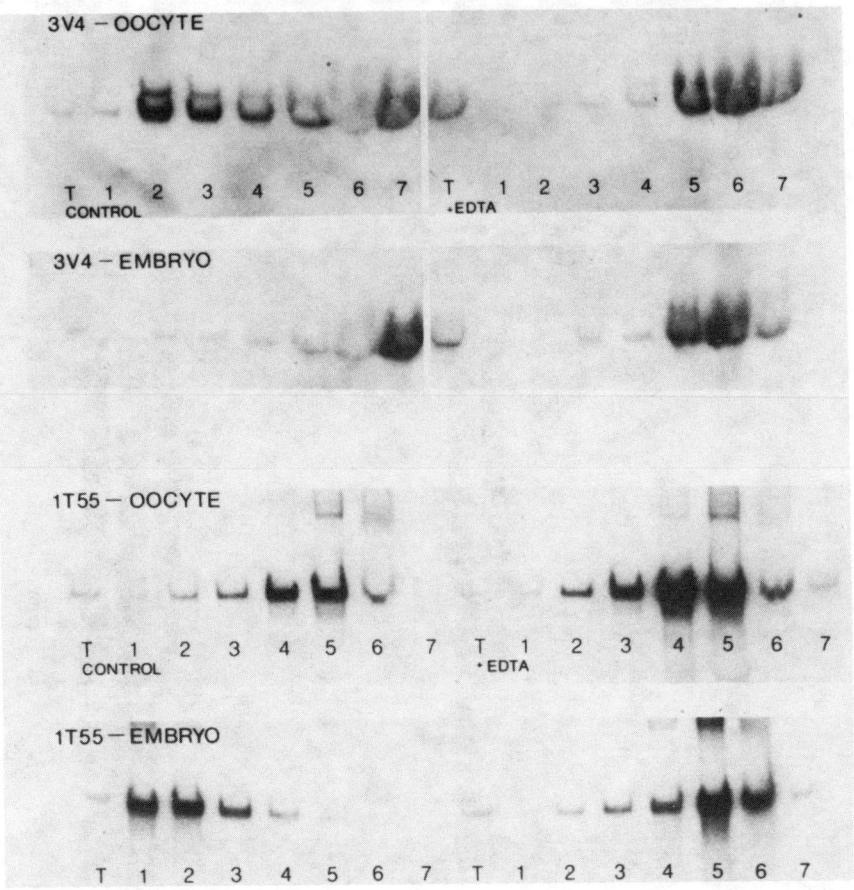

Fig. 3. Translational status of α-tubulin mRNA (3V4) and mRNA (1T55) in oocytes and embryos. Oocytes or embryos were homogenized and fractionated into polysomal and non-polysomal components as described in Figure 2. Sedimentation was from right (fraction 7) to left (fraction 1). "T" indicates an aliquot of total homogenate. "+EDTA" indicates that homogenates were treated with EDTA, as in Figure 2. RNA was extracted from each fraction, electrophoresed in a 1% agarose gel, blotted to nitrocellulose and hybridized with ^{32}P-labelled DNA from clone 1T55 (complementary to mRNA A) or clone 3V4 (complementary to α-tubulin mRNA). The autoradiogram is shown. (Photograph courtesy of E. Rosenthal and T. Tansey.)

Second, many mRNAs that are stored as inactive sequences in the oocyte become active soon after fertilization. For example, mRNA C in oocytes sediments entirely in the free RNP region (Fig. 3C), whereas virtually all of mRNA C is on polysomes in embryos (Fig. 3D). At least eight other mRNAs that follow this same pattern of translational activation have been identified (Rosenthal & Ruderman, in prep.). With one exception so far, each of the mRNAs examined by this method shows a switch in its translational activity. (In this single exception, this RNA is not found on polysomes either before or after fertilization, so it is not yet clear whether the cloned cDNA probe used to monitor the RNA in question is actually complementary to a bona fide messenger RNA or corresponds to some non-mRNA sequence.) Thus, it seems that fertilization of Spisula oocytes is accompanied by large changes in mRNA translation.

I wish to emphasize that this switch in the kinds of proteins made at fertilization is not peculiar to clams. In all organisms studied so far, including Xenopus (Ballantine et al., 1979), sea urchin (Evans et al., 1983), starfish (Rosenthal et al., 1982), mouse (Schultz & Wassarman, 1979), and Urechis (Rosenthal & Wilt, pers. comm.), translational switches in the patterns of protein synthesis are seen at activation or fertilization. While the exact details vary among the various species, the general phenomenon is widespread and, thus, presumably significant.

Messenger RNAs that Become Active After Fertilization Gain Long Poly A Tails, Whereas Those that Become Inactive Use Their Poly A Tails

What are the molecular mechanisms responsible for the stage-specific repression of one set of mRNAs and the activation of another set? Evidence from other systems, most notably the sea urchin embryo, indicates that both mRNA-associated proteins (Ilan & Ilan, 1978; Jenkins et al., 1978; see also Moon et al., 1982) and ribosomes (Danilchik & Hille, 1981; Hille, this volume) play significant roles in the overall rise in the rate of protein synthesis. In Spisula embryos, there is some indirect evidence that mRNA-associated proteins or other phenol-soluble components modulate the translational availability of individual mRNAs (Rosenthal et al., 1981). Pursuing the roles of such putative mRNP-proteins will require developing a cell-free protein synthesis lysate from Spisula oocytes, a task which is not trivial.

A second line of inquiry asks the question: Are there any changes in the primary structure of mRNAs as they go from the inactive to the active state, and vice versa? To answer this question we compared the pre- and post-fertilization sizes of several mRNAs from both categories. Messenger RNA A (Fig. 4, top panel), and several others not shown here, show no evidence of being stored as a larger, precursor sequence that gets processed down to become a mature, active mRNA. The oocyte and embryo versions of this RNA sequence are roughly the same size. If anything, the activated mRNAs are larger than the versions stored in the oocyte. We now know that this is due to the post-fertilization addition of poly A to the oligo A tails of the stored mRNA (Rosenthal et al., 1983). In contrast, the mRNAs, such as α-tubulin mRNA, which become inactive after fertilization possess a poly A tail in the oocyte. They lose this tail at the same time that they become inactive. Thus there is a strong correspondence between possession of a poly A tail and translational activity in vivo.

Size comparisons of oocyte and embryos mRNAs from which poly A tails have been removed show that there are no obvious

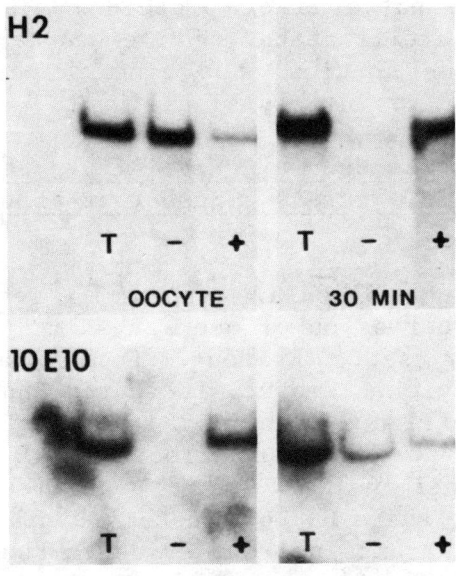

Fig. 4. Comparisons of the sizes and polyadenylation status of mRNA A in oocytes and embryos. Total RNA (T) was extracted from oocytes or 30-min embryos and fractionated into poly A-deficient RNA (-) and poly A-containing RNA (+) on oligo dT-cellulose. RNAs were electrophoresed, blotted and hybridized with ^{32}P-labelled clone H2 (complementary to mRNA) or 10E10 (complementary to α-tubulin mRNA). (Photograph courtesy of E. Rosenthal and T. Tansey.) The autoradiogram is shown.

differences. However, these methods are fairly crude and would not be capable of detecting small (>30 nt) differences. Thus, a final answer awaits other kinds of comparisons.

Primary Structure of Stored, Maternal mRNAs: Signal Sequences for Translational Control

It seems very likely that the translational behavior of individual maternal mRNAs will be in some way specified by their nucleotide sequences, much as the transcriptional activities of individual genes are controlled using specific DNA sequences. To test this idea, we have started to determine the RNA sequences of several mRNAs from each of the two classes of maternal mRNA--those that are activated by fertilization and those that are repressed. The nucleotide sequence for mRNA C, which is activated after fertilization, was determined by isolating restriction enzyme fragments of DNA from four overlapping partial length cDNA clones, subcloning those fragments into the phage M13, and sequencing the inserted DNAs (Standart et al., in press). The overall features of this mRNA are shown in Figure 5. mRNA C has a 900 nucleotide (nt) coding region which codes for a 34,000 dalton polypeptide. It is flanked by a 5' noncoding region of ~300 nt and a 3' noncoding region of ~500 nt. The 3' noncoding region contains a sequence that can form a perfectly matched 77 base pair stem with a 300 nt loop. The significance of this structure is intriguing--could it act as a translational control sequence, a polyadenylation signal

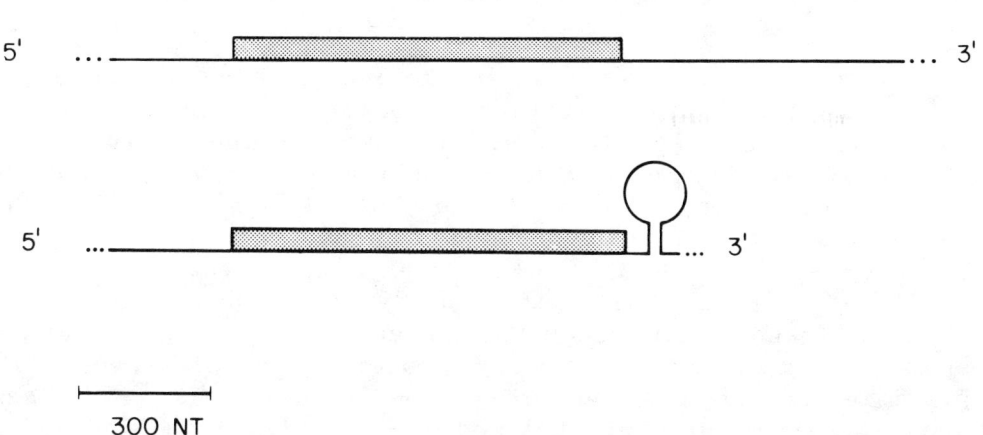

Fig. 5. Diagrammatic representation of the sequence of mRNA C. (Courtesy of N. Standart.)

sequence. The ~300 nt 5' noncoding region has no such peculiar features, or at least none that are obvious at this point.

The sequence of mRNA A, another messenger RNA that is activated at fertilization, contains a 1264 nt coding region which specifies a 48,000 dalton protein (Fig. 6). Its 5' noncoding region is about 300 nt. No obvious homologies have been seen between the 5' noncoding regions of mRNA A and C, but it should be emphasized that neither 5' sequence is complete at this time. The 3' noncoding region of mRNA A is about 1500 nt and, of the portions sequenced so far, none show any significant homology with the 3' end of mRNA C. Definitive comparisons await completion of the two sequences and a consideration of their predicted secondary structures as well as the primary nucleotide sequences.

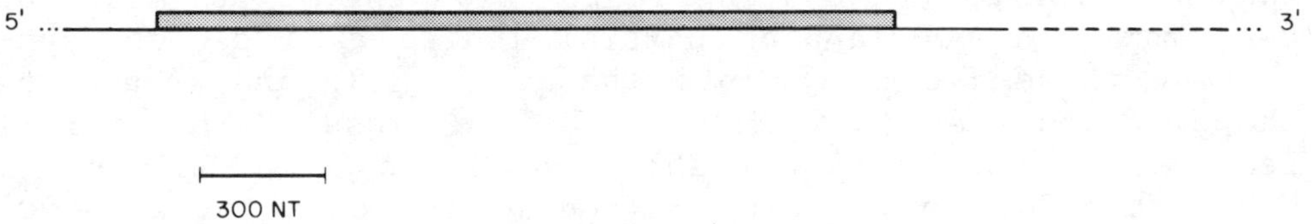

Fig. 6. Diagrammatic representation of mRNA A. For explanation, see Figure 5. Dashed line indicates unconfirmed sequence data. (Courtesy of N. Standart.) Coding region is indicated by shaded area, noncoding regions by solid lines, unsequenced regions by dots. (Courtesy of K. Farrell.)

Stored Maternal mRNAs Encode Proteins with Important Roles in the Cell Cycle

Many of the proteins stockpiled in the egg provide materials for DNA replication, chromatin assembly, spindle formation, and cell cleavage. Why should the early embryo bother to synthesize any new proteins at all? One idea is that, for reasons of space, the oocyte cannot provide the entire pool of proteins needed to sustain the rapid cell division cycles of the early embryo, and that storing a single mRNA rather than the hundreds or thousands of its protein translation products represents a very efficient, space-saving way of providing proteins for cleavage. The provisioning of histones might be such a case. In sea urchins, the mature unfertilized egg contains a small pool of histone proteins, sufficient to complex with about 30-50 nuclear equivalents of DNA (Poccia et al., 1981). The egg also stores a modest amount of histone messenger RNA (curiously, in the nucleus) that is recruited onto polysomes about 90 min after fertilization, or about the time of first cleavage (Gross et al., 1973; Venezky et al., 1981; Wells et al., 1981). Finally, transcription of significant amounts of new histone mRNA occurs throughout cleavage (Kedes, 1979).

A second idea is that the proteins (like A, B and C) encoded by mRNAs that are completely inactive in the oocyte and activated within minutes after fertilization represent key regulatory proteins, ones that control progress through the cell cycle and other aspects of early development. Indeed, recent work has shown that two of these proteins, namely A and C, fit this description.

Evans et al. (1983) found that sea urchin eggs, like those of the marine clam Spisula, contain large amounts of stored mRNAs which are active only after fertilization. They discovered that urchin protein A (55,000 daltons) is synthesized continuously throughout early cleavage, but that it is specifically degraded at the end of each mitosis. Because of this cyclical pattern of accumulation and destruction, they called protein A "cyclin". Protein A in Spisula shows this same pattern. (Spisula protein B also cycles but with a different time course: it accumulates to a peak and then disappears slightly later than protein A.) If cleavage is prevented, cyclin fails to be degraded. Thus cyclin is in some way tied into progress through the cell cycle. Comparisons of the complete amino acid sequence of Spisula protein A (Farrell & Ruderman, unpubl.) with other known protein sequences in the NEWAT Sequence Bank (Doolittle, 1981) showed that regions of protein A are homologous with several kinds of protein including viral capsid proteins, DNA binding proteins, membrane proteins and basic proteins. This suggests that protein A may be a membrane-associated DNA-binding protein, possibly involved in chromosome condensation or nuclear membrane breakdown.

Comparison of the amino acid sequence of protein C with those in the NEWAT database was even more revealing. The C-terminal half of protein C shows a 33% homology with a Herpes virus 39K protein. The N-terminal halves show a lower, but still significant, degree of homology. The function of this Herpes 39K protein was not known, but the protein was suspected to be associated with a 140K protein which was implicated in ribonucleotide reductase activity (Dutia, 1983). Comparisons of the sequences of clam protein C, Herpes 39K (McLaughlin & Clements, 1983a & b) and the recently obtained E. coli ribonucleotide reductase sequences

(J. Fuchs, pers. comm.) suggested that clam C and Herpes 39K represented the smaller of the two subunits of reductase (Standart et al., in press).

What is the role of ribonucleotide reductase? This enzyme catalyzes the reduction of ribonucleotides to deoxyribonucleotides, usually as rNDP to dNDP (Thelander & Reichard, 1979). The dNDP's are then phosphorylated and incorporated into DNA. Thus, reductase is the first enzyme in the de novo pathway of DNA synthesis and controls the concentration of substrates for DNA replication. The presence of reductase is essential in early embryos since oocytes have only very small pools of deoxynucleotides (Brachet, 1972).

Early work by Noronha et al. (1972) had in fact shown that there is no detectable reductase activity in sea urchin oocytes (this remains to be tested for clams). They found that reductase activity appeared soon after fertilization and that this appearance was blocked by puromycin, an inhibitor of protein synthesis, but not by Actinomycin D, an inhibitor of RNA transcription. This result is what would be predicted if some component of reductase activity were encoded by a stored, maternal mRNA that is activated after fertilization.

In summary, mature oocytes awaiting fertilization translate only a fraction of their total available mRNA pool. Presumably these proteins represent maternal components that are still being accumulated, or are needed for maintenance of the oocyte's pre-fertilization status. Within 10 min after fertilization, many--perhaps all--of these mRNAs stop being translated and the activated zygote begins to translate a very different group of mRNAs. Of the two examples studied carefully so far (proteins A and C), both of these newly made proteins are involved in progression through the cell cycle. Protein A, also known as cyclin, is specifically degraded at the end of each mitosis, suggesting a role in chromosome decondensation or nuclear membrane breakdown. The synthesis of protein C, the small subunit of ribonucleotide reductase, is essential for the activation of reductase activity which provides deoxynucleotides for DNA synthesis.

ACKNOWLEDGEMENTS

The work discussed here was done with Nancy M. Standart, Eric T. Rosenthal, Terese R. Tansey, Kevin M. Farrell, and Tim Hunt. It was supported by NSF grant PCM 82-16917.

LITERATURE CITED

Ballantine, J.E.M., H.R. Woodland, and E.A. Sturgess. 1979. Changes in protein synthesis during the development of Xenopus laevis. J. Embryol. Exp. Morph. 51: 137-153.
Brachet, J. 1967. Effects of hydroxyurea on development and regeneration. Nature 214: 1132.
Danilchik, M.V. and M.B. Hille. 1981. Sea urchin egg and embryo ribosomes: Differences in translational activity in a cell-free system. Dev. Biol. 84: 291-198.
Davidson, E.H. 1976. Gene Activity in Early Development. Academic Press, New York.
Doolitle, R.F. 1981. Similar amino acid sequences: Chance or common ancestry. Science 214: 149-159.
Dutia, B.M. 1983. Ribonucleotide reductase induced by Herpes Simplex virus has a virus specified constituent. J. Gen. Viol. 64: 513-521.
Evans, T., E. Rosenthal, J. Youngbloom, D. Distel and T. Hung. 1983. Cyclin: A protein specific maternal mRNA in sea urchin eggs that is destroyed at each cleavage division. Cell 33: 389-396.
Gross, K.W., M. Jacobs-Lorena, C. Baglioni, and P.R. Gross. 1973. Cell-free translation of maternal mRNA from sea urchin eggs. Proc. Nat. Acad. Sci. USA 71: 2614-2618.
Ilan, J. and J. Ilan. 1978. Translation of maternal mRNP particles from sea urchin in a cell-free system from unfertilized eggs and product analysis. Dev. Biol. 66: 375-385.
Jenkins, N., J. Kaumeyer, E. Young and R.A. Ratt. 1978. A test for masked mRNA. Devel. Biol. 63: 9-28.
Kane, R.E. 1980. Induction of either contractile or structural actin-based gels in sea urchin cytoplasmic extract. J. Cell Biol. 86: 813-819.
Kedes, L.H. 1979. Histone genes and histone messengers. Ann. Rev. Biochem. 48: 847-871.
Kuhn, O. and F.H. Wilt. 1981. Double labeling of chromatin proteins, in vivo and in vitro, and their two dimensional electrophoretic resolution. Dev. Biol. 85: 416-424.
Mabuchi, I. and J.A. Spudich. 1980. Purification and properties of soluble actin from sea urchin eggs. J. Biochem. (Tokyo) 87: 785-812.
McLaughlan, J. and J.B. Clements. 1983a. DNA sequence homology between two co-linear loci on the HSV genome which have different transforming abilities. EMBO J. 2: 1953-1961.

McLaughlan, J. and J.B. Clements. 1983b. Organization of the Herpes Simplex virus type 1 transcription unit encoding two early proteins with molecular weights of 141,111 and 41,111. J. Gen. Virol. 64: 997-1016.

Moon, R.T., M.V. Danilchik, and M.B. Hille. 1982. An assessment of the masked message hypothesis: Sea urchin egg mRNP complexes are efficient templates for in vitro protein synthesis. Dev. Biol. 93: 389-413.

Noronha, J.M., G.H. Sheys, and J.M. Buchanan. 1972. Induction of a reductive pathway for deoxyribonucleotide synthesis during early embryogenesis of the sea urchin. Proc. Nat. Acad. Sci. USA 69: 2006-2010.

Poccia, D., J. Salik, and G. Krystal. 1981. Transitions in histone variants of the male pronucleus following fertilization and evidence for a maternal store of cleavage-stage histones in the sea urchin egg. Dev. Biol. 85: 416-424.

Racine, F.M. and P.W. Morris. 1978. DNA polymerase α and β in the California urchin. Nucleic Acids Res. 5: 3945-3957.

Raff, R.A. 1975. Regulation of microtubule synthesis and utilization during early embryonic development of the sea urchin. Amer. Zool. 15: 661-678.

Rosenthal, E., T. Hunt, and J.V. Ruderman. 1980. Selective translation of mRNA controls the pattern of protein synthesis during early development of the surf clam Spisula solidissima embryos. Cell 21: 487-496.

Rosenthal, E.T., B.P. Brandhorst, and J.V. Ruderman. 1982. Translationally mediated changes in patterns of protein synthesis during maturation of starfish oocytes. Dev. Biol. 91: 215-221.

Rosenthal, E.T., T.R. Tansey, and J.V. Ruderman. 1983. Sequence-specific adenylations and deadenylations accompany changes in the translation of maternal mRNA after fertilization of Spisula oocytes. J. Mol. Biol. 166: 319-327.

Sanger, F., S. Nicklen, and A.R. Coulson. 1977. DNA sequencing with chain termination inhibitors. Proc. Natl. Acad. Sci. USA 74: 5463-5467.

Schultz, R. and P.M. Wassarman. 1979. Specific changes in the pattern of protein synthesis during meiotic maturation of mammalian oocytes in vitro. Proc. Nat. Acad. Sci. USA 74: 538-541.

Stardaft, N., S. Bray, E. George, T. Hung and J.V. Ruderman. In Press. The small subunit of ribonucleoside reductase is encoded by one of the most abundant translationally regulated maternal mRNAs in clam and sea urchin eggs. J. Cell Biol.

Thelander, L. and P. Reichard. 1979. Reduction of ribonucleotides. Ann. Rev. Biochem. 48: 158-188.

Venezky, D.L., L.M. Angerer, and R.C. Angerer. 1981. Accumulation of histone repeat transcripts in the sea urchin egg pronucleus. Cell 24: 385-391.

Wells, D.E., R.M. Showman, W.H. Klein, and R.A. Raff. 1981. Delayed recruitment of maternal histone H3 mRNA in sea urchin embryos. Nature 292: 477-479.

THE SEA URCHIN SPEC FAMILY OF CALCIUM-BINDING PROTEINS: CHARACTERIZATION AND CONSIDERATION OF POSSIBLE ROLE IN LARVAL DEVELOPMENT

William H. Klein, Clifford D. Carpenter, Liane E. Philpotts, and Bruce P. Brandhorst

ABSTRACT

This paper describes the structural and functional properties of the Spec gene family. The Spec mRNAs accumulate specifically in presumptive dorsal ectoderm cells of Strongylocentrotus purpuratus blastulae. They encode a family of 10 to 12 proteins having molecular weights ranging from 14,000 to 17,000 daltons. Isolation and partial sequencing of several Spec cDNA clones demonstrate that the Spec proteins are members of the troponin C superfamily of calcium-binding proteins. The proteins are shown to have calcium-binding properties characteristic of troponin C-related proteins in vitro. We suggest that the Spec proteins are part of a contractile system found in dorsal ectoderm cells and are involved in the changes in the shape of the dorsal ectoderm cells which occur during larval development and metamorphosis.

INTRODUCTION

We have been investigating a small family of genes, the Spec genes, active during embryonic development of the sea urchin, Strongylocentrotus purpuratus (Bruskin et al., 1982). The mRNAs of this gene family are rare or undetectable in the unfertilized egg but accumulate specifically in presumptive dorsal ectoderm cells of the blastula. They code for a group of about ten closely related proteins having molecular weights ranging from 14,000 to 17,000 (Bruskin et al., 1982; Lynn et al., 1983). cDNA clones representing several of these mRNAs have been isolated and partially sequenced. A comparison of the translational reading frames of these sequences with known protein sequences demonstrated that the sea urchin proteins are related to the troponin C superfamily (Carpenter et al., 1984).

Proteins belonging to the troponin C superfamily are calcium-binding proteins, usually of molecular weights 15,000-20,000, that include troponin C, a component of the thin filament responsible for the regulation of muscle contractions by calcium ions; parvalbumin, a soluble protein in muscle cells thought to modulate intracellular calcium ion concentrations; myosin heavy chains; and calmodulin, an ubiquitous protein involved in the regulation of various calcium-dependent events such as secretion, cell motility, cell division and cyclic nucleotide metabolism.

The sequence homology of the Spec proteins with troponin C-related proteins is especially strong in the four calcium-binding domains of the molecule. We have recently shown that the Spec proteins bind calcium and have suggested that they perform roles in sea urchin larvae analogous to the roles of some troponin C-related proteins in other systems. Specifically, we have proposed that the Spec proteins are part of a contractile system found in dorsal ectoderm cells and are involved in the changes in shape of the dorsal ectoderm cells which occur during larval development and metamorphosis (Carpenter et al., 1984).

In this report we summarize the information about the structural characteristics of the Spec genes and proteins. In addition, we discuss their possible roles in larval development.

Spec 1 and Spec 2 mRNAs

Large numbers of sea urchin plutei can be conveniently fractionated into endoderm, mesoderm, and ectoderm (McClay & Chambers, 1978). We used this procedure and a library of pluteus cDNA clones to isolate tissue-specific sequences from S. purpuratus plutei (Bruskin et al., 1981). Two of the cDNA clones we originally isolated, Spec 1 and Spec 2, code for two closely related mRNAs present in the pluteus ectoderm. The cloned sea urchin sequences are approximately 80% homologous and represent the 3' untranslated region of a 1.5 kb mRNA (Spec 1) and a 2.2 kb mRNA (Spec 2). Spec 1 and Spec 2 are developmentally regulated mRNAs (Bruskin et al., 1982). As monitored during embryogenesis by RNA gel blot analysis, the mRNAs are present in low or undetectable levels in the unfertilized egg and early cleaving embryo. By hatching blastula stage (20 h) the Spec 1 mRNAs have accumulated significantly in mass, increasing over 100 fold from early cleavage to gastrula stage. The Spec 2 mRNAs also accumulate during embryogenesis, their accumulation becoming detectable about 10 h later than for Spec 1 mRNAs. Quantitation of the amount of Spec 1 mRNAs by solution hybridization suggests they constitute about 0.6% of the embryo mRNA at 50 h (Lynn et al., 1983). The Spec 2 mRNAs are about one-tenth as prevalent at their maximum level, also at 50 h (Bruskin et al.,

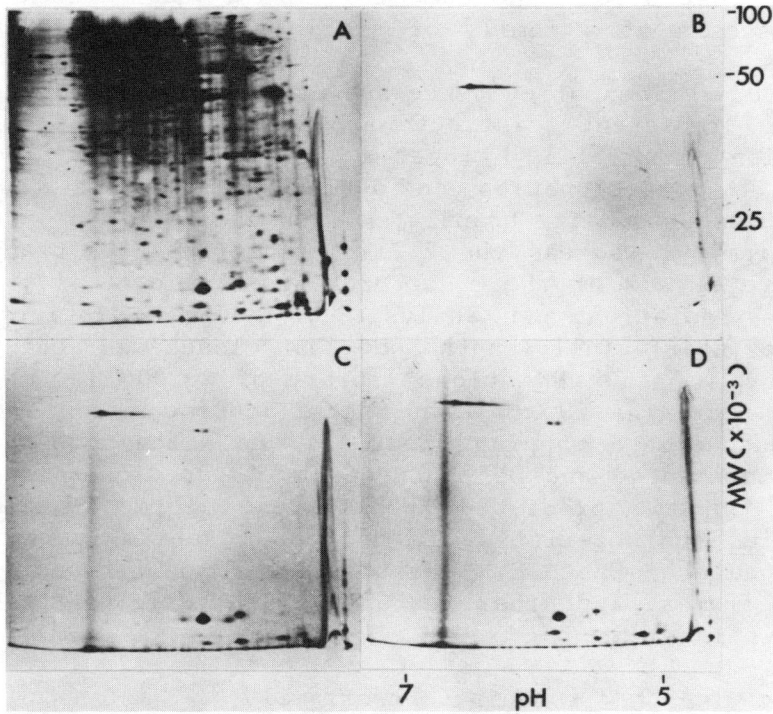

Fig. 1. Two-dimensional electrophoresis of products of translation of RNA selected by hybridization to Spec 1 and Spec 2 DNAs. One hundred micrograms of total cellular RNA from gastrula-stage embryos were hybridized to 15 μg of filter-bound Spec 1 or Spec 2 DNA. The hybridized RNA was eluted and translated in a rabbit reticulocyte lysate cell-free system supplemented with [^{35}S] methionine. The radiolabeled proteins were electrophoresed in two dimensions. The second dimension uses a 10% polyacrylamide gel. The gels were prepared for fluorography with Enhance (New England Nuclear) and exposed to X-Omat RP X-ray film for 2 wk at -70°C. (A) Translation of 10 μg of total cellular RNA from gastrula-stage embryos. (B) Translation with no exogenous RNA. (C) Translation with gastrula RNA hybrid selected by Spec 1 DNA. (D) Translation with gastrula RNA hybrid selected by Spec 2 DNA.

1981). Thus, while they are two very closely related mRNAs, Spec 1 and Spec 2 show distinct patterns of accumulation both in terms of time of appearance and absolute quantity.

We were able to demonstrate a family of proteins corresponding to the Spec 1 and Spec 2 genes by using the technique of hybrid-selected translation (Fig. 1 and Bruskin et al., 1982). RNA selected by either clone codes for approximately ten polypeptides having molecular weights ranging from 14,000 to 17,000 and isoelectric points ranging from pH 5.0 to pH 6.0. A comparison of the in vitro translation products of Spec 1 and Spec 2 mRNAs with ectoderm proteins synthesized in vivo indicates that at least seven of the polypeptides co-migrate on two-dimensional polyacrylamide gels. We conclude that Spec 1 and Spec 2 code for mRNAs which direct the synthesis of a family of similar but distinct ectoderm proteins.

Several lines of evidence suggested that Spec 1 and Spec 2 mRNAs represent distinct subfamilies of Spec genes. Separation of the 1.5 kb (Spec 1) and 2.2 kb (Spec 2) transcripts by fractionation on agarose gels shows that the 1.5 kb transcripts are translated into the two major proteins of the family, whereas the 2.2 kb transcripts are translated into five or six of the minor proteins (Klein et al., 1984).

The isolation and analysis of several distinct full length or nearly full length Spec cDNA clones was achieved by screening a λgt10 cDNA clone library of 20,000 recombinants using the original Spec 1 and Spec 2 sequences. By restriction endonuclease mapping, Southern and Northern hybridization experiments, and DNA sequencing, these clones fall into one of the two predicted subfamilies. Three clones have properties characteristic of the Spec 1 subfamily and three correspond to Spec 2. Restriction maps, translational reading frames, and other features characterizing the clones are shown in Figure 2.

Homology with the Troponin C Superfamily

Protein sequences derived from the translation of the reading frames of five of the cDNA clones, three from the 1.5 kb transcript subfamily and two from the 2.2 kb transcript subfamily, are shown in Figure 3. Spec 1a, 1b, and 1c, members of the 1.5 kb transcript subfamily, appear to encode identical proteins 159 residues in length (1c protein). We cannot state with absolute certainty that all three proteins are identical. Spec 1a is not a full length cDNA clone and does not contain a complete translational

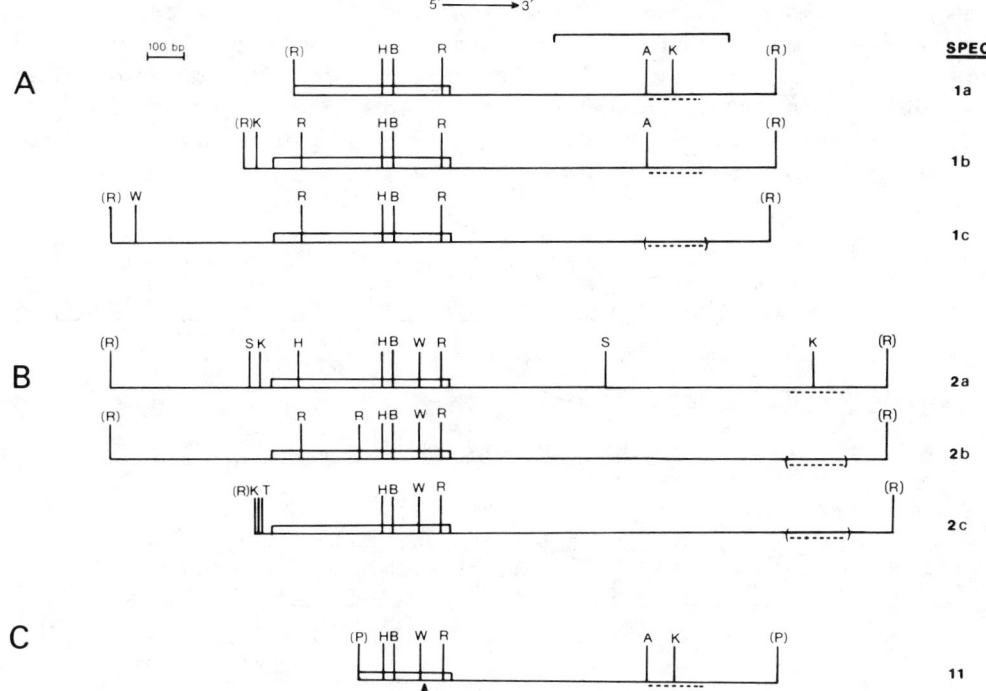

Fig. 2. Restriction maps of Spec cDNA clones. (A) Spec 1 subfamily. (B) Spec 2 subfamily. (C) A hybrid clone of the two subfamilies. The open bars are the translational reading frames determined by sequencing Spec 1a, 1b, 1c, 2a, 2c, and 11. The translational reading frame for Spec 2b is inferred. The dashed lines indicate the position of a 0.15 kb repetitive element. In Spec 1c, 2b, and 2c, the exact position of the element within the 3' Eco RI fragment has not been determined and is therefore enclosed in parentheses. The bracket above Spec 1a indicates the position of the Spec 1 cDNA clone. The vertical arrow beneath Spec 1 indicates the position where the sequence changes from Spec 2a-like to Spec 1a. Restriction enzymes are R, Eco RI; H, Hind III; B, Bam HI; A, Kba I; K, Kpn I; S, Sal I; W, Cla I; T, Sst I; and P, Pst I. The sites at the ends of each insert, generated by the cloning procedure, are shown in parenthesis.

reading frame. The open frame begins with an asparagine (indicated by a dark arrow in Fig. 3) and continues for another 131 amino acid residues until a translational termination signal is reached. Spec 1c, which does have a complete reading frame, contains these identical 132 residues. A 0.4 kb Eco RI fragment of Spec 1b has been partially sequenced and the 44 amino acid residues derived from this sequence are also identical with the corresponding amino acids of Spec 1a and 1c (shown with dashed arrows in Fig. 3).

Spec 2a and 2c, members of the 2.2 kb transcript subfamily, encode two related proteins 150 and 151 residues in length (2a and 2c proteins). Both the Spec 1- and Spec 2-type proteins are highly charged with one-third of their amino acid residues acidic or basic. The size and net negative charge of the proteins is consistent with the migration seen in vivo for the Spec proteins on two-dimensional polyacrylamide gels (Fig. 6 and Bruskin et al., 1982). The proteins for Spec 2a and 2c are 79% homologous to each other but are only 54% and 52% homologous to the 1c protein.

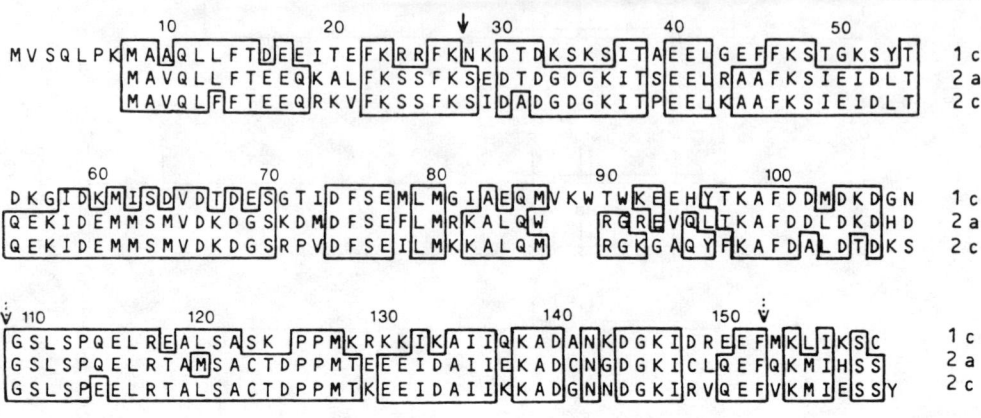

Fig. 3. Putative Spec protein sequences deduced from the translational reading frames of Spec 1a, 1b, 1c, 2a, and 2c. The one-letter notation for the amino acid residue is taken from Dayhoff (1978). Boxed areas are residues homologous to all proteins. Complete sequences are shown for Spec 1c, 2a, and 2c. The solid vertical arrow denotes the beginning of the Spec 1a sequence, which continues to the end of the reading frame. The dashed arrows denote the region of Spec 1b, which has been sequenced. The amino acid residues corresponding to the 1c protein are numbered above the sequence.

A search of the protein sequence data base from the Atlas of Protein Sequence and Structure (Dayhoff, 1978) demonstrated that significant homology exists between troponin C-related proteins and the Spec proteins described here. Figure 4 shows a comparison of the 1c protein sequence with frog skeletal muscle troponin C. When the sequences are aligned colinearly, 32% of the residues are held in common. More dramatic is the location of the homologous residues. Troponin C is divided into four calcium-binding domains (Kretsinger, 1980), indicated by vertical lines in Figure 4. The residues most highly conserved in each domain among the troponin C-related proteins are those that interact with calcium ions. They are indicated by the letters X, Y, Z, -X, and -Z above the protein sequences. Glycine (G) and isoleucine (I) are other residues usually conserved in these regions. Of these 28 residues, 21 are conserved between the 1c protein and troponin C. Three other positions, indicated by asterisks, are amino acids that are found in other troponin C-like proteins. Domains II and IV of the Spec protein have perfect calcium-binding sites based on this sequence analysis. One of the other features characteristic of the proteins of this superfamily is the placement of hydrophobic residues at particular locations within the calcium-binding domains. The 1c protein also meets this criterion in 24 of 32 positions.

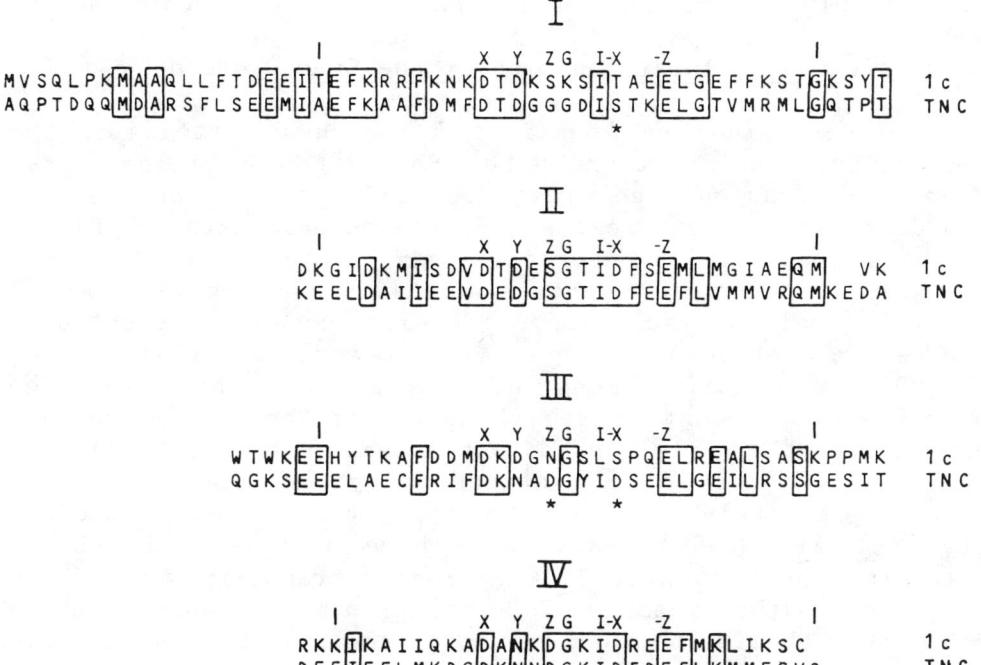

Fig. 4. Sequence comparison between the 1c protein and frog skeletal muscle troponin C. Boxed areas are regions of homology between the two proteins. The sequences are displayed to indicate the four calcium-binding domains of troponin C. The Roman numerals and vertical lines indicate the four domains. The letters above the sequences are residues either directly involved in binding calcium ion (X, Y, Z, -X, -Z) or conserved residues thought to be important for maintaining the E-F hand helical structure of the protein domain (G & I). The asterisks indicate residues of the 1c protein not present in troponin C but present in other troponin C-related proteins.

Similar comparisons can be made for the other Spec protein sequences obtained from Spec 2a and 2c. The consideration of the calcium-binding domains of troponin C-related proteins is at least as strong with these proteins as with the 1c protein. We conclude that the Spec proteins are related to the troponin C superfamily.

It is unlikely that the Spec proteins are troponin C (a highly conserved protein thought to be restricted to muscle cells) or calmodulin, which is considerably more acidic (see below). Myosin light chains, which share more homology with calmodulin than with troponin C (Dayhoff, 1978), are heterogeneous in size and isoelectric point and have been identified in nonmuscle contractile cells (Weeds, 1978; Strohman et al., 1983). It is not yet clear whether the members of the Spec family are functionally related to any of these or other defined members of the troponin C superfamily, but they do not closely correspond in sequence to any known protein.

Embryonic and Cellular Location of the Spec Proteins

To determine the location of the Spec proteins in the embryo, Spec antiserum (raised to Spec 1-type proteins eluted from two-dimensional gels) was reacted with sectioned plutei and detected by the peroxidase-antiperoxidase method. Figure 5 shows that the Spec proteins are localized to the epithelial cells on the surface of the larva away from the ventral (oral) surface. Spec proteins cannot be detected in the ventral ectoderm, the gut, or in mesenchyme cells.

Higher magnifications of the embryos show that the proteins are present in the cytoplasm of these cells but are absent from the nuclei and do not appear to be extracellular; they are not part of a distinctive cytoskeletal structure. We have previously termed the reactive tissue dorsal ectoderm (Lynn et al., 1983), but others have designated it aboral ectoderm (Shott et al., 1984). Spec 1 transcripts are found localized in these same cells. Electron micrographs of these cells show them to be thin, squamous cells with large nuclei.

Characterization of the Spec Proteins

Some of the polypeptides identified among the translation products of the Spec genes may represent different modifications of the same primary gene product. The analysis of gene structure indicates that there are at least two gene subfamilies, Spec 1 and 2, each coding for distinct proteins, and that each subfamily may include several nonallelic members. The proteins of the two subfamilies have distinctive properties, suggesting the possibility that they have different functions. As shown in Figure 6a-c, the two identified major Spec 1-type polypeptides, 1 and 2, are preferentially precipitated from aqueous solution by 50% ethanol, while several polypeptides of Spec 2-type are preferentially precipitated by 80% ethanol, as is the putative sea urchin calmodulin discussed below. Some members of the troponin C superfamily have higher electrophoretic

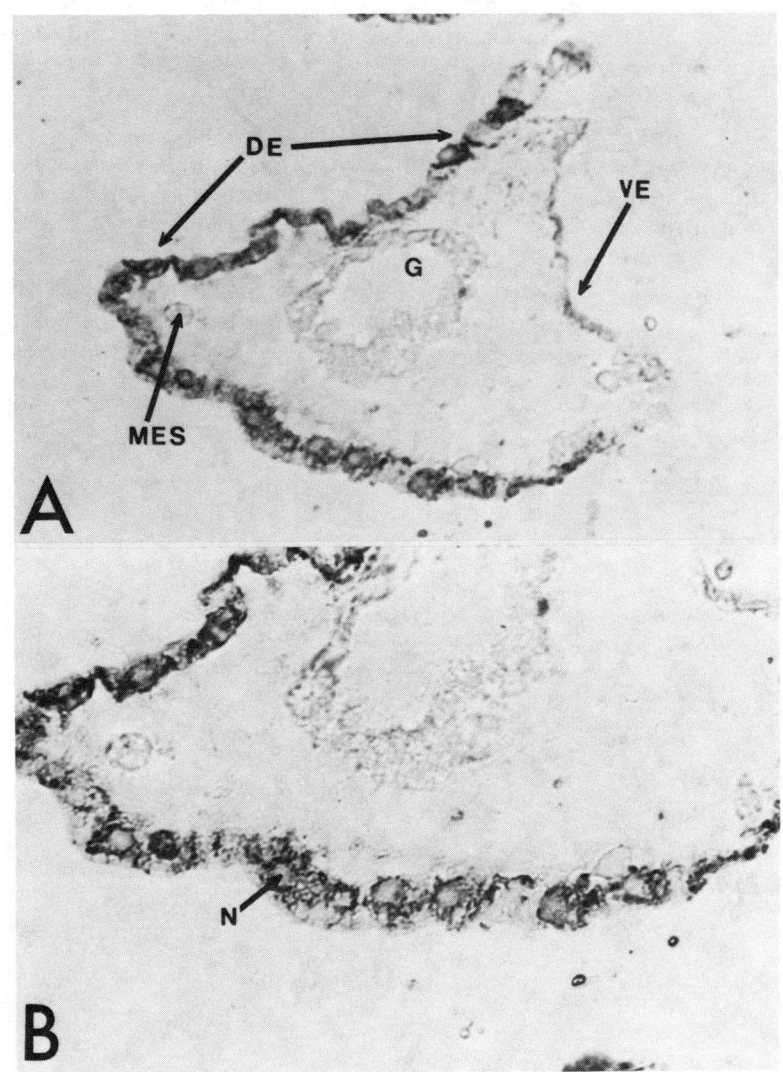

Fig. 5. Immunostaining of S. purpuratus plutei with antibody to the Spec proteins. (A) and (B) are different magnifications of 3-μm sections stained by the peroxidase-antiperoxidase method. DE, dorsal ectoderm; VE, ventral ectoderm; G, gut; MES, mesenchyme; N, nucleus.

mobilities on polyacrylamide gels containing SDS when bound to calcium ions than in the presence of the calcium chelator EGTA (Burgess et al., 1980). As shown in Figure 6a, the Spec 1-type proteins precipitated by 50% ethanol have slightly different mobilities in the presence of calcium ions or EGTA, while proteins precipitated by 80% ethanol include some which shift substantially. When isoelectric focusing gels were equilibrated with SDS buffer containing either calcium ions or EGTA before their application to the second dimension (SDS electrophoresis), differences in mobility of some spots were observed, as shown in Figure 6d and e. Pronounced differences were seen in the mobility of the spot(s) C_m migrating similarly with bovine calmodulin (tentatively identified then as sea urchin calmodulin) and complex z, which includes a Spec 2-type protein, as well as one not enriched in ectoderm (Bruskin et al., 1982). Calmodulin is considerably more acidic than the Spec proteins and does not

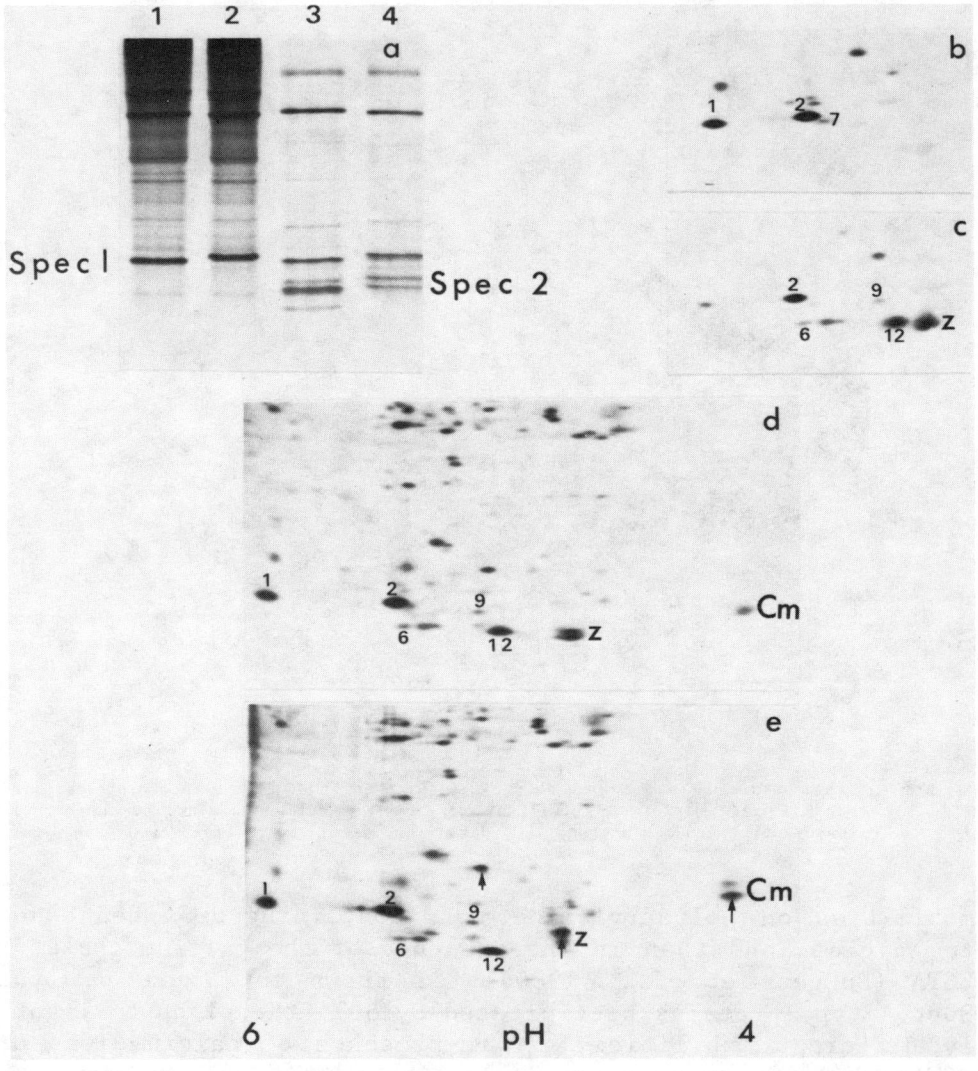

Fig. 6. Characterization of Spec proteins by electrophoresis. Aqueous extracts of gastrulae of S. purpuratus labeled with ^{35}S-methionine were brought to 50% ethanol. Precipitated proteins were collected by centrifugation and the supernatant was brought to 80% ethanol and centrifuged. Panel a: Proteins precipitated with 50% ethanol (Lanes 1 & 2) or 80% ethanol (Lanes 3 & 4) were dissolved by heating the SDS sample buffer (Laemmli, 1970) containing either 5 mM Ca^{++} (Lanes 1 & 3) or 5 mM EGTA and 2 mM EDTA (Lanes 2 & 4), and subjected to discontinuous electrophoresis on polyacrylamide gels containing SDS (Laemmli, 1970). A band corresponding to Spec 1-type proteins in Lanes 1 and 2 is indicated by an arrow. A group of proteins, including some of Spec 2-type and calmodulin, are indicated in Lanes 3 and 4; some of these undergo considerable change in mobility when calcium ions are chelated by EGTA. Panel b: The proteins precipitated with 50% ethanol were subjected to two-dimensional electrophoresis (O'Farrell, 1975) as described by Bedard and Brandhorst (1983). The major Spec 1-type proteins labeled one and two were predominantly precipitated.

Panel c: The proteins precipitated with 80% ethanol were analyzed by two-dimensional electrophoresis. Spot numbers in Panel b and c correspond to those identified as members of the Spec family (Bruskin et al., 1982). The complex of spots labeled z includes protein of the Spec family as well as protein synthesized in endoderm/mesoderm as well. Panel d and e: Extracts of gastrulae were subjected to isoelectric focusing using conditions in which ampholytes having a pH range of 3-5 were substituted for those having a range of 5-7. The isofocusing gels were equilibrated in SDS sample buffer containing 5 mM Ca^{++} (Panel d) or 5 mM EGTA before being electrophoresed in a gel containing SDS. Bovine calmodulin co-migrated with the spot(s) marked Cm. Other Spec-type proteins are identified as in Panels b and c. Proteins showing obvious shifts in the presence of EGTA are shown by arrows, the base of each indicating the center of the spot when electrophoresed in the presence of calcium ions.

focus in the pH range (about 5-7) normally used for our two-dimensional electrophoretic investigations. The shifts in mobility observed between lanes 3 and 4 of panel 6a can probably be accounted for by the slight difference in mobility of protein 2 and the more pronounced differences for calmodulin and complex z. These observations indicate that there are several forms of Spec 2 proteins having different responses to calcium binding (although this does not prove that the forms are distinct proteins; see Plancke & Lazarides, 1983), that Spec 1 and 2 proteins have distinctive properties, and that the Spec proteins are not a sea urchin form of calmodulin, but at least some bind calcium.

We recently obtained additional evidence that Spec 1-type proteins undergo conformational changes upon binding calcium. In these experiments a fusion protein, made in E. coli between the 1c protein and the rop gene product of the ColE1 plasmid (Cesareni et al., 1982), was electrophoresed on SDS polyacrylamide gels in the presence of calcium ions or EGTA. A distinct shift in mobility of the 1c-fusion protein was observed in the presence of EGTA (Fig. 7). It is not clear why the shift in mobility of the fusion protein in the presence of EGTA is greater than the shift for proteins 1 and 2 under similar conditions of electrophoresis. It may be because Spec 1c, from which the fusion gene was constructed,

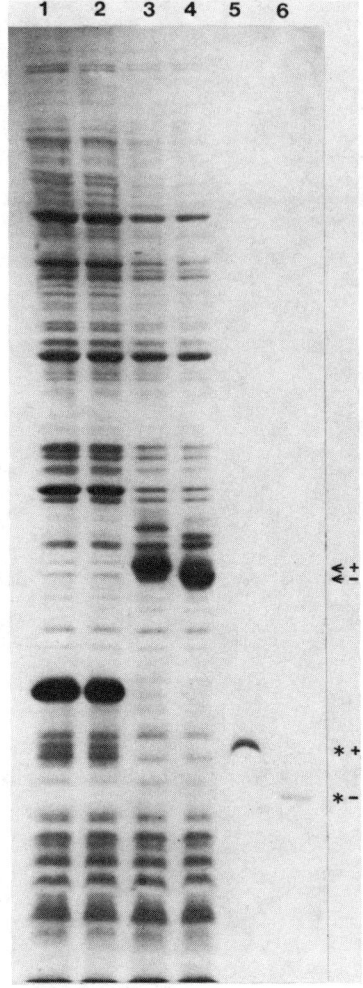

Fig. 7. Comparison of the electrophoretic mobility of the rop-Spec fusion protein in the presence and absence of 2 mM EGTA. Crude extracts of temperature-induced E. coli cells harboring pMAM19, a plasmid expressing a rop-fusion protein unrelated to the Spec protein (Lanes 1 & 2), or pMAM20, a plasmid expressing the rop-Spec fusion protein (Lanes 3 & 4), or bovine brain calmodulin (Sigma Chemicals) (Lanes 5 & 6) were electrophoresed on a 12% polyacrylamide/SDS gel. Lanes 1, 3, and 5, pretreatment with 2 mM EGTA; lanes 2, 4, and 6, no pretreatment. Following electrophoresis, the gel was stained with Coomassie brilliant blue. Arrows and asterisks denote the mobility of the pMAM20 rop-Spec fusion protein and calmodulin, respectively, in the presence (+) or absence (-) of EGTA.

codes for a minor variant not co-migrating with proteins 1 and 2.

It is possible that the Spec proteins represent a newly discovered family of proteins functionally or structurally conserved in other organisms. However, nucleic acid sequence comparisons indicate that they are quite distinct in amino acid sequence from any known protein. Moreover, the Spec proteins are not detectable in testis, intestinal tissue, or coelomocytes of adult S. purpuratus. A distinctive set of major proteins having similar electrophoretic properties does not appear in the ectoderm of embryos of the distantly related echinoid species Lytechinus pictus and Arbacia punctulata (Brandhorst, 1976; Bedard, 1984). Thus, the Spec proteins are not ubiquitous in sea urchin embryos, though homologues may exist having different electrophoretic properties or developmental patterns of expression.

Possible Functions of Spec Proteins

The existence of several Spec proteins belonging to two distinct subfamilies suggests that these proteins may have more than one role in sea urchins. The Spec 1 mRNAs accumulate extensively in presumptive dorsal ectoderm cells of the blastula before these cells are morphologically distinct from other epithelial cells (Lynn et al., 1983). These mRNAs and their translation products accumulate rapidly in cells which have ceased division. Shortly later, gastrulation begins and the morphology of the dorsal (as well as ventral) ectoderm cells is transformed from cuboidal to squamous. It is possible that the Spec 1 proteins are involved in this change in cell shape, perhaps providing a force or structural element to counteract the stretching of the epithelium by the invaginating archenteron. Alternatively, these proteins might have a role in mesenchymal/ectodermal interactions or skeletal formation which begin during the period of extensive accumulation of the Spec proteins.

We presently favor the hypothesis that the Spec proteins function during later larval development. While the fraction of cellular RNA corresponding to Spec 1 and 2 reaches a maximum at gastrula stage and then declines, substantial amounts of the mRNAs persist throughout larval development. This conclusion was drawn from the observed persistence of synthesis of Spec 1 and 2 types of proteins in larvae. Figure 8 shows the synthesis of Spec proteins in gastrulae and in nine-week-old feeding larvae which have formed adult rudiments but have not metamorphosed. Spot 2 disappears before plutei begin to feed and is never observed in growing larvae by staining or labeling with methionine. It is likely that this polypeptide is entirely modified to migrate as spot 1 in larvae, but the functional significance of this change is not understood; neither spots 1 nor 2 can be labeled with

^{32}P-orthophosphate, making it unlikely that the state of phosphorylation is involved in this conversion. Both protein 1 and 2 are observed among the products of cell-free translation of embryonic mRNA in rabbit reticulocyte lysates, though the relative amounts are dependent on the method of sample preparation (Bedard, 1984). Some Spec 2 polypeptides are also actively synthesized during larval development. The restriction of Spec proteins to dorsal ectoderm during larval development has not been demonstrated.

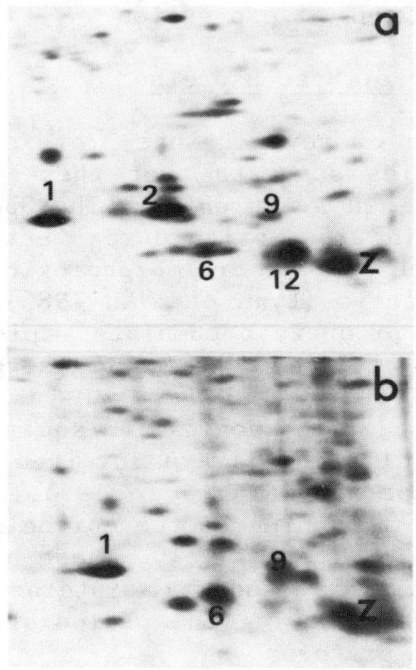

Fig. 8. Synthesis of Spec proteins in gastrulae and advanced larvae. Proteins labeled with ^{35}S were extracted from 48-h gastrulae and 9-wk feeding larvae, using identical extraction conditions. They were analyzed by two-dimensional electrophoresis. Proteins co-migrating with Spec proteins are labeled as in Figure 6.

Unlike embryonic development, when the Spec proteins increase in mass relative to other proteins, their mass does not change noticeably relative to other proteins during larval development. However, there is extensive larval growth resulting in substantial increases in mass of all larval protein, including the Spec proteins. Ectodermal proteins having electrophoretic mobilities characteristic of the Spec proteins are not appreciably synthesized during embryonic development of L. pictus (Brandhorst, 1976; Bedard,

1984). During larval growth, however, some proteins having identical or quite similar mobilities are very actively synthesized and may correspond to the Spec proteins, though they have not been shown to be structurally homologous. These observations suggest that the Spec proteins function during larval rather than (or in addition to) embryonic development.

Other investigators have recently found that the mRNA of an actin gene, Cy IIIa, also accumulates in the dorsal ectoderm cells of the embryo with kinetics very similar to those of Spec 1 transcripts (Shott et al., 1984). Moreover, the Cy IIIa mRNA also persists throughout larval development but abruptly disappears during metamorphosis (Shott et al., 1984). These results suggest that the Spec proteins and Cy IIIa actin (and possibly other as yet unidentified gene products) constitute a similarly regulated battery of proteins which perform related specialized functions.

One of the most important related functions of actin and several characterized members of the troponin C superfamily (particularly troponin C, myosin light chains, parvalbumin, and calmodulin) is in the formation and regulation of contractile systems of both muscle and nonmuscle cells. We suggest that the Spec family of proteins participates in a contraction of the larval dorsal epithelium during adult metamorphosis. During larval growth and development an adult rudiment is formed within the larva. Metamorphosis consists of several rapid and complex changes leading to the eversion and release of the adult rudiment from within the larval epithelium. A change in the shape of the dorsal epithelial cells from squamous to cuboidal appears to be the result of an active contraction of these cells (Cameron & Hinegardner, 1978). This contraction leads to the splitting open of the larval epithelium and its collapse onto the aboral surface of the juvenile urchin; eventually the residue of the dorsal epithelium is consumed by the juvenile urchin. During contraction the dorsal epithelial cells of L. pictus have subapical bundles of microfilaments oriented parallel to the plane of contraction (Cameron & Hinegardner, 1978). The inhibition of this contraction by cytochalasin B suggests that the bundles of microfilaments include actin filaments; other contractile proteins are undoubtedly involved as well, perhaps including the Spec proteins. The multiplicity of Spec proteins may correspond to the heterogeneity in form and function of some characterized members of the troponin C superfamily, such as the myosin light chains. Moreover, as for the characterized members of the troponin C family involved in contraction, some Spec proteins may be elements of the contractile apparatus while others may serve to regulate the contraction. We have not yet established when during larval development bundles of microfilaments are formed (they are not observed in embryos) and whether Spec 1 proteins are part of the bundles. Metamorphosis of ascidian larvae is also accompanied by a contraction of ectodermal epithelial cells of the tail which is mediated by actin microfilaments (Cloney, 1982). In some ascidian species this contractile apparatus is assembled only after the induction of metamorphosis by external cues (Cloney, 1982); a similar situation might apply in sea urchin larvae since the bundles of filaments are not observed in dorsal ectoderm cells of early larvae.

If calcium-binding Spec proteins serve in the epithelial contraction during metamorphosis, experimentally induced elevations of the intracellular concentrations of free calcium ions might induce metamorphosis in competent larvae. Larvae of S. purpuratus having completely formed adult rudiments tend to undergo metamorphosis spontaneously upon settling onto a solid substrate or even when suspended in sea water, making investigations of the induction of metamorphosis difficult. Mature larvae of L. pictus have a period of several weeks during which they normally metamorphose only in response to appropriate external stimuli; in particular, the surface films of bacteria on the substrate to which they attach upon settling (Cameron & Hinegardner, 1974). We have subjected these larvae to a variety of treatments expected to alter intracellular free calcium levels.

As shown in Figure 9a, the calcium ionophore A23187 in sea water induces complete metamorphosis at an optimum concentration of 5 µM; higher concentrations lead to epithelial contraction but also, ultimately, larval disintegration. As shown in Figure 9b, quercetin, an inhibitor of the $[Ca^{++},Mg^{++}]$-ATPase thought to act as a cellular calcium pump, induces metamorphosis at moderate concentrations but epithelial contraction followed by disintegration at higher concentrations. The responses to both ionophore and quercetin are slow. In the case of A23187, no juvenile urchins were observed 24 hours after the beginning of treatment with the optimum dose, and the completion of metamorphosis (the collapse of the larval epithelium on the everted juvenile) frequently did not occur until several days after the initiation of treatment. The response to quercetin was more rapid but still required up to a day. On the other hand, sea water lacking Mg^{++} ions (which might inhibit the exchange of intracellular free Ca^{++} for external Mg^{++} via the calcium pump) and natural stimuli (bacterial surface films) sometimes results in much more rapid responses; completion within two hours was observed sometimes.

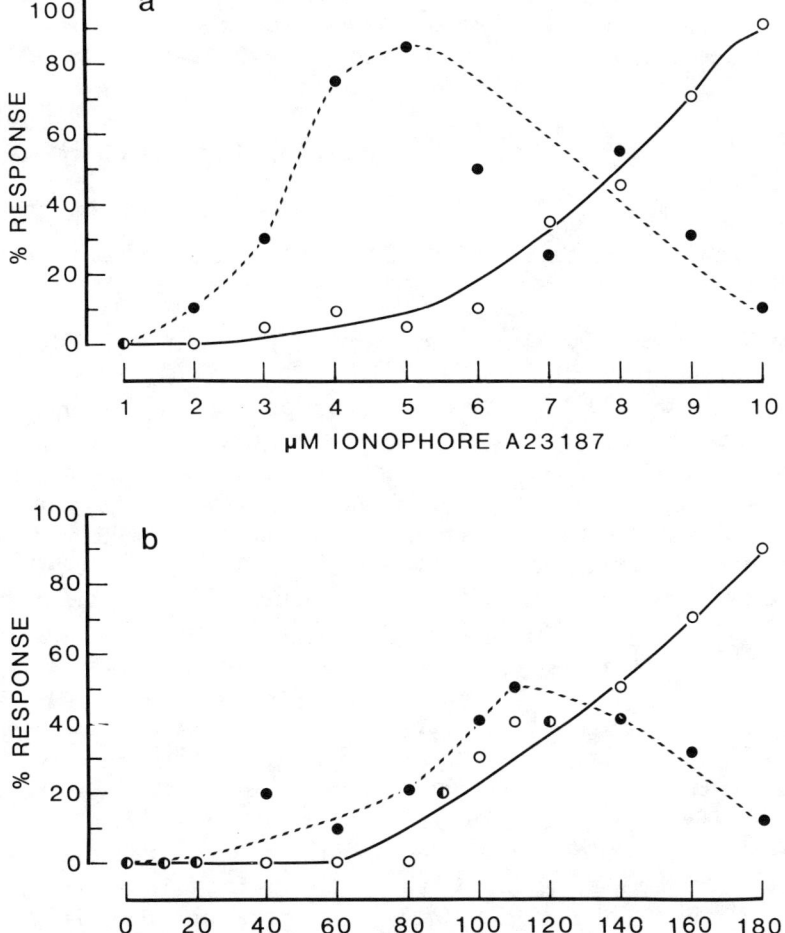

Fig. 9. The induction of metamorphosis by ionophore A23187 and quercetin. Groups of 20 mature larvae of L. pictus were placed in dishes containing sea water and various concentrations of ionophore (a), or quercetin (b). The percentage which underwent metamorphosis and formed essentially normal juveniles is plotted (closed circles). At higher ionophore concentrations, epithelial contraction was usually followed by disintegration of the larvae (and adult rudiments) rather than the completion of metamorphosis; the percentage of larvae showing this response is also plotted (open circles). None of the controls (no drug added) underwent either metamorphosis or contraction.

Burke (1983a & b) has provided some evidence for a model whereby a sensory response to an external stimulus leads to an induction of metamorphosis via the release of an endocrine signal or a local neurotransmitter. Contraction of isolated larval arms (epithelia) can be stimulated electrically and by application of several catecholamines in some echinoids; again the response was very slow. In our investigations the ionophore A23187 and quercetin may be acting indirectly, perhaps by stimulating secretions of neurotransmitters from neurons which may underlie the larval epithelium (Burke, 1983a). Further investigations will be required to establish the role of Ca^{++} ions and Spec proteins in metamorphosis, but we have found a convenient method for inducing metamorphosis in L. pictus larvae.

Delayed Functions of Proteins Synthesized During Embryogenesis

It is commonly expected that a protein actively synthesized and accumulating only in a particular cell lineage during a restricted period of embryonic development serves a specialized function in those embryonic cells. We have proposed that the Spec proteins, as well as other contractile proteins yet to be identified, perform their specialized functions only after weeks of larval development, though we do not exclude the possibility that they carry out functions in embryos as well. We contend that rather than being a special case this may be expected of many lineage-specific proteins accumulating during embryonic development; that is, such proteins may appear long before their specialized functions are activated. Most proteins of plutei are present in eggs in similar amounts, and many of these proteins are not detectably synthesized during embryonic development (Bedard & Brandhorst, 1983); such proteins accumulate during oogenesis and are retained (and presumably utilized) by embryos. Examples of proteins synthesized and stored during oogenesis but utilized during embryogenesis include the histones, tubulins, actins, and a variety of chromosomal proteins. The storage of proteins during one period of development for use during a subsequent period is thus a common developmental strategy, one which may ensure sufficient levels and coordinate interactions of proteins within a cell and appropriate activity of the tissue constituted by such cells.

We propose as a useful generalization that proteins functioning during embryonic development, which is characterized by cellular proliferation and morphogenesis (but little or no growth), are mostly synthesized during oogenesis. Proteins whose accumulation is initiated during embryonic development may mostly function during larval development, characterized by growth and cellular specialization. Finally, the many changes in protein synthesis which occur during echinoid larval development (Brandhorst, unpubl. obs.) may be in preparation for adult metamorphosis and juvenile development. A corollary to these speculations is that many proteins first appearing during oogenesis, embryogenesis, or larval development continue to be synthesized during later development in order to maintain appropriate cellular levels in spite of growth or protein turnover. Key developmental transitions such as meiotic maturation, initiation of gastrulation, and initiation of larval growth upon feeding may be accompanied not only by extensive changes in patterns of gene expression (Brandhorst et al., 1983) but also by changes in the utilization of stored material, be it RNA, protein, or other molecules.

ACKNOWLEDGMENTS

We thank Mary Bannet and Zhiyuan Gong for assistance with the protein electrophoresis. The research reported was supported in part by a grant from NSERC to B.P. Brandhorst, NSERC summer fellowships to L.E. Philpotts, and a grant from the NIH to W.H. Klein.

LITERATURE CITED

Bedard, P-A., 1984. Regulation of patterns of protein synthesis during sea urchin embryogenesis. Ph.D. Thesis, McGill University.

Bedard, P-A. and B.P. Brandhorst. 1983. Patterns of protein synthesis and metabolism during sea urchin embryogenesis. Dev. Biol. 96: 74-83.

Brandhorst, B.P. 1976. Two-dimensional gel patterns of protein synthesis before and after fertilization of sea urchin eggs. Dev. Biol. 52: 310-317.

Brandhorst, B.P., P-A. Bedard, and F. Tufaro. 1983. Patterns of protein metabolism and the role of maternal RNA in sea urchin embryos. pp. 29-48. In: Time, Space, and Pattern in Embryonic Development, W.R. Jeffery and R.A. Raff (eds.). Alan R. Liss, Inc., New York.

Bruskin, A.M., A.L. Tyner, D.W. Wells, R.M. Showman, and W.H. Klein. 1981. Accumulation in embryogenesis of five mRNAs enriched in the ectoderm of the sea urchin pluteus. Dev. Biol. 87: 308-318.

Bruskin, A.M., P.A. Bedard, A.P. Tyner, R.M. Showman, B.P. Brandhorst, and W.H. Klein. 1982. A family of proteins accumulating in ectoderm of sea urchin embryos specified by two related cDNA clones. Dev. Biol. 91: 317-324.

Burgess, W.H., D.K. Jemiolo, and R.H. Kretsinger. 1980. Interaction of calcium and calmodulin in the presence of sodium dodecyl sulfate. Biochim. Biophys. Acta 623: 257-270.

Burke, R.D. 1983a. Neural control of metamorphosis in Dendraster excentricus. Biol. Bull. 164: 176-188.

Burke, R.D. 1983b. The induction of metamorphosis of marine invertebrate larvae: Stimulus and response. Can. J. Zool. 61: 1701-1719.

Cameron, R.A. and R.T. Hinegardner. 1974. Initiation of metamorphosis in laboratory cultured sea urchins. Biol. Bull. 146: 335-342.

Cameron, R.A. and R.T. Hinegardner. 1978. Early events in sea urchin metamorphosis: Description and analysis. J. Morphol. 151: 21-32.

Carpenter, C.D., A.M. Bruskin, P.E. Hardin, M.J. Keast, J. Anstrom, A.L. Tyner, B.P. Brandhorst, and W.H. Klein. 1984. Novel proteins belonging to the troponin C superfamily are encoded by a set of mRNAs in sea urchin embryos. Cell 36: 663-671.

Cesareni, G., M.A. Muesing, and B. Polisky. 1982. Control of ColEl DNA replication: the rop gene product negatively affects transcription from the replication primer promoter. Proc. Natl. Acad. Sci., USA 79: 6313-6317.

Cloney, R.A. 1982. Ascidian larvae and the events of metamorphosis. Amer. Zool. 22: 817-826.

Dayhoff, M.O. 1978. Atlas of protein sequence and structure. Volume 5, supplement 3, pp. 273-283.

Klein, W.H., L.M. Spain, A.L. Tyner, J. Anstrom, R.M. Showman, C.D. Carpenter, E.D. Eldon, and A.M. Bruskin. 1984. A family of mRNAs expressed in the dorsal ectoderm of sea urchin embryos. pp. 131-140. In: The Molecular Aspects of Early Development. G.M. Malacinski and W.H. Klein (eds.). Plenum Pess, Inc. New York.

Kretsinger, R.H. 1980. Structure and evolution of calcium-modulated proteins. CRC Crit. Rev. Biochem. 8: 119-174.

Laemmli, U.K. 1970. Cleavage of structural proteins during assembly of the head of bacteriophage T4. Nature 227: 680-685.

Lynn, D.A., L.M. Angerer, A.M. Bruskin, W.H. Klein, and R.C. Angerer. 1983. Localization of a family of mRNAs in a single cell type and its precursors in sea urchin embryos. Proc. Natl. Acad. Sci., USA 80: 2656-2660.

McClay, D.R. and A.F. Chambers. 1978. Identification of four classes of cell surface antigens appearing at gastrulation in sea urchin embryos. Dev. Biol. 63: 179-186.

O'Farrell, P.H. 1975. High-resolution two-dimensional electrophoresis of proteins. J. Biol. Chem. 250: 4007-4021.

Plancke, Y.D. and E. Lazarides. 1983. Evidence for a phosphorylation form of calmodulin in chicken brain and muscle. Mol. Cell. Biol. 3: 1412-1420.

Shott, R.J., J.J. Lee, R.J. Britten, and E.H. Davidson. 1984. Differential expression of the actin gene family of Strongylocentrotus purpuratus. Dev. Biol. 101: 295-306.

Strohman, R.C., J. Micou-Eastwood, C.A. Glass, and R. Matsuda. 1983. Human fetal muscle and cultured myotubules derived from it contain a fetal-specific myosin light chain. Science 221: 955-957.

Weeds, A. 1978. Myosin: Polymorphism and promiscuity. Nature 274: 417-418.

THE ORIGIN OF THE MICROMERES AND FORMATION OF THE SKELETAL SPICULES IN DEVELOPING SEA URCHIN EMBRYOS

F.H. Wilt, S. Benson, and J.A. Uzman

ABSTRACT

There are a number of events involved in the formation of the endoskeletal calcareous spicule during sea urchin embryogenesis. We enumerate many of them here in an attempt to demonstrate the large variety of things we must understand in order to learn how a new phenotype arises in an embryo. Consequently, a variety of methods and different experimental attacks must be brought to bear on the developing embryo if one is to learn how its different parts arise. Micromeres arise at the fourth cell division because of the plane of that cell division. The formed micromeres undergo a program of cell divisions and of changes in adhesivity and motility. At the time of the fourth cell division there are large changes in the rate of RNA synthesis; experiments are described which implicate the loading of RNA polymerase as a controlling factor in that increase. After emigration into the blastocoel, the cells of the primary mesenchyme, which is the tissue formed from the micromeres, put forth protrusions which fuse with one another. Within this syncytial cable, the calcareous spicule is deposited on an organic matrix. The matrix has been isolated and characterized, and is composed of a small number of glycoproteins rich in acidic amino acids. The over-all shape and architecture of the spicule is species-specific, and the shape may be the result of factors indigenous to the primary mesenchyme cells and to their interaction with the blastocoel environment.

INTRODUCTION

"Nowadays, when investigators consider the problem of differentiation, they bypass intermediate steps and immediately try to prove the formation of new specific proteins. Certainly synthesis of new proteins is the most definite proof of differentiation, but in another sense, it is the end result of differentiation. Unequal division is the initial step for differentiation because of its selectivity in segregating special cytoplasm and placing it in an appropriate orientation" (Dan et al., 1983).

The study of the origin and fate of the micromeres of sea urchin embryos illustrates well the aptness of Dan's statement. This quartet of cells that arises at the fourth cell division has been studied for a very long time by many embryologists. The fascinating and profound role(s) these cells perform in pattern orientation and gastrulation is well known, and many excellent reviews exist (Giudice, 1973; Horstadius, 1973; Czihak, 1975; Davidson, 1976; Angerer & Davidson, submitted). The micromeres offer a splendid opportunity to study many different facets of determination and differentiation, and to learn how these different aspects relate to one another. As Dan points out, both the origin of the cells in question and the subsequent program of differential gene expression must be understood if we are to grasp the principles by which a new phenotype arises at the proper time and place in the developing embryo. In this article, we propose to enumerate the different known steps involved in the formation of the endoskeletal spicules of the pluteus larva of sea urchins. We shall comment upon two of the steps in some detail, using experimental approaches currently employed by us.

The Timing and Orientation of Cleavage

The distribution of daughter nuclei during cleavage is partly a function of the timing of the cell cycle which may influence the plane of cell division. Even a cursory review of the control of cell cycle timing is outside our purview, but one can mention the well known fact that early sea urchin embryos have rhythms of DNA synthesis, centriole replication, and changes in the cytoplasm (e.g., the balance of reduced and oxidized sulfhydryl). These different aspects of the cell cycle may be uncoupled to some extent. Of special interest are manipulations that alter the timing of micromere formation. In 1915, Painter used phenylurethane (ethyl carbamate) to interfere with the relative timing of nuclear divisions, and obtained clear examples of alterations of cleavage planes and precocious formation of micromeres. More recently Dan (1972) and his colleagues have employed several ways to obtain micromeres at the third cell division rather than the fourth. Using appropriate timing of administration of antimitotic agents, one may obtain a third cell division at a time near when the fourth cell division occurs in the controls, and in such instances micromeres will form at the third cell division. Under such circumstances, the normal plane of the third equatorial division is displaced toward the vegetal pole, resulting in an animal quartet of very large cells and a vegetal quartet of very small ones. These small cells look like authentic micromeres. Kitajima and Okazaki (1980) have shown these precocious micromeres may be isolated, cultured in vitro, and give rise to the skeletal spicule elements in vitro. On the other hand, Tanaka (1976) has shown that brief treatments of cleaving eggs near the time of the second cell division with some ionic detergents will alter the normally asymmetric fourth cleavage so that micromeres never form. The embryos are viable, and differentiate, but never form primary mesenchyme nor an endoskeleton. Hence, the plane of cell division, and by influence, the timing and placement of the centrioles, is crucial to the origin of the micromere lineage. The unequal fourth cell division in the vegetal pole is preceded by a movement of the vegetal cell nuclei of the eight-cell stage, and the forming asymmetric spindle interacts at the vegetal end with a pigment granule-free zone of the vegetal cortex (Dan, 1979; Dan et al., 1983). Somehow these interactions of the nuclei and their spindles with the superficial cytoplasmic region of this area must be important, but the nature of the interactions is unknown. Tanaka's (1976) experiments show that the formation of a cell containing vegetal cytoplasm is not sufficient; small cells that arise from progenitors with spindles oriented parallel to the animal-vegetal axis seem to be necessary for a functional primary mesenchyme. On the other hand, there is a special quality to the cytoplasm of the vegetal pole of the egg, a point thoroughly documented by the classic work of Horstadius (1973) and recently substantiated by Schroeder (1980a & b).

The Endowment of the Micromeres

From the outset the newly formed micromeres have different behaviors and properties from their neighbors and siblings. The program of cell division is different for the micromeres and the timing of this program has been followed in some detail (Dan et al., 1980). The micromeres divide to give one small and one large cell. The former only divides once more, while the quartet of larger cells divides three times more to give 32 cells; the rhythms of these divisions are not synchronous with those occurring in the progeny formed by mesomeres and macromeres. Sano (1977) has shown by electrophoresis of dissociated cells that the micromeres acquire a greater surface electronegativity as they mature, and the pattern of agglutinability by concanavalin A also shows distinct patterns of change (Sano, 1980).

Micromeres also differed from their congeners in the complexity of the nucleic acids which they contain. Early experiments by Mizuno et al. (1974) employing RNA competition-hybridization studies with repetitive DNA sequences indicated substantial differences between the RNA of micromeres and the other cells. More recently, Rodgers and Gross (1978) showed the total sequence complexity of the RNA of micromeres was substantially less than that of the RNA from mesomeres and macromeres. Ernst et al. (1980) showed that micromeres apparently lack high complexity nuclear RNA, and that in contrast with the other cells, all the sequence complexity in the cytoplasm is present in the polysomal RNA. A thorough investigation using 2 dimensional gel electrophoresis has shown that the proteins synthesized by micromeres are apparently identical to those made in mesomeres and macromeres (Tufaro & Brandhorst, 1979); later in development the descendants of these cells do diverge in the patterns of proteins that they synthesize.

At approximately the same time that the micromeres normally form at the fourth cell division, there is a substantial upturn in the total rate of RNA synthesis by the embryo (Wilt, 1970). When calculated on a per cell basis, it has been estimated that there is a four-fold or greater increase in RNA accumulation, presumably due to the transcription rate (Maxson & Wilt, 1981). One may prepare nucleated merogons by conventional procedures first devised by E. B. Harvey. After stratification in sucrose density gradients, centrifuged eggs will separate into nucleated and enucleated halves. The nucleated half may be fertilized and it will divide at exactly the same time as the full size control embryos. The nucleated merogons give rise to normal but miniature plutei. The incorporation of ^{3}H-guanosine into RNA was followed by continuous incorporation, and there was, as reported earlier, a substantial increase in incorporation about the time of the fourth cell division. Surprisingly, the nucleated merogons, with an average volume roughly half of the controls, showed this same increase in RNA accumulation one cell division earlier (Uzman, 1983). It is possible that the increase in the rate of RNA accumulation is sensitive to some cytoplasmic component. A similar suggestion was made by Newport and Kirschner (1982) to explain the precise timing of the onset of RNA synthesis at the 12th cell division in Xenopus laevis embryos.

Table 1. Transcript elongation in isolated nuclei.

Development Stage (Number of Cells)	4	8	16	100	400 hatching	600 gastrula
pM of UTP per mg of DNA	0.11	0.14	0.34	0.55	0.29	0.32

Nuclei were isolated from embryos at the different stages, and the pM of UTP per µg of DNA in the nuclei incorporated in 5 min is shown above. Each number is the average of 3 or more experiments. Details may be found in Uzman and Wilt (1984).

The increase in RNA synthesis at the fourth cell division also may be studied by examination of RNA synthesis in isolated nuclei (Uzman & Wilt, 1984). Some of the early work on multiple forms of RNA polymerase was carried out with nuclei obtained from sea urchin embryos, and very active incorporation of ribonucleoside triphosphates into RNA occurred in isolated nuclei. Several groups of investigators (Roeder & Rutter, 1970; Levy et al., 1978) have shown that the incorporation that occurs is an elongation of already initiated transcripts. Thus far a convincing demonstration of initiation of new transcripts in isolated nuclei of sea urchin embryos has not been forthcoming. Since already initiated polymerases with nascent chains are the only sites of incorporation, the activity in isolated nuclei may reflect the number of nascent elongating chains present at the time the nuclei are isolated. We decided to exploit this in order to look at the level of overall RNA synthesis during sea urchin development. Table 1 presents some representative findings. The level of transcript elongation in isolated nuclei shows a marked increase after the eight-cell stage, just as is found in whole embryos. This increase is on a "per nucleus" basis; since only elongation is measured, this increase means that either the elongation rate increases, or the number of nascent chains that can elongate has increased. We favor the latter alternative because measurements of RNA chain elongation rates, in vivo, by Aronson and Chen (1977) do not show such large changes.

It is also possible that there are polymerase molecules present that have formed a ternary complex that do not, for some unknown reason, elongate. This situation has been reported for polymerases in the nuclei of hen erythrocytes; Gariglio et al. (1981) showed that chain elongation could occur in these nuclei if sarkosyl was included in the assay. Sarkosyl is a detergent that may remove much of the protein from chromatin without removing initiated polymerases. We found sarkosyl had little effect on transcript elongation in nuclei from the early stages, and hence, it is unlikely there are any cryptic polymerases present at the time when RNA synthesis shows an upsurge near the time of the fourth cell division. These findings suggest that there is an increase in transcription near the time of micromere formation because more polymerases are engaged in transcription. We do not know if there are differences between the different blastomeres in this respect, nor do we know the biological significance of this upsurge in the synthesis of RNA.

Changes in the level of RNA synthesis are not the only changes that become noticeable when micromeres arise; several other changes in biosynthetic capacities and cell behaviors become apparent after the cell division. Elsewhere in this volume Harkey describes his studies on the kinds of proteins synthesized by the descendants of micromeres, mesomeres and macromeres (Harkey & Whiteley, 1983). Kitajima and Matsuda (1982) have also shown that micromere progeny and the descendants of mesomeres and macromeres synthesize different proteins.

The Primary Mesenchyme

The descendants of the micromeres become the primary mesenchyme. The behavior of these cells is profoundly different way from their immediate neighbors. According to Horstadius (1973), the primary mesenchyme has an influence on archenteron formation. There is a dramatic change in the adhesivity of the primary mesenchyme cells to the hyalin layer, and to each other, as outlined by McClay (this volume). In addition, the motility of the mesenchyme cells soon becomes pronounced. In a sense the primary mesenchyme cells detach from the epithelium and become a population of single migratory cells, apparently exploring the blastocoel. It is significant that the primary mesenchyme only ingresses into the blastocoel. Even in exogastrulae the prospective primary mesenchyme cells detach from the epithelium and wander into the blastocoel; this ingression never occurs in the other direction (into the sea water). There is a polarity to the change in adhesivity and ingression that is not understood. Does the hyalin form a barrier to outward migration? Is the change in motility polarized?

It is also not clear whether the locomotory activity is random. The blastocoel contains considerable fibrillar material; it is not an empty space (Solursh & Katow, 1982; Crise-Benson & Benson, 1979). Do the primary mesenchyme cells migrate on or around these fibers as a kind of scaffolding? How much of the migration is innate and how much determined by the environment? A close examination of primary mesenchyme cells, in vitro, as already begun by Okazaki (1975), Harkey and Whiteley (1980) and McClay (1982) will be of interest. The movement of the cells in the blastocoel is not random. The cells move up the blastocoel wall, then return to near the vegetal pole where they align into a ring.

There are indications in the literature that the morphogenesis and differentiation of the primary mesenchyme cells are influenced by collagen formation and deposition. This is fascinating because of the involvement of collagen in vertebrate bone formation, and collagen has been reported to be a constituent of the sea urchin larval spicule as well (Pucci-Minafra et al., 1972). Some recent studies of the influence of collagen metabolism on the primary mesenchyme differentiation, in vitro, shed some light on this. Blankenship and Benson (1984) have examined the role of inhibitors of collagen metabolism on the formation of spicules in culture. These inhibitors have different primary modes of action, but all are known to inhibit collagen formation. α, α, dipyridyl inhibits prolyl hydroxylase, an enzyme necessary for post-translational hydroxylation of proline in the collagen chains. L-azetidine-2-carboxylic acid and L-3, 4 dehydroproline are proline analogs. Blankenship and Benson found that these agents stopped spicule formation, in vitro, from cultured micromeres (cf. also, Mintz et al., 1981). On the other hand, if the culture dishes were pre-coated with a layer of Type I collagen before the micromeres are cultured on them, the formation of spicules takes place almost normally in the presence of the same inhibitors. The clear implication is that while collagen may be involved in micromere differentiation to form spicules, collagen only creates a set of permissive conditions for terminal differentiation. Is it possible that the primary mesenchyme elaborates an extracellular collagen, which in turn allows the cells that secrete it in the first place to complete their final phases of differentiation?

Earlier studies of the primary mesenchyme activity in embryos describe a unique behavior of these cells. After the completion of the migratory phase and assumption of the appropriate placement in the blastocoel, cellular extrusions (rather broad-lobed filopodia) extend from the cell bodies, and these cellular protrusions fuse so that the primary mesenchyme cells acquire a syncytial character. First described with the light microscope by Okazaki (1960), this was confirmed with electron microscopy by Gibbins et al. (1969), who also described the appearance of a vacuole in the syncytial cables of cytoplasm. It is within this vacuole that the calcareous skeleton is elaborated. Few details of the structure of the vacuole and its relationship to the mineralization are available because of the propensity of the hard $CaCO_3$ deposits to pop out of the embedding material during sectioning. High resolution morphological techniques have not yet been applied to study syncytium formation, vacuole formation, and the onset of biomineralization.

The Organic Matrix of the Spicule

Okazaki (1960) studied the demineralization of purified spicules under the light microscope and noticed the presence of a thin strand of material that might constitute a matrix; she favored the idea that the matrix was an internal ground substance permeating the calcite crystal of the spicule. In order to pursue this idea, we set out to purify spicules from larvae to study their structure and characterize the components of the structure (Benson et al., 1983). The isolation of spicules by repeated exposure to ionic and non-ionic detergents according to published methods (Pucci-Minafra et al., 1972) resulted, in our hands, in spicules contaminated with basal laminae and blastocoel contents (Fig. 1). Accordingly, we resorted to somewhat different treatments; either extensive treatment with 4 M urea, 2% sodium dodecylsulfate, or brief exposure to 1-2% sodium hypochlorite. Both procedures resulted in spicules free of contamination when viewed by a scanning electron microscope. The morphological evidence (Benson et al., 1983) has supported the view that the calcareous spicule does contain a ground substance of organic material permeating the mineral phase. In surface view the matrix seems lamellar, while it seems more reticular in nature when viewed by transmission electron microscopy. The material is soluble in water unless it is fixed, and the reticular material is not cross banded. When fixed material is viewed in cross section after partial demineralization of the spicule, concentric lamellae are observed.

Our more recent studies have concentrated on a chemical characterization of the water-soluble matrix material. It is chiefly glycoprotein. Amino acid analysis has shown a glycine content of 16%, glutamic acid 13%, aspartic acid 9%, alanine 8%, and serine 8%. Hence, five amino acids constitute about 54% of the residues in the protein, and this composition is very reminiscent of some other proteins associated with calcareous skeletal elements found among the invertebrates, e.g., the matrix proteins of shells of molluscs and cuticles of arthropods (Weiner et al., 1983).

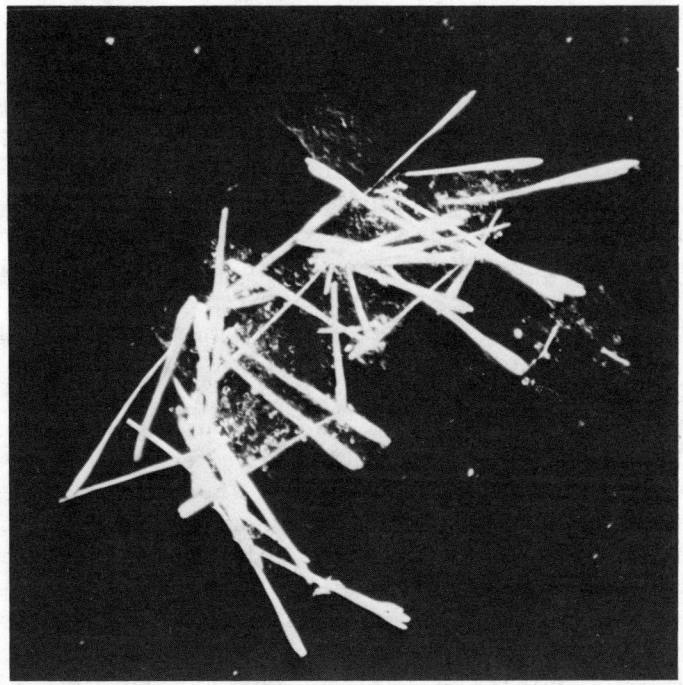

Fig. 1. Spicules isolated with detergent. Plutei were lysed and washed repeatedly (8x) in 2% deoxycholate, 1% Triton-X-100, Tris buffer, pH 8.0. Each washing involved centrifugation and resuspension by homogenization. This preparation is viewed by dark field microscopy (x220) and is still visibly contaminated with basal laminae and blastocoel contents.

Antibodies have been raised against these protein(s) and show a good specificity when tested by immunocytochemistry or "western" blotting procedures. We are currently using immunochemical methods to define the time course of appearance of the spicule matrix components and to help characterize them. Protein blotting experiments reveal there are several proteins that react with the polyclonal antibodies. Among the detectable components are those with apparent molecular weights of 117,000, ~51,000 and ~39,000. When the proteins are subjected to extensive treatment with the endoglycosidase, "endo F", prior to electrophoresis, the apparent molecular weight of the 51,000 and 39,000 components is reduced about 5,000 for each, confirming the glycoprotein nature of these components. The amino acid composition, solubility, and structure of the matrix components are not consistent with collagen being a prominent component of the spicule matrix.

The Pattern of the Spicule

The different parts of the spicule form and are aligned in precise positions with respect to the rest of the embryo. Since the spicule is an intracellular organelle, the overall morphology and shape are a result of the position and alignment of the primary mesenchyme cells and the fusion of the cell protrusions. The possible role of extracellular material in this phenomenon has been mentioned previously. There may also be cues issuing from the blastocoel wall and/or its associated basal lamina. The primary mesenchyme cells occupy final positions along the junction of the descendants of "animal 2" and "vegetal 1" tiers of the 32-cell stage. Unfortunately, detailed fate mapping using marked cells has not yet been carried out with these cells, and it is not clear how precise are the earlier studies of lineage. The primary mesenchyme cells show autonomous behavior in culture that may reflect a role in spicule formation. The cells arrange themselves linearly on the culture substrate and form linear needle-like spicules. This morphology is similar to that of a single arm of the basic triradiate spicule. Is there some cell polarity that aligns the fusing cell protrusions? Or does the formation of one fused element produce, or induce, a subsequent polarity in the next forming fusion? The morphology produced in culture is not what one would expect from random collision and adhesion. Finally, the triradiate spicule has an elaborate microarchitecture that is species-specific. S. purpuratus spicules possess numerous hooks and spurs at reproducible places. D. excentricus has spicules with a beautifully fenestrated appearance. No two species produce spicules that are exactly alike.

CONCLUSION

The purpose of this review is to underline with a single, "simple" example the incredible richness of different phenomena that are involved in the origin of a new phenotype. Even a sketch of the differentiation of the micromeres to form a spicule requires a deep knowledge of regulation of mitosis in time and space, at least a notion of what is involved in cell type determination at the fourth cell division, some information on how a program of mitosis is bestowed on a lineage of cells, and some knowledge of how cell type gene expression is regulated. Furthermore, against this background at the cellular level, knowledge of cues from other cells and the extracellular matrix, and how they impinge on the autonomous behavior of the primary mesenchyme is necessary to appreciate how the final form of the spicule is achieved.

LITERATURE CITED

Aronson, A.I. and K. Chen. 1977. Rates of RNA chain growth in developing sea urchin embryos. Dev. Biol. 59: 39-48.
Benson, S., E.M.E. Jones, N. Crise-Benson, and F. Wilt. 1983. Morphology of the organic matrix of the spicule of sea urchin larva. Exp. Cell Res. 148: 249-253.
Blankenship, J. and S. Benson. 1984. Collagen metabolism and spicule formation in sea urchin micromeres. Exp. Cell Res. 152: 98-104.
Crise-Benson, N. and S. Benson. 1979. Ultrastructure of sea urchin embryo collagen. Wilh. Roux. Archiv. Develop. Biol. 186: 65-70.
Czihak, G. 1975. The Sea Urchin Embryo. Springer Verlag Berlin.
Dan, K. 1972. Modified cleavage pattern after suppression of one mitotic division. Exp. Cell Res. 72: 69-73.
Dan, K. 1979. Studies on unequal cleavage in sea urchins. I. Migration of the nuclei to the vegetal pole. Dev. Growth Differ. 21: 527-535.
Dan, K., S. Tanaka, K. Yamazaki, and Y. Kato. 1980. Cell cycle study up to the time of hatching in the embryos of the sea urchin Hemicentrotus pulcherrimus. Dev. Growth Differ. 22: 589-598.
Dan, K., S. Endo, and I. Uemura. 1983. Studies of unequal cleavage in sea urchins. II. Surface differentiation and the direction of nuclear migration. Dev. Growth Differ. 25: 227-237.

Davidson, E.H. 1976. Gene Activity in Early Development. 2nd Edition, Academic Press, N.Y.
Ernst, S.G., B.R. Hough-Evans, R.J. Britten, and E.H. Davidson. 1980. Limited complexity of the RNA in micromeres of 16 cell sea urchin embryos. Dev. Biol. 79: 119-127.
Gariglio, P., M. Bellard, and P. Chambon. 1981. Clustering of RNA polymerase B molecules in the 5' moiety of the adult β globin gene of hen erythrocytes. Nuc. Acid. Res. 9: 2589-2598.
Gibbins, J.R., L.G. Tilney, and K.R. Porter. 1969. Microtubules in the formation and development of the primary mesenchyme in Arbacia punctulata. J. Cell Biol. 41: 201.
Giudice, G. 1973. Developmental Biology of the Sea Urchin Embryo. Academic Press, N.Y.
Harkey, M.A. and A.H. Whiteley. 1980. Isolation, culture and differentiation of echinoid primary mesenchyme cells. Wilh. Roux. Archiv. Develop. Biol. 189: 111-122.
Harkey, M. and A.H. Whiteley. 1983. The program of protein synthesis during the development of the micromere-primary mesenchyme cell line in the sea urchin embryo. Dev. Biol. 100: 12-28.
Horstadius, S. 1973. Experimental Embryology of Echinoderms. Clarendon, Oxford.
Kitajima, T. and K. Okazaki. 1980. Spicule formation in vitro by the descendants of precocious micromere formed at the 8 cell stage of sea urchin embryo. Dev. Growth Differ. 22: 265-279.
Kitajima, T. and R. Matsuda. 1982. Specific protein synthesis of sea urchin micromeres during differentiation. Zoological Magazine 91: 200-205.
Levy, S., G. Childs, and L. Kedes. 1978. Sea urchin nuclei use RNA polymerase II to transcribe discrete histone RNAs larger than messengers. Cell 15: 151-163.
Maxson, Jr., R.E. and F.H. Wilt. 1981. The rate of synthesis of histone mRNA during development of sea urchin embryos. Dev. Biol. 83: 380-386.
McClay, D.R. 1982. Cell recognition during gastrulation in the sea urchin. Cell Differ. 11: 341-344.
Mizuno, S., Y.R. Lee, A.H. Whiteley, and H.R. Whiteley. 1974. Cellular distribution of RNA populations in 16 cell stage embryos of the sand dollar, Dendraster excentricus. Dev. Biol. 37: 18-27.
Mintz, G.R., S. De Francesco, and W.J. Lennarz. 1981. Spicule formation by cultured embryonic cells from the sea urchin. J. Biol. Chem. 256: 13205-13111.

Newport, J. and M. Kirschner. 1982. A major developmental transition in early Xenopus embryos. II. Control of the onset of transcription. Cell 30: 687-695.

Okazaki, K. 1975. Spicule formation by isolated micromeres of the sea urchin embryo. Amer. Zool. 15: 567-581.

Okazaki, K. 1960. Skeleton formation of sea urchin larvae. II. Organic matrix of the spicule. Embryologica 5: 283-320.

Painter, T. 1915. An experimental study in cleavage. Exp. Zool. 18: 299-322.

Pucci-Minafra, I., C. Casano, and C. LaRosa. 1972. Collagen synthesis and spicule formation in sea urchin embryos. Cell Differ. 1: 157-165.

Rodgers, W.H. and P.R. Gross. 1978. In homogeneous distribution of egg RNA sequences in the early embryos. Cell 14: 279-288.

Roeder, R.G. and W.J. Rutter. 1970. Multiple RNA polymerases and ribonucleic acid synthesis during sea urchin development. Biochem. 9: 2543-2553.

Sano, K. 1977. Changes in cell surface charges during differentiation of isolated micromeres and mesomeres from sea urchin embryos. Dev. Biol. 60: 404-415.

Sano, K. 1980. Changes in Concanavalin A mediated cell agglutinability during differentiation of micromere and mesomere derived cells of the sea urchin embryo. Zoological Magazine, 89: 321-325.

Schroeder, T.E. 1980a. Expressions of the prefertilization polar axis in sea urchin eggs. Dev. Biol. 79: 428-443.

Schroeder, T.E. 1980b. The jelly canal marker of polarity for sea urchin oocytes, eggs and embryos. Exp. Cell Res. 128: 490.

Solursh, M. and H. Katow. 1982. Initial characterization of sulfated macromolecules in the blastocoels of mesenchyme blatulae of S. purpuratus and L. pictus. Dev. Biol. 94: 326-336.

Tanaka, Y. 1976. Effects of the surfactants on the cleavage and further development of the sea urchin embryos. I. The inhibition of micromere formation at the 4th cleavage. Dev. Growth Differ. 18: 113-122.

Tufaro, F. and B.P. Brandhorst. 1979. Similarity of proteins synthesized by isolated blastomees of early sea urchin embryos. Dev. Biol. 71: 390-397.

Uzman, J.A. 1983. An investigation into the acceleration of the rate of RNA synthesis at the 4th cleavage of the sea urchin zygote. Ph.D. Dissertation. University of California, Berkeley.

Uzman, J.A. and F.H. Wilt. 1984. The role of RNA polymerase initiation and elongation in control of total RNA and histone mRNA synthesis in sea urchin embryos. Dev. Biol. 106: 174-180.

Weiner, S., W. Traub, and H.A. Lowenstam. 1983. Organic matrix in calcified exoskeletons. pp. 205-224. In: Biomineralization and Biological Metal Accumulation. P. Westbroek and E.W. DeJong (eds.). Reidel Publishing Co., Dordrecht.

Wilt, F.H. 1970. The acceleration of RNA synthesis in cleaving sea urchin embryos. Dev. Biol. 23: 444-455.

ASPECTS OF GENE EXPRESSION IN THE SEA URCHIN MICROMERE-PRIMARY MESENCHYME CELL LINE
Michael A. Harkey

ABSTRACT

In the 16-cell stage sea urchin embryo, four micromeres are committed to form the primary mesenchyme, which ultimately differentiates the larval skeleton. This chapter describes methods for isolation and culture of the micromere-primary mesenchyme cell line, and then examines aspects of gene expression during their development. The program of protein synthesis in developing micromeres is highly cell-specific. The cells undergo a major qualitative shift in protein synthesis at about the time of primary mesenchyme ingression. Preliminary analysis of primary mesenchyme mRNAs, using cloned cDNAs as probes, suggests that at least some of the changes in protein synthesis are regulated at the level of mRNA abundance rather than by selective translation or localization of mRNAs.

The sea urchin embryo has provided a wealth of information in the field of developmental biology at both the morphological and molecular levels. While the clarity of the embryo allows direct and detailed observation of morphogenesis, the synchrony of embryo cultures facilitates examination of the underlying molecular events.

Traditionally, molecular analysis of early development for most organisms, including the sea urchin, has relied on homogenates of whole embryos. While this approach has yielded much valuable information regarding temporal relationships between morphological and molecular events, it has not addressed the spatial relationships. Thus, even though the morphological data clearly demonstrate that development involves a branching pattern of phenotypic pathways, and although molecular analysis of adult tissues shows that each of these pathways culminates in a specialized pattern of gene expression, we have been unable to distinguish the early molecular events of cell specialization.

Recently, investigators have begun to dissect the sea urchin embryo into its component cell-types. Two approaches have been used successfully to examine early molecular events of development. First, techniques have been developed to localize specific antigens (McClay et al., 1983) and RNAs (Angerer & Angerer, 1983) in biological sections of embryos. With the appropriate probe in hand, these techniques require very little embryo material, and are capable of detecting sub-populations of cells that are otherwise indistinguishable. The powerful utility of this approach is demonstrated elsewhere in this volume (Klein; Showman).

The second approach involves the physical isolation of individual cell-types from the embryo. In addition to its analytical utility, this approach facilitates the purification of cell-specific macromolecules. These, in turn, can be used to generate antibody or nucleic acid probes to study the expression of individual genes. Isolation protocols now exist for (1) separation of the three types of blastomeres from 16-cell stage embryos (Spiegel & Tyler, 1966; Hynes & Gross, 1970; Mizuno et al., 1974); (2) isolation of primary mesenchyme, and of presumptive ecto-endodermal epithelium from early gastrulae (Harkey & Whiteley, 1980); and (3) isolation of ectoderm and of endoderm plus mesoderm from prism and pluteus-stage larvae (McClay & Marchase, 1979; Harkey & Whiteley, 1980, 1983). In another chapter of this volume, W. Klein describes the use of this strategy to isolate and characterize genes whose expression is restricted to the larval ectoderm. The data that are emerging from the application of both of these approaches clearly demonstrate that each cell lineage undergoes a distinct program of gene expression, and that these programs begin to diverge very early in development (Bruskin et al., 1981; Harkey & Whiteley, 1982, 1983; Lynn et al., 1983; Cox et al., 1984).

The micromere-primary mesenchyme cell line is particularly tractable for the second approach to the dissection of sea urchin embryogenesis. As I will describe below, this cell line is accessible en mass in pure form at any stage of development. In addition, its early determination and very specialized morphogenesis allow us to follow its entire program of development in culture. I have been studying gene expression in sea urchin micromeres, at the protein and nucleic acid levels, in the laboratories of Arthur H. Whiteley and Helen R. Whiteley. This chapter describes some of our results concerning the isolation and culture of this cell-line from Strongylocentrotus purpuratus, and the program of gene expression that accompanies its development.

Normal Development of Micromeres

At the fourth cleavage of the sea urchin embryo, the vegetal quartet of blastomeres divides unequally, producing four micromeres. These cells are fully committed at this early stage to differentiate the skeletogenic primary mesenchyme of the larva (Okazaki, 1975). The micromeres divide 4-5 times during the next few hours, and become incorporated along with all of the blastomere descendents into the ciliated epithelium of the blastula.

Real-time observation of the blastula indicates a uniformly ciliated epithelium with no distinguishable sub-populations of cells. However, time-lapse video analysis demonstrates that the micromere descendents are a behaviorally distinct group of cells (R. Langelan, pers. comm.). They exhibit a lower mitotic rate and a higher level of pulsatory activity than do other cells in the embryo.

Shortly after the blastula hatches, the majority of micromere descendents (excluding a specific sub-population) disengage from the epithelium and emerge as the primary mesenchyme into the blastocoel. They migrate within this space, initially without apparent organization, but soon accumulate as a sub-equatorial ring. Through multiple filopodial interactions, the primary mesenchyme cells form a syncytial cable of cytoplasm along the ring (Gibbins et al., 1969; Millonig, 1970), and the skeletal spicules of the pluteus larva are deposited within this syncytial network. For a more detailed review of micromere development see Harkey (1983).

Isolation of Micromeres and Primary Mesenchyme Cells

We employ three types of procedures to isolate the micromere-primary mesenchyme cell line from early embryos. Each is optimally useful during a limited period of development, but the combination of methods yields access to these cells throughout development. Prerequisite to all of the isolation procedures is the dissociation of embryos into single cells by removal of divalent cations from the medium.

1: <u>Micromeres</u>. At the 16-cell stage, micromeres are sufficiently smaller than the other blastomeres to allow separation by differential sedimentation. Several protocols exist for the isolation of micromeres on gradients of Ficol (Hynes & Gross, 1970), BSA (Chamberlain, 1977), or sucrose (e.g., Spiegel & Tyler, 1966). These methods, which can yield preparations of 99% or higher purity, are strictly limited to the 16-cell stage of development.

In order to facilitate the generation of cell-specific cDNA libraries, we recently developed a scaled-up and partially automated procedure for micromere isolations (Harkey & Whiteley, In press). Illustrated in Figure 1, this procedure typically yields 1-2 grams of >99% pure viable micromeres from about 60 ml of embryos.

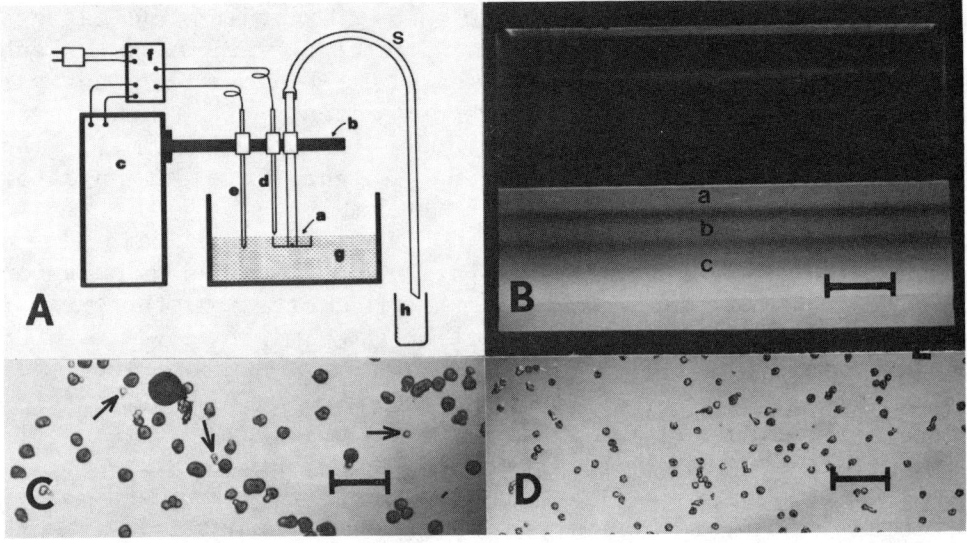

Fig. 1. Mass isolation of micromeres. (A) Schematic diagram of cell-separating equipment. 1.5 L of cell suspension derived from 30 ml of 16-cell stage embryos are layered over a 5 L sucrose gradient (g). The suspension is introduced using a siphon (s) equipped with a baffled orifice (a) that facilitates rapid fluid transfer without disturbing the gradient. After 1-2 h of sedimentation, the gradient is fractionated from the top using the same baffled siphon device. The siphon orifice follows the descending gradient surface powered by a syringe pump motor (c), and controlled by a switching circuit (f) that detects the surface with a pair of electrodes (e & d). (B) Stratification of cells on a gradient after 2 h of sedimentation. Under oblique illumination, the regions of high cell density are visible as bright bands. The micromere band (b) is fully resolved from the cell debris in the original sample layer (a), and from the faster sedimenting cells and aggregates (c). Bar = 5 cm. (C) Total cell suspension prior to loading onto gradient. Arrows indicate micromeres. The large sphere is an unfertilized egg. Bar = 100 µm. (D) Isolated micromeres derived from band (b) of the gradient. Bar = 100 µm.

2: <u>Gastrula Primary Mesenchyme</u>. At the early gastrula stage, primary mesenchyme cells can be obtained in high purity (95%) by exploiting their unique position inside the blastocoel (Harkey & Whiteley, 1980). Gentle dissociation of early gastrulae with 1 M glycine disperses the presumptive ectoderm and endoderm cells, leaving the primary mesenchyme cells trapped in "bags" of basal lamina. These bags are then isolated by buoyant density centrifugation. Because of the rapid morphogenetic rearrangements taking place in the gastrula, this method is optimally useful only during the early phase of gastrulation.

3: <u>Late Primary Mesenchyme</u>. Primary mesenchyme-enriched cell preparations are easily obtained from late gastrula or older embryos using a modification of the above procedure. Basal laminar bags, from which the ectoderm has been dispersed with glycine, are collected either by filtering with Nitex netting (McClay & Marchase, 1979), or by differential sedimentation (Harkey & Whiteley, 1980). These bag preparations contain secondary mesenchyme and endodermal structures; however, by eliminating the major cell mass, they are highly enriched in primary mesenchyme cells.

Differentiation in Culture

Okazaki (1975) discovered that isolated micromeres from several sea urchin species are capable of normal synchronous differentiation in culture if they are grown in sea water supplemented with horse serum. We have observed the same phenomenon in <u>S</u>. <u>purpuratus</u> micromeres prepared by both small-scale (Harkey & Whiteley, 1980) and large-scale (Harkey & Whiteley, In press) procedures.

The development of isolated micromeres can be considered in three phases. During the first phase, the cells are mitotically active, adherent to each other, and non-adherent to the substratum. Thus, they drift in the medium as irregular clumps of dividing cells. The second phase begins at the time of ingression in control embryos. The now mitotically quiescent cells reverse their affinities, preferring the substratum to each other, and begin a period of active migration (Fig. 2a). During this phase the cells interact with multiple filopodia, forming a complex network of branching cytoplasmic cables (Fig. 2b). In addition, groups of cells often appear to contribute to large cytoplasmic processes as shown in Figure 2c. These observations suggest a syncytial interaction among cultured micromeres similar to that described for primary mesenchyme cells in the intact embryo (Gibbins et al., 1969; Millonig, 1970). The third phase of development of isolated micromeres is the differentiation of spicules.

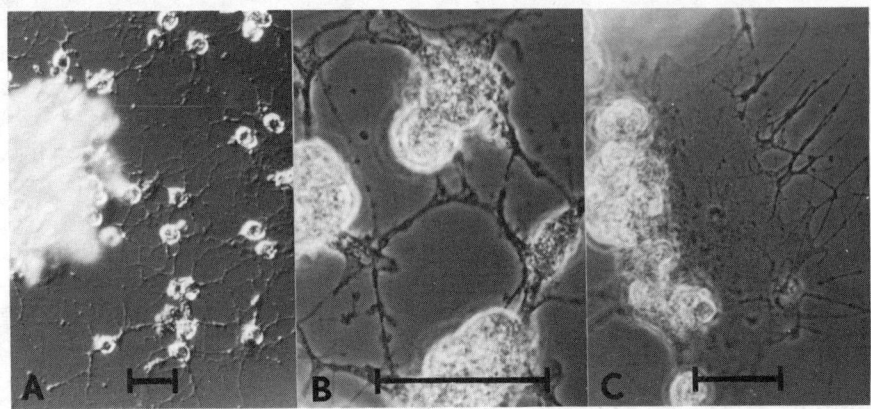

Fig. 2. Aspects of the development of micromeres in culture. (A) The initial spreading of cells from an aggregate at 35 h of culture. This is coincident with primary mesenchyme ingression in control embryos. (B) A filipodial network at 50 h. (C) A large cytoplasmic sheet, apparently derived from the combined contributions of several cells. This culture is also 50 h old. Bars = 10 µm.

Under our culture conditions, only a small and variable fraction of micromeres appears to become directly associated with growing spicules (Harkey & Whiteley, 1983). It is this observation that led us to develop alternate methods for obtaining primary mesenchyme cells from older embryos. In contrast to isolated micromeres, the cells isolated from early gastrulae consistently exhibit 90-100% efficiency of spicule formation (Harkey & Whiteley, 1980).

It is increasingly evident that the low efficiency of spicule formation in micromere cultures is not the result of an aberration or abortion of the program of gene expression leading to spiculogenesis. Rather, it probably reflects a more direct inhibition of the process of spicule deposition. That these events can be separated was demonstrated by Okazaki (1971a & b, 1975). She showed that the latter, but not the former, is dependent upon the presence of horse serum in the medium.

At the level of protein synthesis, micromeres differentiate normally in culture. They undergo the same changes in protein synthesis during post-gastrula development, and differentiate the same terminal pattern of protein synthesis as do the more uniformly spiculogenic primary mesenchyme cells isolated from early gastrulae (Harkey & Whiteley,

1983). Furthermore, many of the stage- and cell-specific events in the program of protein synthesis observed in cultured micromeres have been corroborated in vivo (Harkey & Whiteley, 1982, 1983). Thus, micromere cultures appear to undergo a consistent program of gene expression that is independent of the method or time of isolation, or the apparent efficiency of spicule formation.

The available data suggest that the low spicule-forming efficiency in our micromere cultures reflects a normal process of "contact" inhibition exerted by dominant members of the primary mesenchyme syncytium. Okazaki (1960) reported that in normal embryos spiculogenesis begins with the formation of numerous spicule rudiments by individual cells or small groups of cells in the mesenchymal ring. The growth of all, but two dominant, rudiments is inhibited as the cells become linked into a continuous syncytial cable. However if this linkage is prevented by a properly timed exposure to hypotonic or low-calcium media, the dominance effect is blocked and all of the rudiments grow. Similar transient spicule rudiments appear during spiculogenesis in micromere cultures (Okazaki, 1975; Harkey & Whiteley, 1983). The frequency of stable spicules that continue to grow in culture is generally highest in the most sparsely populated regions of the dish (unpubl. obs.). In particular, spicules are almost always present in small groups of cells that become dislodged from the cellular network and drift freely in the medium.

It is of interest in this context that primary mesenchyme cells from early gastrulae, which are isolated as discrete groups contained within basal laminar "bags", can be cultured at high density with >95% spicule-forming efficiency (Harkey & Whiteley, 1980). Therefore, the inhibitory effect is not strictly dependent on cell density. Rather, it would appear that physical parameters of cell-cell interaction may be important. We suggest that overcrowding in micromere cultures results in syncytial linkage of multiple centers of "dominance" or inhibition, and that by mutual inhibition, these centers produce large fields in which spicule formation does not occur.

The Micromere Program of Protein Synthesis

The high degree of morphological specialization of the micromere-primary mesenchyme cell line suggests an equally specialized program of gene expression. We have characterized this program at the level of protein synthesis using two-dimensional polyacrylamide gel electrophoresis (Harkey & Whiteley, 1982, 1983). Micromere cultures, gastrula primary mesenchyme cultures and mesenchyme-enriched "bags" from older embryos were employed to obtain data throughout the development of this cell line. The following general observations were made.

1) The pattern of protein synthesis in micromeres changes dramatically during development. At least 50% of the detectable proteins synthesized by these cells exhibit de novo appearances, disappearances, increases, or decreases in their rates of accumulation. The differentiated pattern of protein synthesis is dominated by a small group of rapidly accumulating proteins that are not detectable at the 16-cell stage.

2) These changes are mostly cell-specific. At the 16-cell stage, the patterns of protein synthesis are virtually identical in all of the cell types (Tufaro & Brandhorst, 1979; Harkey & Whiteley, 1983). However, by the time of spicule differentiation, the micromere-primary mesenchyme pattern is very different from that of ectoderm. As a result, the dominant group of actively synthesized proteins in differentiated micromeres is highly cell-specific as well as stage-specific.

3) Most changes cluster around the time of ingression. We followed the labeling histories of 161 individual proteins through 8 arbitrarily designated stages of micromere development (Harkey & Whiteley, 1983). Several proteins appeared de novo within a few hours of culture, and a small group of proteins did not appear until the onset of spicule formation. However, the majority of changes occurred during the period between hatching and gastrulation.

4) The micromere program of protein synthesis is essentially bimodal. Of those proteins that exhibit changes during development, only 10% show transient changes. The rest exhibit maximal rates of synthesis either at the beginning or at the end of development, and minimal rates at the opposite time. Thus it appears that most of the changes reflect a simple transition from an early mode to a late mode of gene expression.

A particularly interesting case of cell-specific gene expression in micromeres is that of the tubulins. We have identified α- and β-tubulin on two-dimensional gels (Figs. 3a & b) by co-electrophoresis of total gastrula proteins with vinblastine sulfate-precipitated tubulin from eggs (Bryan,

GENE EXPRESSION 319

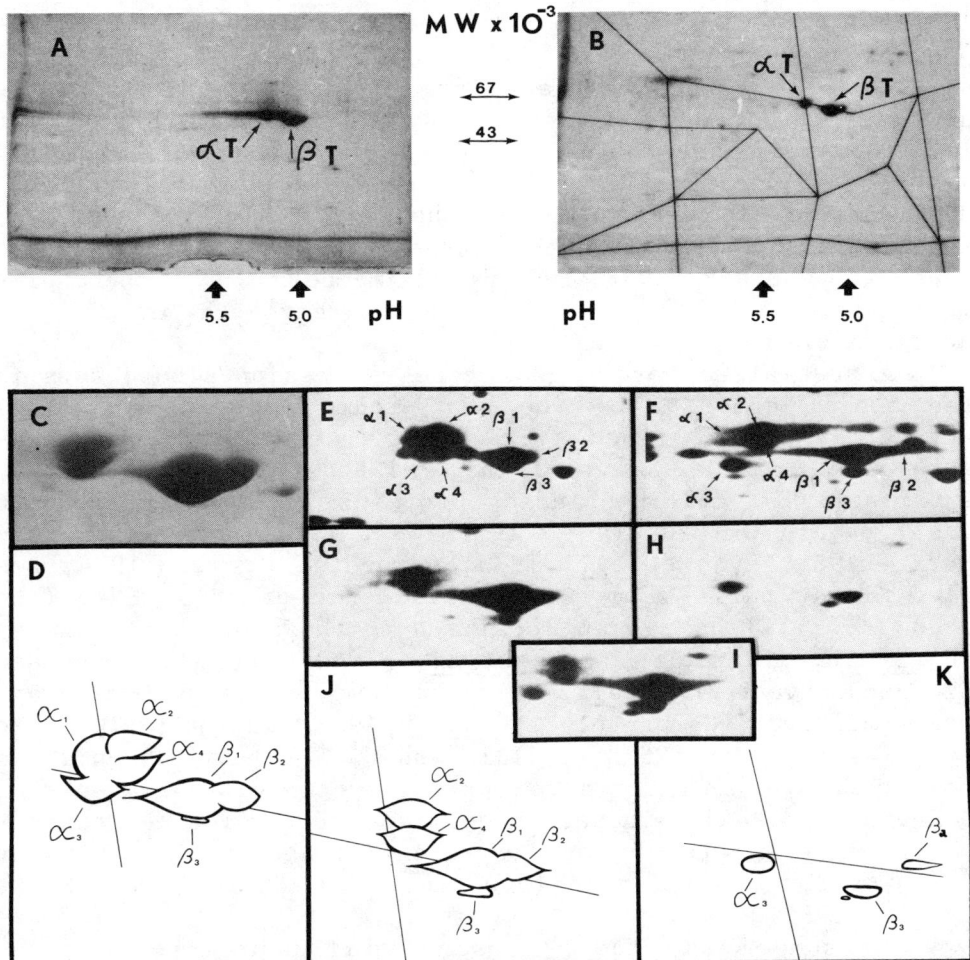

Fig. 3. Analysis of tubulin synthesis by two-dimensional electrophoresis. (A) Coomassie blue-stained gel of purified tubulin from unfertilized eggs. (αT) α tubulin, (βT) β tubulin. (B) Preparation similar to (A), except that ^{3}H-labeled gastrula proteins were included as reference proteins to construct the standard reference grid (Harkey & Whiteley, 1982). (C) Magnified view of the tubulin region in (B). (D) Interpretive drawing of (C) indicating the individual electrophoretic variants of the tubulins. The numbering of tubulin variants is arbitrary. The crossed lines are components of the reference grid in (B). (E & F) The tubulin regions of fluorographs of newly synthesized pluteus proteins. The second example (F) was produced using a new batch of ampholines for isoelectric focusing (horizontal dimension of electrophoresis: O'Farrell, 1975) which affected better resolution of the tubulins (Harkey, 1982). The following fluorographs are derived as in (F) using the new ampholines. (G-K) Comparison of tubulin synthesis by the two distinguishable cell-types of the early gastrula. The tubulin regions on fluorographs of ^{3}H-valine labeled proteins are shown for (G) epithelium, and (H) primary mesenchyme isolated from early gastrulae. Superposition of these images (I) demonstrates that the two cell fractions account for the normal embryonic complement of tubulin variants, with the possible exception of $α_1$T. Interpretive drawings of (J) epithelial and (K) mesenchymal tubulins are shown below their respective fluorographs. It can be seen that $α_2$T, $α_4$T and $β_1$T are specific to the epithelium, while $α_3$T is specific to the mesenchyme.

1971). The tubulin preparations exhibit several electrophoretic variants in both the α and β positions (Figs. 3c & d), and all of these variants are distinguishable among the newly synthesized proteins of advanced embryos (Figs. 3e & f). However, separation of the cell-types at the early gastrula stage reveals that primary mesenchyme cells (Figs. 3h & k) and ecto-endodermal epithelial cells (Figs. 3g & j) synthesize distinct subsets of the α- and β-tubulin groups. Thus it appears that individual members of the tubulin multigene family are differentially regulated in the sea urchin embryo.

A specialized program of gene expression also occurs in the aboral ectoderm of the sea urchin embryo. A set of ectoderm-specific (Spec) mRNAs has been described, which encodes a family of proteins related to the vertebrate superfamily of calcium-binding proteins including calmodulin and troponin C (Bruskin et al., 1981, 1982; Bedard & Brandhorst, 1983; Lynn et al., 1983; Carpenter et al., 1984; see also Klein, this volume). At least some of these RNAs (the Spec 1 family) have been localized in the aboral ectoderm by in situ hybridization (Lynn et al., 1983), and their encoded proteins have been localized in the same tissue by immunological methods (Carpenter et al., 1984). Similarly, two cytoplasmic actin mRNAs (Cy IIIa and Cy IIIb) have been shown to accumulate with the same spatial specificity (see Angerer & Davidson, 1984). These ectoderm-specific molecules exhibit their highest rates of syntheses and/or accumulation between hatching and gastrulation (Bruskin et al., 1981, 1982; Shott et al., 1984).

In some ways, these tissue-specific programs of gene expression resemble that described for whole embryos (Brandhorst, 1976; Bedard & Brandhorst, 1983). In each case the program involves a peak of changes between hatching and gastrulation, and very few of these changes are transient. Two conclusions can be drawn from these observations. First, it is clear that a profound transition in gene expression occurs embryo-wide at about the mesenchyme blastula stage. Indeed, a large body of work characterizes this period as one of transition from predominantly maternal, to predominantly embryonic gene expression (for reviews see Whiteley & Whiteley, 1975; Davidson, 1976). Second, this transition involves the appearance and accumulation of major tissue-specific gene products; products that are characteristic of the differentiated states of the various tissues of the larva. Thus, regional specialization of gene expression begins well before any morphological specialization is detectable in the embryo.

Regulation of the Micromere Program

The protein synthesis data provide a foundation from which to examine the mechanisms of regulation of gene expression during development. In general, we want to know how the observed changes in protein synthesis are coordinated temporaly, and how they are restricted to certain cells. Our first approach has been to ask where and when the mRNAs encoding primary mesenchyme-specific proteins first occur. If these mRNAs exist in the maternal message pool, then the cell-specific expression could be regulated either by cytoplasmic localization of messages during or before early cleavage, or by a selective translation mechanism. Temporal regulation would require the latter type of mechanism. On the other hand, if the early embryo does not contain these mRNAs, then any number of transcriptional, processing and translational regulatory mechanisms may be important. Therefore we have undertaken to obtain cDNA clones representing major mRNAs that are specific to differentiated micromeres, and to use the clones to examine the metabolism of these RNAs during development.

In order to determine whether differentiated micromeres contain abundant cell- and stage-specific mRNAs suitable for this study, we translated poly A^+ RNA from differentiated micromere cultures in a rabbit reticulocyte lysate and examined the products on fluorographs of SDS acrylamide gels (Fig. 4). By comparison with the translation products of poly A^+ RNAs from an earlier stage (16-cell embryos) (not shown) and from another cell-type of the same stage (pluteus ectoderm), it was possible to detect several major stage- and cell-specific bands (arrows). At least some of these bands corresponded closely in apparent molecular weight to previously characterized (Harkey & Whiteley, 1983) stage- and cell-specific products of living differentiated micromeres. These results indicated that at least some of the specificity in protein synthesis observed in differentiated micromeres reflects specificity in the abundance of translatable poly A^+ RNAs. They further suggested that abundant RNAs would be available in these cells for use in generating stage- and cell-specific cDNA clones.

We constructed a cDNA library in the E. coli plasmid, PBR322, from the RNA described previously and screened 1000 clones for cell- and stage-specificity by three rounds of colony hybridization. ^{32}P-cDNA, generated from the same RNA, was used as a "positive" probe. Polysomal poly A^+ RNA from

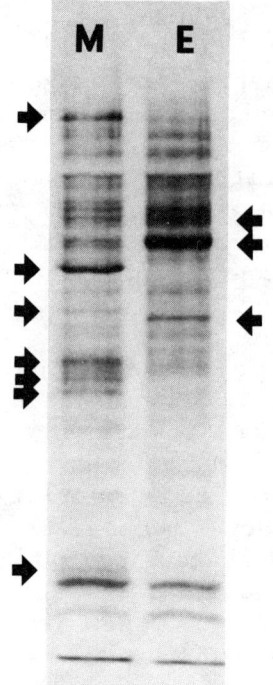

Fig. 4. In vitro translation products (rabbit reticulocyte lysate) of poly A⁺ RNA obtained from (E) pluteus ectoderm, and (M) a pluteus-stage micromere culture. Arrows indicate bands that are unique or enriched in each preparation.

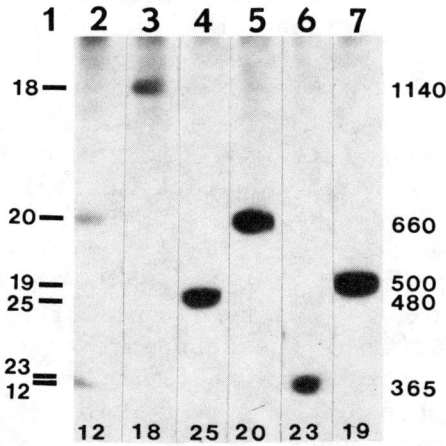

Fig. 5. A test of homology among six SpLM clones. Lane 1 illustrates the DNA banding pattern observed on agarose gels after Pst 1 digestion and coelectrophoresis of the six SpLM plasmid preparations. Each excised SpLM cDNA insert is identified to the left, and its size in base pairs is indicated to the right. Lanes 2-7 show the results of hybridizations of replicate blots of this same pattern of DNA fragments to ^{32}P probes made from each of the clones. The SpLM probe used in each case is indicated at the bottom of the lane.

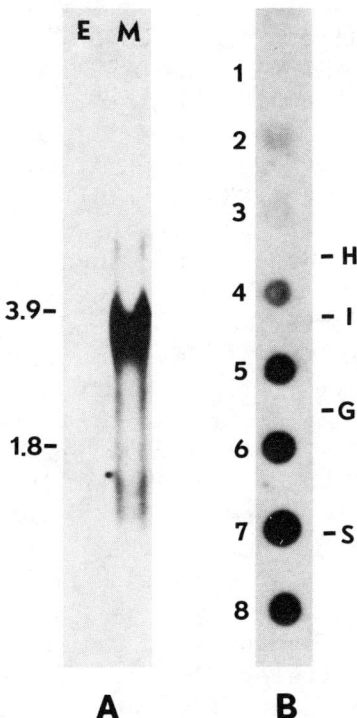

Fig. 6. Analysis of SpLM 18 RNA. (A) Northern analysis of cell specificity of the RNA. Poly A$^+$ RNA (2 µg) from differentiated micromeres (M) or contemporary pluteus ectoderm (E) were electrophoresed on agarose/formaldehyde gels, blotted onto nitrocellulose, hybridized to the SpLM18 probe, and autoradiographed. The positions of the 3.9 and 1.8 kilobase ribosomal RNAs are indicated to the left. (B) Changes in the abundance of SpLM18 RNA during micromere development. Total RNA (10 µg) from each of the indicated stages of developing micromere cultures was spotted onto nitrocellulose, hybridized to the SpLM18 probe, and autoradiographed. (Row 1) RNA from 1-h-old zygotes. (Rows 2-8) RNA from micromeres at 0 h (2), 10 h (3), 22 h (4), 35 h (5), 50 h (6), 65 h (7), and 89 h (8) of culture. The progress of development of control embryos is indicated to the right. (H) Hatching, (I) the start of ingression of the primary mesenchyme, (G) the start of gastrulation, and (S) the start of spicule formation.

16-cell embryos was used to make a stage-specific "negative" probe, while poly A$^+$ RNA from pluteus ectoderm was used to make a cell-specific "negative" probe. Several clones showed reproducible strong specificity for the positive probe. This set of clones will be referred to as SpLM (<u>Strongylocentrotus purpuratus</u> Late Micromeres).

An examination of sequence homology among six of these clones by Southern blot analysis is shown in Figure 5. In this case, SpLM12 and SpLM20 appeared to share at least some homology, while the others did not cross-hybridize. Similar analysis of the remaining SpLM clones revealed additional distinct sequences; however, the present discussion will focus on SpLM18.

The cell-specificity of SpLM18 RNA was demonstrated by Northern blot analysis of poly A^+ RNAs from differentiated micromeres and from contemporary pluteus ectoderm. Hybridization of the RNA blots with a ^{32}P-probe generated from the SpLM18 cDNA (Fig. 6a) showed that this clone binds predominantly to a 3.8 kilobase RNA from differentiated micromeres. Ectoderm does not contain detectable SpLM18 RNA.

Developmental changes in the abundance of the SpLM18 RNA were examined using dot blots of total RNA from seven stages of developing micromere cultures. Total nucleic acid was prepared from each of the stages and treated with RNAase-free DNAase under conditions that eliminated all hybridization of the probe to sperm DNA, but completely preserved the hybridization to poly A^+ RNA. The resulting RNA preparations were dot blotted and hybridized to the SpLM18 probe. As shown in Figure 6b, SpLM18 RNA increased in abundance during development, showing an initial rise at the time of primary mesenchyme ingression, and a maximum abundance at the start of spicule formation. This is the same pattern of change shown by many of the newly synthesized primary mesenchyme-specific proteins (Harkey & Whiteley, 1983).

We have not yet identified the protein encoded by SpLM18 RNA. However, its high abundance and its stage- and cell-specificity suggest that it encodes one of the primary mesenchyme-specific proteins. Since little or no SpLM18 RNA is detectable either in the egg, or in 16-cell stage micromeres, it follows that synthesis of the SpLM18 protein later in development involves regulation at the transcriptional or RNA-processing levels. We are currently examining the other SpLM clones to determine whether this level of regulation is generally representative of primary mesenchyme-specific gene expression. Preliminary data indicate that it is.

ACKNOWLEDGMENTS

This work was supported in part by the following grants from the National Institutes of Health: Grant HL-10312 to Arthur H. Whiteley, Grant GM-20784 to Helen R. Whiteley, Grant ES-02190 to N. Karle Mottet, and Training Grants ES-07032 and HD-00266.

LITERATURE CITED

Angerer, R.C. and L.M. Angerer. 1983. RNA localization in sea urchin embryos. pp. 101-129. In: Time, Space and

Patterns in Embryonic Development. W.R. Jeffery and
R.A. Raff (eds.). Liss, N.Y.
Angerer, R.C. and E.H. Davidson. 1984. Molecular indices of
cell lineage specification in sea urchin embryos.
Science 226: 1153-1160.
Bedard, P.-A. and B.P. Brandhorst. 1983. Patterns of protein synthesis and metabolism during sea urchin embryogenesis. Develop. Biol. 96: 74-83.
Brandhorst, B.P. 1976. Two dimensional gel patterns of protein synthesis before and after fertilization of sea
urchin eggs. Develop. Biol. 52: 310-317.
Bruskin, A.M., P.-A. Bedard, A.L. Tyner, R.M. Showman, B.P.
Brandhorst, and W.L. Klein. 1982. A family of proteins
accumulating in ectoderm of sea urchin embryos specified
by two related cDNA clones. Develop. Biol. 91: 317-324.
Bruskin, A.M., A.L. Tyner, D.E. Wells, R.M. Showman, and W.H.
Klein. 1981. Accumulation in embryogenesis of five
mRNAs enriched in the ectoderm of the sea urchin pluteus. Develop. Biol. 87: 308-313.
Bryan, J. 1971. Vinblastine and microtubules. Induction and
isolation of crystals from sea urchin oocytes. Exp.
Cell Res. 66: 129-136.
Carpenter, C.D., A.M. Bruskin, P.E. Hardin, M.J. Keast, J.
Anstrom, A.L. Tyner, B.P. Brandhorst, and W.H. Klein.
1984. Novel proteins belonging to the troponin C superfamily are encoded by a set of mRNAs in sea urchin
embryos. Cell 36: 663-671.
Chamberlain, J.P. 1977. Protein synthesis by separated
blastomeres of sixteen-cell sea urchin embryos. J. Cell
Biol. 75: 33a (Abstract).
Cox, K.H., D.V. DeLeon, L.M. Angerer, and R.C. Angerer.
1984. Detection of mRNAs in sea urchin embryos by in
situ hybridization using asymmetric probes. Develop.
Biol. 101: 485-502.
Davidson, E.H. 1976. Gene Activity in Early Development,
2nd Ed. Academic Press, N.Y.
Gibbins, J.R., L.G. Tilney, and K.R. Porter. 1969. Microtubules in the formation and development of the primary
mesenchyme in Arbacia punctulata. I. The distribution
of microtubules. J. Cell Biol. 41: 201-227.
Harkey, M.A. 1982. The Program of Protein Synthesis in the
Development of the Echinoid Primary Mesenchyme. Ph.D.
Thesis. Dept. of Zoology, University of Washington.
Harkey, M.A. 1983. Determination and differentiation of
micromeres in the sea urchin embryo. pp. 131-155 In:

Time, Space and Patterns in Embryonic Development. W.R. Jeffery and R.A. Raff (eds.). Liss, N.Y.

Harkey, M.A. and A.H. Whiteley. 1980. Isolation, culture and differentiation of echinoid primary mesenchyme cells. Wilhelm Roux's Arch. 189: 111-122.

Harkey, M.A. and A.H. Whiteley. 1982. Cell-specific regulation of protein synthesis in the sea urchin gastrula: A two-dimensional electrophoretic study. Develop. Biol. 93: 453-462.

Harkey, M.A. and A.H. Whiteley. 1983. The program of protein synthesis during the development of the micromere-primary mesenchyme cell line in the sea urchin. Develop. Biol. 100: 12-28.

Harkey, M.A. and A.H. Whiteley. In Press. Mass isolation and culture of sea urchin micromeres. In Vitro.

Hynes, R.O. and P.R. Gross. 1970. A method of separating cells from early sea urchin embryos. Develop. Biol. 21: 383-402.

Lynn, D.A., L.M. Angerer, A.M. Bruskin, W.H. Klein, and R.C. Angerer. 1983. Localization of a family of mRNAs in a single cell type and its precursors in sea urchin embryos. Proc. Natl. Acad. Sci. U.S.A. 80: 2656-2660.

McClay, D.R., G.W. Cannon, G.M. Wessel, R.D. Fink, and R.B. Marchase. 1983. Patterns of antigenic expression in early sea urchin development. pp. 157. In: Time, Space and Patterns in Embryonic Development. W.R. Jeffery and R.A. Raff, (eds.). Liss, N.Y.

McClay, D.R. and R.B. Marchase. 1979. Separation of ectoderm and endoderm from sea urchin pluteus larvae and demonstration of germ layer-specific antigens. Develop. Biol. 71: 289-296.

Millonig, G. 1970. A study on the formation of the sea urchin spicule. J. Submicrosc. Cytol. 2: 157-165.

Mizuno, S., Y.R. Lee, A.H. Whiteley, and H.R. Whiteley. 1974. Cellular distribution of RNA populations in 16-cell stage embryos of the sand dollar Dendraster excentricus. Develop. Biol. 37: 18-27.

O'Farrell, P.H. 1975. High resolution two-dimensional electrophoresis of proteins. J. Biol. Chem. 250: 4007-4021.

Okazaki, K. 1960. Skeleton formation of sea urchin larvae. II. Organic matrix of the spicule. Embryologia 5: 283-320.

Okazaki, K. 1971a. Spicule formation in sea urchin larvae; observations in vivo and in vitro. Symp. Cell Biol. 22: 163-171 (in Japanese).

Okazaki, K. 1971b. In vitro culture of micromeres and primary mesenchyme cells isolated from sea urchin embryos and larvae. In: Cells in Early Stages of Development, Jpn. Soc. Dev. Biol. (eds.). Tokyo: Iwanami Shoten (in Japanese).

Okazaki, K. 1975. Spicule formation by isolated micromeres of the sea urchin embryo. Am. Zool. 15: 567-581.

Shott, R.J., J.J. Lee, R.J. Britten, and E.H. Davidson. 1984. Differential expression of the actin gene family of Strongylocentrotus purpuratus. Develop. Biol. 101: 295-306.

Spiegel, M. and A. Tyler. 1966. Protein synthesis in micromeres of the sea urchin egg. Science 151: 1233-1234.

Tufaro, F. and B.P. Brandhorst. 1979. Similarity of proteins synthesized by isolated blastomeres of early sea urchin embryos. Develop. Biol. 72: 390-397.

Whiteley, H.R. and A.H. Whiteley. 1975. Changing populations of reiterated DNA transcripts during early echinoderm development. pp. 39-88. In: Current Topics in Developmental Biology, Vol. 9. A.A. Moscona and A. Monroy (eds.). Academic Press, N.Y.

Index

Adhesion:
    Of Cells..198
Actin Filaments:
    Of Microvilli...49
Activation:
    Calcium Flux..50
    Change In Membrane Fluidity...........................40,45
    pH Changes..50
    Proteolytic Processing....................................40
    Transport Systems...40
Aeguipecten..209
Animal (AN) Hemisphere (Pole).......................9,130,132
    AN1..9,10
    AN2...9,10,17
Animalization..157
Antibodies:
    Monoclonal...171
Antigens:
    De Novo..183
    Localized In Eggs..173
    Localized To Basal Lamina................................175
    Localized To Blastula....................................182
    Localized To Ectoderm....................................173
    Localized To Endoderm................................176,180
    Localized To Mesoderm....................................177
    Localized To Stomodael Opening...........................180
    Localized To Vegetal Pole................................182
Arbacia..165
Archenteron...10,19
Ascidians..126
Asters..69
    Assymetry Of..75
    Aster-Cortex Interactions.................................85
    Diaster...72
    "Interphase" Aster..72
    Monaster..72
    Pre-Meiotic Aster...73
    Sperm Aster...71
Astral Rays:
    Discussion Of...85
Attachment, Of Bryozoan Larvae...............................202
Axes:
    Animal-Vegetal......................11,29,126,157,183
    Dorsal-Ventral...155
    Of Bilateral Symmetry....................................126

Basement Membrane..4
    Basal Lamina...........................4,11,17,19,175
Benthic Marine Invertebrates..............................198
Bilateral Symmetry...15
Blastocoel..23,177
Blastoderm..240
    Specific Genes Of....................................252
Blastomeres:
    AB...134,145
    CD...134,145
Blastula...3,182
    Premesenchyme...13
Blastulation...3,11
Body Column, Maintaining Proportions In Polyps............235
Bowerbankia gracilis......................................200
Bryzoans...197,199
Bugula neritina:
    Metamorphosis Of.....................................210

Calcium-Binding Proteins..............................275,285
Calcium Ionophore A23187..................................290
Cells:
    Differential Adhesion................................198
    Lineage Of...126
    Totipotency Of.......................................126
    Surface Antigens.................................171,182
Centrifuged Eggs..............................137,140,173
    Centrifugal Fragments................................140
    Centripetal Fragments................................140
    Development Of.......................................145
Centrosomes...71
Chaetopterus..125,137
Chromatin...159
Cilia:
    Membrane Proteins Of.................................209
    Reversal of Beat.....................................209
Cleavage...7,298
    Cleavage Furrow.......................................79
Clypeaster japonicus......................................78
Coated Vesicles...51
Coelenterates...221
Collagen..303
Colloidal Gold..30
Concanavalin A......................................17,42,48
Coronal Involution..207

Cortex:
    Of Eggs....................................69,125
Cortical Granule Reaction.........................36
    and Pinocytosis......................................51
    In *Arbacia*.....................................37,42,44
    In Ascidians..37
    In Molluscs...37
    In *Sabellaria*......................................36
    In Sea Urchins......................................37
    In *Strongylocentrotus*..............................42
    In *Urechis*...36
Cortical Granules:
    Membrane Of...35
        Vesicles From...................................39
    Peroxidase Activity Of..............................39
    Surface Area Of.....................................49
Crenation...79
Critical Point Drying..............................6
Cyclin.......................................260,269
Cytochalasin B..........................83,202,214
Cytokinesis.......................................72
Cytoskeletal Recognition Sites...................136
Cytoskeleton.....................42,45,125,134,138,197

Dendraster excentricus......................96,179
Determination....................................126
    In *Drosophila melanogaster*..........239,242,244,253
    Of Germ Cells...................................187
Development:
    Sea Urchin..7
    Translational Control Of...........................91
Diptera..155
Drosophila.....................................155
Drosophila melanogaster........................187
    Blastoderm-Specific Genes Of......................239
    Polar Granules Of.................................188

Echinophores......................................10
Ectoderm..............19,129,173,224,226,228,275,276
Ectoplasm....................................128,129
Eggs..35
    Activation Of.......................................35
    Centrifugation Of..................................145
    Crenation of..77
    Cytochalasin-Inhibition Of..........................83

            Endoplasm Of..................................70,129
            Exocytosis In.....................................43
            Fragments Of.....................................140
            Of Annelids......................................125
            Of Ascidians.....................................126
            Of Asteroids.....................................100
            Of Chaetognath...................................131
            Of Chaetopterus..................................140
            Of Echinoderms........................69,91,96,100
            Of Xenopus..93
            Polarity Of......................................126
            Protein Synthesis In:
                Cellular Compartments........................101
                Cytoplasmic Inhibitory Factors...............101
                Recruitment Factors..........................101
Embryogenesis:
            Of Drosophila....................................240
Embryonic Induction..184
Embryonic Lethal Loci..244
Embryos:
            Translational Control In..........................91
Endocytosis...51
Endoderm.............................19,173,224,226,227,276
            Radial organization In...........................228
Epithelial Cells, Role In Polyp Morphogenesis........225,229
Eucidaris tribuloides............................80,165,179
Exogastrulae:
            Lithium (Li)-Induced.......................11,27,29
Extracellular:
            In Spicule Formation.............................306
            Matrix (ECM).........................6,11,17,23,27,173
            Substance..3
Euechinoids..165

Fascin..50
Fate Maps:
            In Drosophila....................................242
            Of Micromeres....................................298
Fertilization..35,91
            Ironic Events Of..................................92
            Post-Fertilization Changes.......................108
            Translation of mRNA Following................98,259
Filipin..42,52
            Filipin-Sterol Complexes..........................45
Filopodia..11,19,25

Fixatives..5
    Cetylpyridinium Chloride..............................3
    Gluteraldehyde..3
    Osmium..3
    Ruthenium Red...3
Fluorescence:
    Photobleaching.......................................41
Freeze Fracture...6,42

Gametes..35
Gastrula:
    Mid..21,23
Gastrulation..................................3,15,171,242
    Antigen Appearance At...............................177
    Glycosaminoglycans (GAGs).............................7
Gene Expression:
    Tissue-Specific Expression In Sea Urchin............320
Genes:
    Antennapedia..253
    Blastoderm-Specific.............................239,248
    Embryonic Lethals...................................243
    Notch...253
    Spec Family...275
    Tudor (tud)...187
    Ultrabithorax.......................................253
Germ Cells:
    Determination Of....................................187
Germ Layers:
    Specific Antigens Of................................171
Germ Plasm..188
Germinal Vesicle.....................................73,128
    Breakdown Of....................................132,136
    Presence of Alpha Histone mRNA......................160

Heavy Bodies..113
Histones..157
    Alpha Histone mRNAs.................................159
    Cleavage Stage......................................159
    Pronuclear Localization.............................157
Hyaline Layer.......................................7,19,29
Hybridomas:
    Screening Of..173
Hydroids..221
Hydrostatic Pressure.....................................23
Hypostome...235

In Situ Hybridization..................127,131,137,142,160
Intracellular Membranes.....................................134
Intra Membranous Particles:
    Of Cortical Granule Membrane....................38,42,44
    Of Plasma Membrane...........................38,42,44,45
Involution, Force Driving...................................207

Just, Ernest Everett.................................vii,xii,1

Lamellopidia...13,19
Lytechinus variegatus..................................3,27,179
Lytechnicus carolinus...4
Lytechinus pictus..76,99
Lectins:
    Concanavalin A..42
Lectithotrophic Larvae......................................199
Lipids:
    In Lytechnius Eggs......................................46
Lysosomes..53
Lineages:
    Germ Layer...171
    Of Cells...126,139
    Of Mesoderm..134
    Of Micromeres..299
    Lineage-Specific Proteins..............................292
Larva:
    Free-Swimming..199
Locomotory Traction:
    In Morphogenesis.......................................223

Macromeres..9
Manning, Kenneth R..1
Maternal mRNA..........................91,93,125,153,243
    Distribution Of....................125,134,145,146,147
    Role Of..125,139
    Synthesis and Utilization..............................155
Mesenchymal Cells.......................15,17,23,29,177,184
    En Mass Isolation Of...............................312,313
    Ingression Of..311
    mRNA Abundance In......................................311
    Primary (PMC)..297
    Secondary (SMC).....................................10,21
    Syncytial Cable..297
Mesoderm..134,276

Mesolamella:
    Of Coelenterate Polyps..............................224
Mesomeres...9
Metamorphosis.......................................197,198
    Role of Spec Proteins..........................275,290
Microfilaments......................................197
    In Metamorphogenesis...............................215
Micromeres........................9,10,19,157,297,311,313
    Culture Of...315
    Endowment Of.......................................299
    En Mass Isolation Of...............................313
    Nucleic Acids Of...................................300
    Program of Protein Synthesis.......................317
Microvilli:
    Actin Filaments Of..................................49
    Elongation Of.......................................48
    Of Activated Eggs...................................48
Mitotic Spindle..72
Molecular Cloning:
    Blastoderm-Specific Clones.........................245
Morphogenesis.......................................172,197
    Ciliary Motility During........................197,205
    Evidence of Force..................................229
    In Hydroids..221
    Mechanical Causes..................................222
    Microfilament Contraction During...............197,200
    Muscle Contraction During......................197,200
Morphogens...187
Morphological Determinants.............................154
Mosaic: Membranes.................................38,39,51
Muscle Processes, As Mediator of Morphogenesis.........231
Mutations:
    Grandchildless (gs) Mutant.........................190
    Maternal Effect....................................242
    Point Mutations....................................248
    Tudor (tud) Gene...................................187
Myoplasm...129
Membranes:
    Internalization By Endocytosis......................52
    Intracellular Systems..............................134

Nereis...125
Nuage-Like Particles...............................132,137

Oblation..82
    Mechanism Of...82
Okazaki Pattern..13,29
Oogenesis...173
    in Spisula...260
Ooplasmic Segregation..130,136
Organogenesis...210

Pallial Epithelium..210
    Cytochalasin B Treatment...212
Paracentrotus lividus..10,99,106
Pattern Formation:
    in Drosophila..244,253
Pelagic Organisms...199
Phosphorylation:
    Of Ribosomal Protein...108
Pinocytosis..51
Pisaster ochraceus..100
Plasma Membrane:
    Binding of Plant Lectins..41
    Filipin "Staining" Of...42
    Insertion Into..38
    Lipids Of...41
    Of Eggs...35
    Of Sperm...35,47
    Of Spisula..45
    Vesicles From...39
Polar Granules...188,189
Pole Cells...155,188
Poly A+ RNA..93
    Association with Cortical Complex................................137
    Association with Cytoskeleton....................................135
    Centrifugation of..125
    Cortical Pattern...132
    Distribution Of......................................125,128,131,142
    GV Plasm-Ectoplasm Pattern.......................................128
    In Annelids..128
    In Ascidians...128
    In Chaetopterus..125,132
    Injection Of...155
    In Nereis..125
    In Spisula...265
    In Sea Urchins..93,128
    In Xenopus..93
    Perinuclear Plasm-Ectoplasm Pattern..............................131

Polyps:
    Of Hydrozoan Coelenterates..........................221
    Structure Of..224
Polyribosomes..92
Polytene Chromosomes....................................248
Pores, Sea Urchin Archenteron............................21
Profilin...51
Pronucleus..157,159
Proteins:
    Calcium-Binding....................................275
    DNA-Binding..260
    Heat-Shock...109
    Related To Cell Cycle..........................259,268
    Spec Family..275
    Storage Of...260
    Troponin C...276
Protein Synthesis..............................98,153,298
    Activation Of.......................................92
    In Micromeres......................................317
    Qualitative Change at Fertilization............261,265
    Rate Of...96
    Regulation By Compartmentalization.................112
    Regulation by Discriminatory Factors...............109
    Regulation by Initiation Factors...................109
    Of Sea Urchins.....................................109
Psammechinus miliaris.....................................4

Radialization..27
Rete Muscularis...201
Ribonucleic Acid (RNA):
    Heterogenous Nuclear...............................155
    Messenger RNA (mRNA):
        Alpha-Histone..............................112,159
        Alpha-Tubulin..................................264
        Blastoderm Specific............................246
        Evolutionary Consideration.....................165
        Histone..153
        Of Micromeres..................................321
        pH Regulation of Translation....................99
            Maternal mRNAs......................91,139,259
                Masking Of.....................93,96,114
                Primary Structure of.................267
                Secondary Structure of...............268
                Synthesis Of.........................155

337

    Message-Containing Ribonucleoprotein:
        Particles (mRNP)........................91
           Extraction Of.....................96
           Isolation Of......................96
      Regulatory Sequences Of.........................259
      Sequestration Of................................164
      Spec Family.....................................276
      Transport Of....................................163
  Poly A+ RNA..93,321
  Polymerase...301
  Primary Structure...................................266,267
  Recruitment Of......................................92,128
  Stage Specific Repression of............................265
Ribonucleotide Reductase....................................260
  Role Of...270
Ribosomes..92
  Associated Factors......................................106
  Compartmentalization Of.................................113
  Core Proteins Of..108
  From Embryos..101
  From Unfertilized Eggs..................................101
  Monoribosomes...103
  Polysome-Derived Ribosomes..............................103
  Regulation Of...101
  Subunit Protein (S6) Of.................................108
    Phosphorylation Of....................................108

Scanning Electron Microscopy....................3,11,13,15,42
Sea Cucumber...74
Sea Urchin..................35,69,91,153,157,171,275,297,311
  Separation of Blastomeres...............................312
Septate Junctions..23
Sessile Organism..199
Smittia...155
Spec Family of Proteins.....................................275
  Calcium-Binding...285
  Functions Of..287
Spec 1...27,282
  Antiserum Against.......................................282
  Cellular Location.......................................282
Spec 2...27,282
Spicules...27,178,297
  And Collagen..303
  Organic Matrix Of.......................................304
    Antibodies Against....................................306
  Patterns In...306

Spisula solidissima..............................45,259,261
    Oogenesis In...260
    Protein A..260
    Protein C..260
Stereoimagery...30
Stomodaeum..10
Stop Signals...156
Strongylocentrotus.....................................42,43
    Stongylocentrotus purpuratus......99,107,155,179,275,313
Strongylocentrotus droebachiensis.........................96
Styela plicata...128
Syncytial Blastoderm Of Drosophila.......................240

Tetrahymena..209
Tissue-Specific Gene Expression..........................320
Totipotent...126
Toxopnuestes.variegatus....................................4
Transcription:
    In Drosophila Embryogenesis..........................241
Transcripts:
    Repeat-Containing....................................156
Transitions, Midblastula/Blastoderm......................242
Translational Control:
    Of Protein Synthesis.............................259,260
Tripneustes esculentus...................................179
Troponin C Superfamily...............................276,278
Tudor (tud) Gene.....................................187,190

Ultrastructure:
    Of Bowerbankia gracilis..............................200
    Of Egg Plasma Membrane................................37
    Of Heavy Bodies......................................113

Vegetal (Veg) Hemisphere (Pole).............9,130,133,177,182
    Plate...19
    Veg_1..9,17
    Veg_2...9
Vegetalization.......................................27,157

Yellow Crescent..130

Zygote:
    Genes of...243